Collins

Maths Frameworking

3rd edition

Kevin Evans, Keith Gordon,
Trevor Senior, Brian Speed,
Chris Pearce

Contents

How to use this book

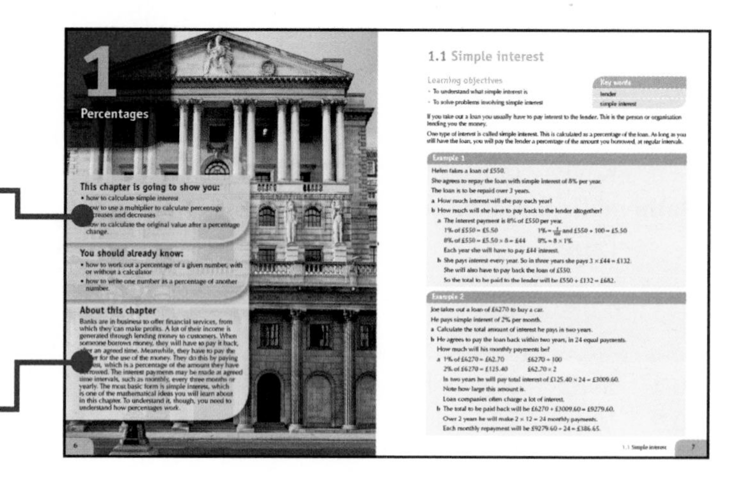

Learning objectives

See what you are going to cover and what you should already know at the start of each chapter.

About this chapter

Find out the history of the maths you are going to learn and how it is used in real-life contexts.

Key words

The main terms used are listed at the start of each topic and highlighted in the text the first time they come up, helping you to master the terminology you need to express yourself fluently about maths. Definitions are provided in the glossary at the back of the book.

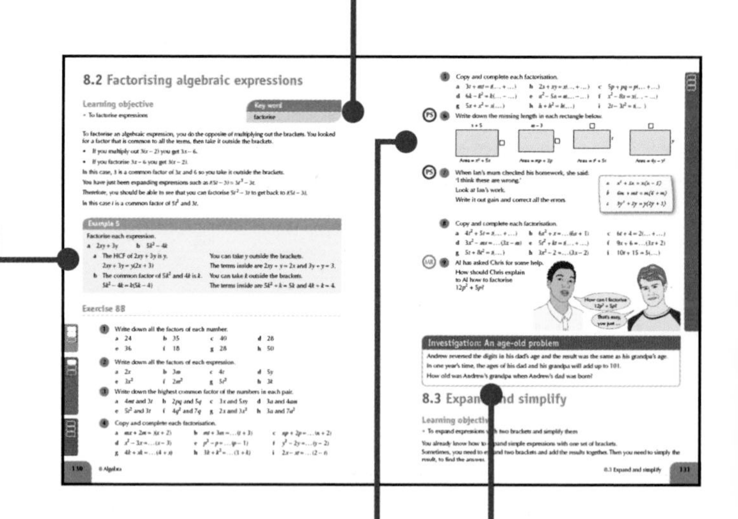

Worked examples

Understand the topic before you start the exercises, by reading the examples in blue boxes. These take you through how to answer a question step by step.

Skills focus

Practise your problem-solving, mathematical reasoning and financial skills.

Take it further

Stretch your thinking by working through the **Investigation**, **Problem solving**, **Challenge** and **Activity** sections. By tackling these you are working at a higher level.

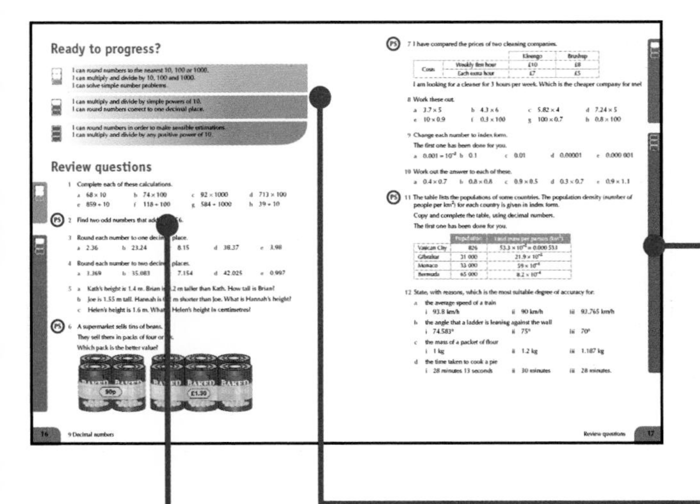

Ready to progress?

Review questions

Progress indicators

Track your progress with indicators that show the difficulty level of each question.

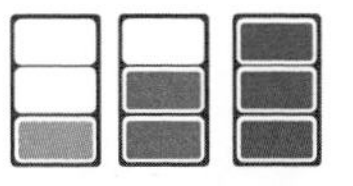

Ready to progress?

Check whether you have achieved the expected level of progress in each chapter. The statements show you what you need to know and how you can improve.

Review questions

The review questions bring together what you've learnt in this and earlier chapters, helping you to develop your mathematical fluency.

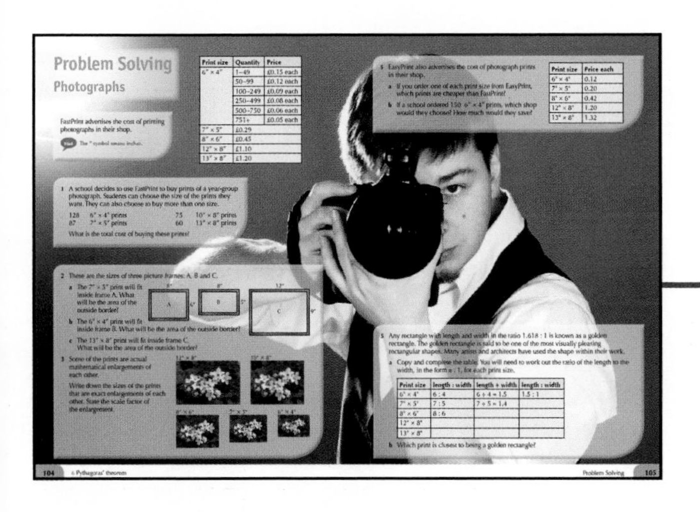

Problem Solving

Photographs

Activity pages

Put maths into context with these colourful pages showing real-world situations involving maths. You are practising your problem-solving, reasoning and financial skills.

Interactive book, digital resources and videos

A digital version of this Pupil Book is available, with interactive classroom and homework activities, assessments, worked examples and tools that have been specially developed to help you improve your maths skills. Also included are engaging video clips that explain essential concepts, and exciting real-life videos and images that bring to life the awe and wonder of maths.

Find out more at www.collins.co.uk/connect

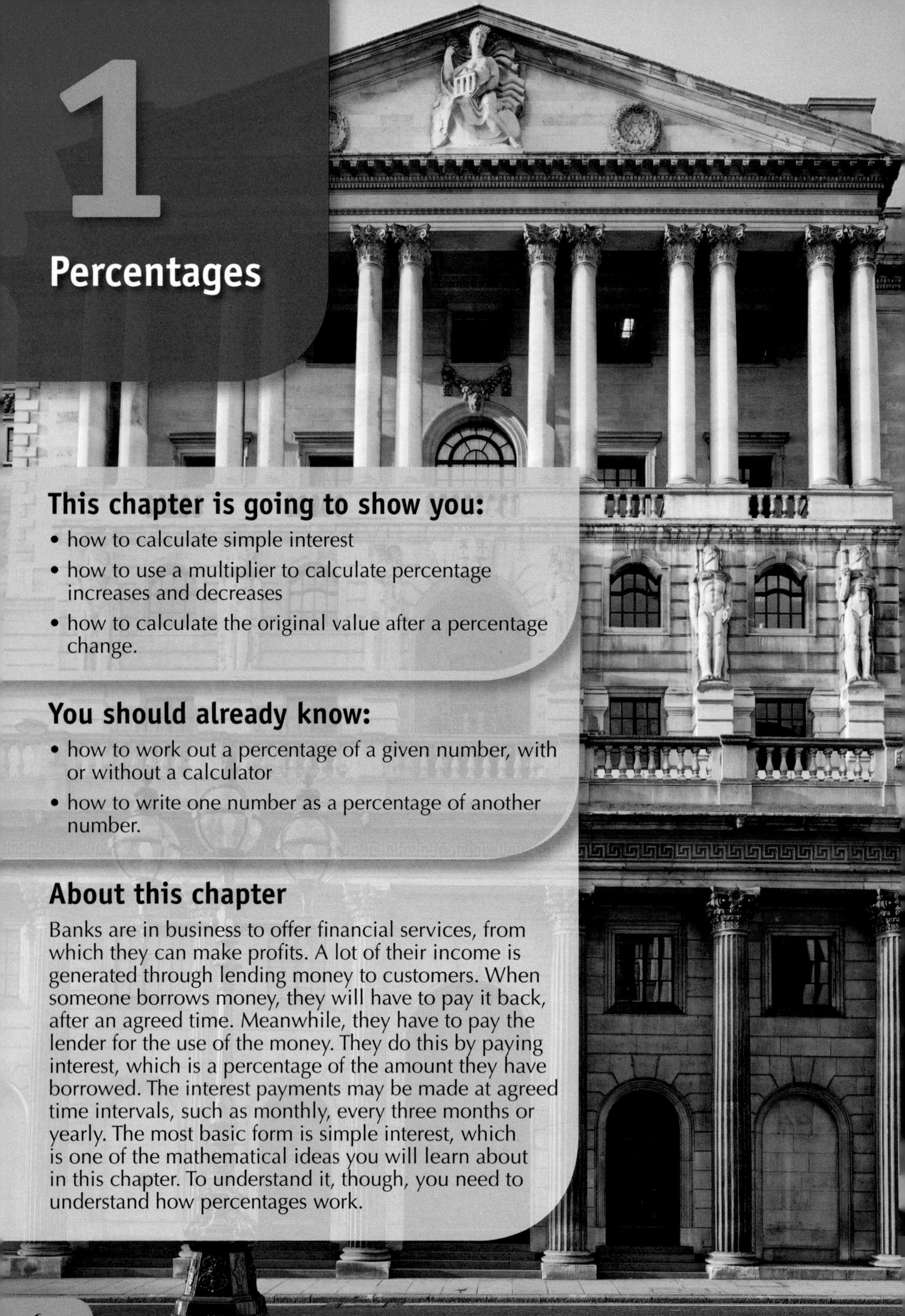

1 Percentages

This chapter is going to show you:

- how to calculate simple interest
- how to use a multiplier to calculate percentage increases and decreases
- how to calculate the original value after a percentage change.

You should already know:

- how to work out a percentage of a given number, with or without a calculator
- how to write one number as a percentage of another number.

About this chapter

Banks are in business to offer financial services, from which they can make profits. A lot of their income is generated through lending money to customers. When someone borrows money, they will have to pay it back, after an agreed time. Meanwhile, they have to pay the lender for the use of the money. They do this by paying interest, which is a percentage of the amount they have borrowed. The interest payments may be made at agreed time intervals, such as monthly, every three months or yearly. The most basic form is simple interest, which is one of the mathematical ideas you will learn about in this chapter. To understand it, though, you need to understand how percentages work.

1.1 Simple interest

Learning objectives

- To understand what simple interest is
- To solve problems involving simple interest

Key words
lender
simple interest

If you take out a loan you usually have to pay interest to the **lender**. This is the person or organisation lending you the money.

One type of interest is called **simple interest**. This is calculated as a percentage of the loan. As long as you still have the loan, you will pay the lender a percentage of the amount you borrowed, at regular intervals.

Example 1

Helen takes a loan of £550.

She agrees to repay the loan with simple interest of 8% per year.

The loan is to be repaid over 3 years.

a How much interest will she pay each year?

b How much will she have to pay back to the lender altogether?

a The interest payment is 8% of £550 per year.

1% of £550 = £5.50 — $1\% = \frac{1}{100}$ and £550 ÷ 100 = £5.50

8% of £550 = £5.50 × 8 = £44 — 8% = 8 × 1%

Each year she will have to pay £44 interest.

b She pays interest every year. So in three years she pays 3 × £44 = £132.

She will also have to pay back the loan of £550.

So the total to be paid to the lender will be £550 + £132 = £682.

Example 2

Joe takes out a loan of £6270 to buy a car.

He pays simple interest of 2% per month.

a Calculate the total amount of interest he pays in two years.

b He agrees to pay the loan back within two years, in 24 equal payments.

How much will his monthly payments be?

a 1% of £6270 = £62.70 — £6270 ÷ 100

2% of £6270 = £125.40 — £62.70 × 2

In two years he will pay total interest of £125.40 × 24 = £3009.60.

Note how large this amount is.

Loan companies often charge a lot of interest.

b The total to be paid back will be £6270 + £3009.60 = £9279.60.

Over 2 years he will make 2 × 12 = 24 monthly payments.

Each monthly repayment will be £9279.60 ÷ 24 = £386.65.

You need to be aware that small monthly interest rates will build up to a large interest rate. The loan companies rely on you thinking a small monthly interest rate is good!

Exercise 1A

1 Work out these percentages. Do not use a calculator.

a 5% of £30 **b** 10% of £35 **c** 4% of £200 **d** 7% of £300

e 3% of £400 **f** 11% of £200 **g** 8% of £300 **h** 10% of £60

2 Work out these percentages. Do not use a calculator.

a 3% of £75 **b** 7% of £25 **c** 3% of £35 **d** 5% of £43

e 4% of £520 **f** 8% of £540 **g** 4% of £120 **h** 6% of £710

3 Tom takes a loan of £600, to be paid back over the year.

He pays simple interest of 5% per year.

a Work out the amount of interest Tom pays.

b Work out the monthly repayment that Tom will pay.

4 James takes a loan of £1500, to be paid back over the year.

He pays simple interest of 4% per year.

a Work out the amount of interest James pays.

b Work out the monthly repayment that James will pay.

5 Hannah takes a loan of £4500, to be repaid over two years.

She pays 3% simple interest per year.

a Calculate how much interest she pays each year.

b Calculate how much she pays back to the lender altogether.

6 Mandy takes a loan of £730, to be paid back over three years.

She pays simple interest of 7% per year.

a Calculate the amount of interest Mandy pays each year.

b Calculate how much Mandy pays back to the lender altogether.

7 David agrees to pay 8% interest per year on a loan of £4200.

He repays the loan over four years.

How much will David have repaid the lender altogether?

8 Kirsty pays 3% interest per month on a loan of £300 over a year.

a How much interest will she have repaid over the year?

b How much will she have repaid to the lender altogether?

c What will her monthly repayments be?

9 Robbie pays 9% interest per month on a loan of £750 over a year.

a How much interest will he have repaid over the year?

b How much will he have repaid to the lender altogether?

c What will his monthly repayments be?

d How would you describe the lender of this loan? Give reasons for your answer.

10 Gabriel takes a loan of £1200 over one year and pays 1% simple interest every month.

Joshua takes a loan of £1200 over one year and pays 10% annual simple interest.

Who pays more interest, Gabriel or Joshua? Explain your answer.

11 Halina wants to take a loan of £900 to repay over three years.

She is offered two loans.

- Loan A with a simple interest rate of 0.5% per month
- Loan B with a simple interest rate of 6% per year.

Halina's friend Anna says they are both the same.

Is Anna correct? Explain your answer.

Challenge: Using a formula

This is a formula you can use to work out simple interest.

$$I = \frac{PRT}{100}$$

where I is the total interest paid

P is the initial loan

R is the annual percentage rate of simple interest

T is the number of years the loan is taken out for.

A If you take out a loan of £1000 over 3 years with an annual simple interest rate of 5%, then $P = 1000$, $R = 5$ and $T = 3$.

Use the formula to work out the total interest paid in this case.

B Work out the total simple interest paid on:

a a loan of £700 at 4% annual simple interest for 2 years

b a loan of £1500 at 6% annual simple interest for 4 years

c a loan of £8000 at 7% annual simple interest for 5 years.

1.2 Percentage increases and decreases

Learning objectives

- To calculate the result of a percentage increase or decrease
- To choose the most appropriate method to calculate a percentage change

Key words	
decrease	increase
multiplier	

A percentage change may be:

- an **increase** if the new value is larger than the original value
- a **decrease** if the new value is smaller that the original value.

There are several methods that you can use to calculate the result of a percentage change.

You could just calculate the percentage change then either add or subtract as necessary.

However, the **multiplier** method is often more efficient.

You change the percentage to a decimal multiplier, then multiply by that decimal, to calculate the result of the percentage change.

Example 3

Tomasz buys a bike for £300. A year later the value of the bicycle has increased by 20%.

Calculate the new value.

The original value was 100%.

You need to find 100% + 20% = 120% of the original price.

$$120\% = \frac{120}{100}$$
$$= 1.2$$

This is the multiplier.

The new value is £300 × 1.2 = £360.

Example 4

Shaun puts £400 in a special savings account.

At the end of each year it will earn 4% interest.

Work out the total amount in the account after 2 years.

The interest must be added to the original amount.

You need to find 100% + 4% = 104%.

104% = 104 ÷ 100 = 1.04 The multiplier is 1.04.

The total after 1 year is £400 × 1.04 = £416.

The total after 2 years is £416 × 1.04 = £432.64.

For this problem, it would be sensible to use a calculator.

Example 5

Pete buys a car for £8000.

He expects the value of the car to decrease by 10% each year.

What would the car be worth after two years?

The change each year is a reduction of 10%.

That will be 90% of the original value. $90 \div 100 = 0.9$, so this is the multiplier.

After the first year, the car will be worth £8000 × 0.9 = £7200.

After the second year, the car will be worth £7200 × 0.9 = £6480.

Peter expects that after two years the value of the car will be £6480.

Example 6

When a tree was planted, it was 100 cm tall.

Alan expects it to increase in height by 5% in each of its early years.

How tall does he expect the tree to be after two years?

After the first year, the height of the tree will have increased by 5%.

1% of 100 cm is 1 cm. $100 \div 100 = 1$

5% will be 5 cm. 5×1 cm

So after 1 year, the tree will be 105 cm tall.

After the second year, the height of the tree will again increase by 5%.

1% of 105 cm is 1.05 cm. $105 \div 100 = 1.05$

5% will be 5.25 cm. 5×1.05 cm

So after 2 years, the tree will be $105 + 5.25 = 110.25$ cm tall.

Exercise 1B

1 Work out the multiplier for each percentage increase.
The first one has been done for you.

a 4% → 100% + 4% = 104% and 104% is $104 \div 100 = 1.04$.

b 6% **c** 9% **d** 10% **e** 12%
f 15% **g** 20% **h** 35% **i** 17%

2 What are the multipliers for a decrease by the percentages below?
The first one has been done for you.

a 2% → 100% – 2% = 98% and 98% is $98 \div 100 = 0.98$.

b 3% **c** 5% **d** 8% **e** 10%
f 15% **g** 20% **h** 30% **i** 25%

3 Work out the result of increasing:

a £12 by 10% **b** £20 by 5% **c** £30 by 4% **d** £100 by 8% **e** £200 by 12%.

4 Find the result of decreasing:

a £18 by 10% **b** £30 by 3% **c** £50 by 6% **d** £150 by 10% **e** £400 by 20%.

5 Carl plants a Leylandii hedge with small plants that are all 80 cm tall.

He is told to expect the hedge to grow taller by 20% each year for the first few years.

How tall will he expect the tree to be after:

a 1 year **b** 2 years?

6 In the first few years of its life, a baby elephant increases its mass by 10% each year.

A baby elephant is born with a mass of 45 kg.

What is the elephant's expected mass after:

a 1 year **b** 2 years?

7 Tobias has a savings account that will give him 6% interest at the end of each year.

Tobias starts this account with £500.

How much is the account worth after:

a 1 year **b** 2 years?

8 A scientist is growing bacteria in a laboratory.

The bacteria will grow at the rate of 20% each day.

She starts with a mass of 2 mg of the bacteria.

How much of the bacteria will she have after:

a 1 day **b** 2 days?

9 In 2014, the population of Melchester was 12 500 people.

The population is expected to increase by 8% each year up to 2020.

What is the expected population in:

a 2015 **b** 2016?

10 In the spring of 2014, a reservoir contained approximately 500 megalitres of water.

During a time of unreliable rainfall, the reservoir was expected to lose 40% of its water each year.

How much water was expected to be in the reservoir in the spring of:

a 2015 **b** 2016?

11 Neil bought a new car at a cost of £12 000.

He was told to expect the value to drop by 15% every 6 months.

How much is the value of the car expected to be after:

a six months **b** a year?

PS **12** In the spring, house prices were expected to increase by 2% each month.

In March, Mayree was looking at a house priced at £120 000.

What might she expect the price to be in May?

MR **13** A shop increased its prices by 10% one day, but found that sales were badly affected.

So they then reduced their prices by 10%.

Read what Maya and Tom are saying.

That means the prices are back to where they were before.
I don't think so, I think they'll be lower now!

Explain why Tom is correct.

Challenge: Population change

The population of a town was increasing by 20% each year.

In 2012 the population was 4800.

A What was the population expected to be in 2013?

B Show that the population of 6912 in 2014 was expected.

C Work out the expected population in 2015.

D In what year is the population expected to be over 10 000 for the first time?

1.3 Calculating the original value

Learning objective

- Given the result of a percentage change, to calculate the original value

Key words

original value

The number of ducks in a park has increased by 10% since last year.

There are now 55 ducks. How many were there last year?

It would be incorrect to reduce 55 by 10% to get $55 \times 0.9 = 49.5$.

This is because the 10% was added to the **original value**, which is the number last year, not the number now.

The correct answer is that last year there were 50 ducks.

This is because if you increase 50 by 10% you get $50 \times 1.1 = 55$.

In this section you will learn how to answer questions like this.

Example 7

Last month, the number of visitors to a cinema increased by 20% to 2160.

What was the number of visitors before the increase?

The multiplier for the increase is 1.2.	$100\% + 20\% = 120\% = 1.2$
Original number $\times 1.2 = 2160$	The result of the increase is 2160.
Original number $= 2160 \div 1.2$	Find the original number, by dividing by 1.2.
$= 1800$	$2160 \div 1.2 = 1800$

Example 8

In a spring sale, a shop took 40% off all prices.

Rachel bought a watch with a sale price of £48.

What was the original price of the watch?

The multiplier for a 40% reduction is 0.6.	$100\% - 40\% = 60\% = 0.6$
Original price $\times 0.6 =$ £48	
Original price $= 48 \div 0.6$	Divide 48 by 0.6.
= £80	

Prices often include a tax called value-added tax (VAT). This is a percentage of the basic cost.

You can use the multiplier method to find the cost before VAT is added.

Example 9

The price of having a carpet cleaned, including 20% VAT, is £108.

Work out the price excluding VAT.

The multiplier for a 20% increase is 1.2. $100\% + 20\% = 120\% = 1.2$

(Price excluding VAT) × 1.2 = £108

Price excluding VAT = £108 ÷ 1.2

= £90

Exercise 1C

1 What are the multipliers for an increase by the given percentages?

The first one has been done for you.

a 5%

> 5% → 100% + 5% = 105%
> 105 ÷ 100 = 1.05

b 7% **c** 11% **d** 13% **e** 18%

f 25% **g** 40% **h** 55% **i** 34%

2 What are the multipliers for a decrease by the percentages below?

The first one has been done for you.

a 4%

> 4% → 100% – 4% = 96%
> 96 ÷ 100 = 0.96

b 3% **c** 5% **d** 8% **e** 10%

f 15% **g** 20% **h** 30% **i** 45%

3 In a sale, a shop reduces its prices by 20%.

Work out the original prices of these articles.

The first one has been done for you.

a £24

> A 20% reduction gives a multiplier of (100 – 20) ÷ 100 = 0.8.
> The original price × 0.8 = £24.
> The original price was £24 ÷ 0.8 = £30.

b £12.80 **c** £28.80 **d** £35.20 **e** £96

4 VAT is a tax of 20% that is added to the basic price of goods.

Work out the pre-VAT price for goods that cost these amounts.

The first one has been done for you.

a £72

> A 20% increase is a multiplier of $(100 + 20) \div 100 = 1.2$.
> The pre-VAT price $\times 1.2 = £72$.
> The pre-VAT price is $£72 \div 1.2 = £60$.

b £84 **c** £96 **d** £60.60 **e** £12.96

5 Last year, Hassan grew taller by 5%. His new height was 126 cm.

a Show that the multiplier for an increase of 5% is 1.05.

b Explain why you divide by 1.05 to get back to Hassan's height at the beginning of the year.

c How tall was Hassan at the beginning of the year?

6 During the first few months of his life, the mass of Alfie the puppy increased by 10% each month.

a Show that the multiplier for an increase of 10% is 1.1.

When Alfie was 2 months old his mass was 6.05 kg.

b Explain why you divide by 1.1 to get back to Alfie's mass when he was one month old.

c What was Alfie's mass when he was one month old?

d What was Alfie's mass when he was born?

Challenge: Trees

A fast-growing tree increases its height by 50% each year.

David planted a tree. After two years it was 2.25 metres tall.

A What height will the tree be after a further two years?

B What was the height of the tree when it was planted?

C If it continues to grow at this same rate, when would the tree be over 10 metres tall?

1.4 Using percentages

Learning objectives

- To revise the links within fractions, decimals and percentages
- To choose the correct calculation to work out a percentage

These examples show the link between fractions and decimals.

$\frac{1}{2} = 1 \div 2 = 0.5$

$\frac{1}{4} = 1 \div 4 = 0.25$ $\quad \frac{3}{4} = 3 \div 4 = 0.75$

$\frac{1}{5} = 1 \div 5 = 0.2$ $\quad \frac{2}{5} = 2 \div 5 = 0.4$ $\quad \frac{3}{5} = 3 \div 5 = 0.6$ $\quad \frac{4}{5} = 4 \div 5 = 0.8$

$\frac{1}{8} = 1 \div 8 = 0.125$ $\quad \frac{3}{8} = 3 \div 8 = 0.375$ $\quad \frac{5}{8} = 5 \div 8 = 0.625$ $\quad \frac{7}{8} = 7 \div 8 = 0.875$

$\frac{1}{10} = 1 \div 10 = 0.1$ $\quad \frac{3}{10} = 3 \div 10 = 0.3$ $\quad \frac{7}{10} = 7 \div 10 = 0.7$ $\quad \frac{9}{10} = 9 \div 10 = 0.9$

> You can change a fraction to a decimal by dividing the numerator (top number) by the denominator (bottom number).

These examples show the link between decimals and percentages.

$0.5 = 0.5 \times 100\% = 50\%$

$0.25 = 0.25 \times 100\% = 25\%$ $\quad 0.75 = 0.75 \times 100\% = 75\%$

$0.2 = 0.2 \times 100\% = 20\%$ $\quad 0.4 = 0.4 \times 100\% = 40\%$ $\quad 0.6 = 0.6 \times 100\% = 60\%$

$0.125 = 0.125 \times 100\% = 12.5\%$ $\quad 0.375 = 0.375 \times 100\% = 37.5\%$

> You can change a decimal to a percentage by multiplying the decimal by 100.

These examples show the link between fractions and percentages.

$\frac{1}{2} = 1 \div 2 \times 100 = 0.5 \times 100 = 50\%$

$\frac{1}{4} = 1 \div 4 \times 100 = 0.25 \times 100 = 25\%$ $\quad \frac{3}{4} = 3 \div 4 \times 100 = 0.75 \times 100 = 75\%$

$\frac{1}{5} = 1 \div 5 \times 100 = 0.2 \times 100 = 20\%$ $\quad \frac{2}{5} = 2 \div 5 \times 100 = 0.4 \times 100 = 40\%$

$\frac{1}{8} = 1 \div 8 \times 100 = 0.125 \times 100 = 12.5\%$ $\quad \frac{3}{8} = 3 \div 8 \times 100 = 0.375 \times 100 = 37.5\%$

$\frac{1}{10} = 1 \div 10 \times 100 = 0.1 \times 100 = 10\%$ $\quad \frac{3}{10} = 3 \div 10 \times 100 = 0.3 \times 100 = 30\%$

> You can change a fraction to a percentage by dividing the numerator (top number) by the denominator (bottom number), then multiplying the decimal by 100.

Quick changes

There is often a quick way to change a fraction to a decimal.

To do this, change the fraction to one that has 100 as its denominator.

- $\frac{1}{2} = \frac{50}{100}$ and so $\frac{1}{2} = 50\%$
- $\frac{9}{10} = \frac{90}{100}$ and so $\frac{9}{10} = 90\%$

You now know how to make calculations involving percentages in a variety of situations. This section will give you practice is choosing the right calculation to use.

Example 10

In a group, the ratio of boys to girls is 7 : 3.

What percentage of the group is made up of boys?

You can represent the situation with a diagram.

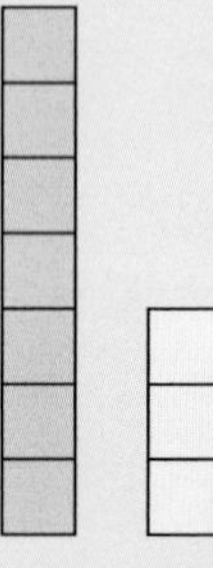

a Seven out of every ten of the group are boys. $7 + 3 = 10$

$\frac{7}{10}$ are boys.

$$\frac{7}{10} = 7 \div 10 \times 100\%$$
$$= 0.7 \times 100$$
$$= 70\%$$

Example 11

Giles reported that the probability that it would rain the next day was 0.3.

What is the probability that it would not rain? Give the answer as a percentage.

Probability of it not raining will be $1 - 0.3 = 0.7$.

Change 0.7 to a percentage.

$0.7 \times 100\% = 70\%$

Example 12

Sophia scored 17 out of 20 in a spelling test.

What percentage score did she get?

You need to change the fraction $\frac{17}{20}$ into a percentage.

If you can change the score to an equivalent fraction out of 100, then you have an easy way to find the percentage.

Change $\frac{17}{20}$ by multiplying top and bottom by 5 to give $\frac{85}{100}$.

Hence the percentage score is 85%.

Exercise 1D

1 Change each fraction to a decimal.

a $\frac{3}{4}$ **b** $\frac{4}{5}$ **c** $\frac{8}{10}$ **d** $\frac{7}{8}$

e $\frac{7}{20}$ **f** $\frac{18}{50}$ **g** $\frac{37}{100}$ **h** $\frac{4}{25}$

2 Change each decimal to a percentage.

a 0.75 **b** 0.85 **c** 0.38 **d** 0.17

e 0.7 **f** 0.9 **g** 0.155 **h** 0.755

3 Change each fraction to a percentage.

The first one has been done for you.

a $\frac{4}{5}$

$$\frac{4}{5} = 4 \div 5 \times 100\%$$
$$= 0.8 \times 100\%$$
$$= 80\%$$

b $\frac{3}{4}$ **c** $\frac{5}{8}$ **d** $\frac{27}{100}$ **e** $\frac{31}{50}$ **f** $\frac{7}{20}$ **g** $\frac{7}{8}$

4 In the boxes there are some fractions, decimals and percentages.

Match up each fraction to the equivalent decimal and percentage.

Fractions	Decimals	Percentages
$\frac{1}{2}$	0.5	50%
$\frac{1}{4}$	0.6	87.5%
$\frac{7}{8}$	0.65	70%
$\frac{3}{5}$	0.7	25%
$\frac{7}{10}$	0.25	60%
$\frac{13}{20}$	0.875	65%

5 A group of pupils took a maths challenge test. Their results are shown in the table.

Andrew	32 out of 50
Khidash	9 out of 10
Chris	12 out of 20
Shireen	18 out of 30
Eve	28 out of 40

Work out the percentage score for each pupil.

The first one has been done for you.

Andrew scored 32 out of 50

$\rightarrow \frac{32}{50} = \frac{64}{100}$ Double both numbers.

$= 64\%$

6 Monty entered a vegetable growing competition and would be happy with any scores of 80% or more.

These are the scores for his vegetables.

Onions	7 out of 10
Beetroot	19 out of 20
Leeks	35 out of 50
Cauliflower	27 out of 40
Marrow	8 out of 10

a Work out Monty's percentage score for each of his vegetables.

b Monty was not happy about some of his scores. Which ones? Explain your answer.

7 One Friday, a school produced its 'absence' figures in a table, like this.

Stage	Number absent	Number of pupils
KS3	20	500
KS4	64	250
Sixth form	3	100

a **i** Express the KS3 absences as a fraction of the number of pupils.

ii Change this fraction to a decimal.

iii Change this decimal to a percentage.

b What percentage of KS4 pupils were absent that day?

c Explain how you can state the percentage absent for the sixth form without doing any calculations at all.

 **8** Sophia came home and told her brother Aran that she had scored 17 out of 20 in a maths test.

Oh dear, that's less than 75%.
No it's not, it's more than 75%.

Who is correct? Explain how you know.

 **9** One week last year, 80 people took a driving test but only 32 passed.

Work out the percentage that passed.

 **10** At a concert there were 70 men, 60 women, 32 girls and 38 boys.

a How many were at the concert altogether?

b **i** What fraction of the audience were women?

ii What percentage of the audience were women?

c What percentage of the audience were girls?

d What percentage of the audience were male?

Challenge: Different representations

Look at this information.

In a class there are 13 boys and 12 girls.

You could show that information in a number of ways.

- 52% of the class are boys.
- The fraction of pupils in the class who are girls is $\frac{48}{100}$.

A Show that each of the statements above is correct.

B In the same school year there are 108 boys and 92 girls.

Write statements like the ones above, to show this information.

Use fractions and percentages.

C In the whole school there are 468 boys and 432 girls.

Write statements like the ones above, to show this information.

Use fractions and percentages.

Ready to progress?

I can calculate percentages.
I can change a value by a given percentage.
I can change fractions to decimals, and decimals to percentages.

I can find an original amount after a percentage change.

Review questions

1 Work out these amounts. Do not use a calculator.

a 3% of £20 b 8% of £25 c 5% of £150 d 10% of £38
e 2% of £360 f 12% of £300 g 7% of £420 h 5% of £820

2 Julie takes a loan of £800, to be paid back over the year.
She pays simple interest of 6% per year.

a Work out the amount of interest Julie pays.
b Work out the monthly repayment that Julie will pay.

3 Work out the multiplier for an increase by each percentage.
The first one has been done for you.

a 3%

3% → 100% + 3% = 103%
103 ÷ 100 = 1.03

b 5% c 8% d 11% e 14%
f 18% g 22% h 45% i 33%

4 Work out the multiplier for a decrease by each percentage.
The first one has been done for you.

a 4%

4% → 100% – 4% = 96%
96 ÷ 100 = 0.96

b 2% c 7% d 9% e 12%
f 18% g 25% h 35% i 33%

5 Work out the final amount after each increase.

a £8 by 4% b £30 by 4% c £20 by 5% d £50 by 6% e £300 by 15%

6 Work out the final amount after each decrease.

a £20 by 5% b £40 by 4% c £60 by 8% d £250 by 10% e £300 by 12%

7 In the first few years of its life, a baby giraffe increases its mass by 20% each year.
A baby giraffe is born with a mass of 25 kg.
What is the giraffe's expected mass after:

a 1 year b 2 years?

8 Change each fraction to a percentage.

The first one has been done for you.

a $\frac{3}{5}$

$\frac{3}{5} = 3 \div 5 \times 100\%$
$= 0.6 \times 100\% = 60\%$

b $\frac{1}{4}$ **c** $\frac{3}{8}$ **d** $\frac{39}{100}$

e $\frac{42}{50}$ **f** $\frac{9}{20}$ **g** $\frac{5}{8}$

9 Joe bought a new car that cost him £14 000.

He was told to expect the value to drop by 20% every 6 months.

How much is the value of the car expected to be after:

a six months **b** a year?

MR **10** Lisa pays 8% interest per month on a loan of £600 over a year.

a How much interest will she have repaid over the year?

b How much will she have repaid to the lender altogether?

c What will her monthly repayments be?

d How would you describe the lender of this loan? Give reasons for your answer.

11 A shop has a sale reduction of 10%.

Work out the original price of the articles with these sale prices.

The first one has been done for you.

a £27

A 10% reduction gives a multiplier of (100 – 10) ÷ 100 = 0.9.
The original price × 0.9 = £27.
The original price = £27 ÷ 0.9 = £30.

b £12.60 **c** £28.80 **d** £35.10 **e** £99

12 In one year, Ivan grew taller by 4%. His height at the end of the year was 130 cm.

a Show that the multiplier for an increase of 4% is 1.04.

b Explain why you divide by 1.04 to get back to Ivan's height at the beginning of the year.

c What was Ivan's height at the beginning of the year?

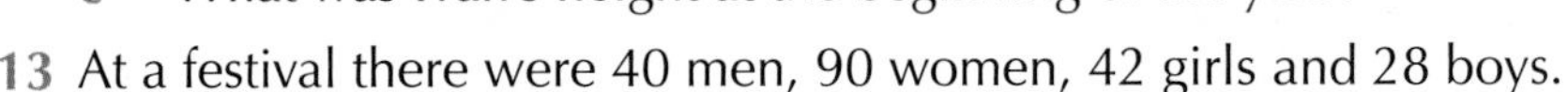

PS **13** At a festival there were 40 men, 90 women, 42 girls and 28 boys.

a How many people were at the festival altogether?

b **i** What fraction of the audience were men?

ii What percentage of the audience were men?

c What percentage of the audience were boys?

d What percentage of the audience were female?

14 VAT is a tax of 20% that is added to the basic price of many goods.

Work out the price of each article, before VAT is added.

The first one has been done for you.

a Shoes costing £60

A 20% increase gives a multiplier of (100 + 20) ÷ 100 = 1.2.
The pre-VAT price × 1.2 = £60.
The pre-VAT price is £60 ÷ 1.2 = £50.

b Trousers costing £36 **c** A coat costing £48

d Boots costing £72 **e** A shirt costing £24.36

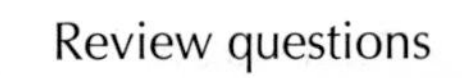

Challenge

The Royal Albert Hall

The Royal Albert Hall is a familiar London landmark. It was founded in 1871.

During 2013, 1 700 000 people attended 390 different events at the Royal Albert Hall.

Their revenue increased by 10% in 2012, to a staggering £18.6 million.

During this year there was the Dr Who Prom as well as concerts by Eric Clapton and Bob Dylan.

During the period from 2012 to 2016 it is being refurbished. The work takes place mainly overnight, between 11 pm and 8 am each day. This is before the event teams need to be there – and after they have gone home.

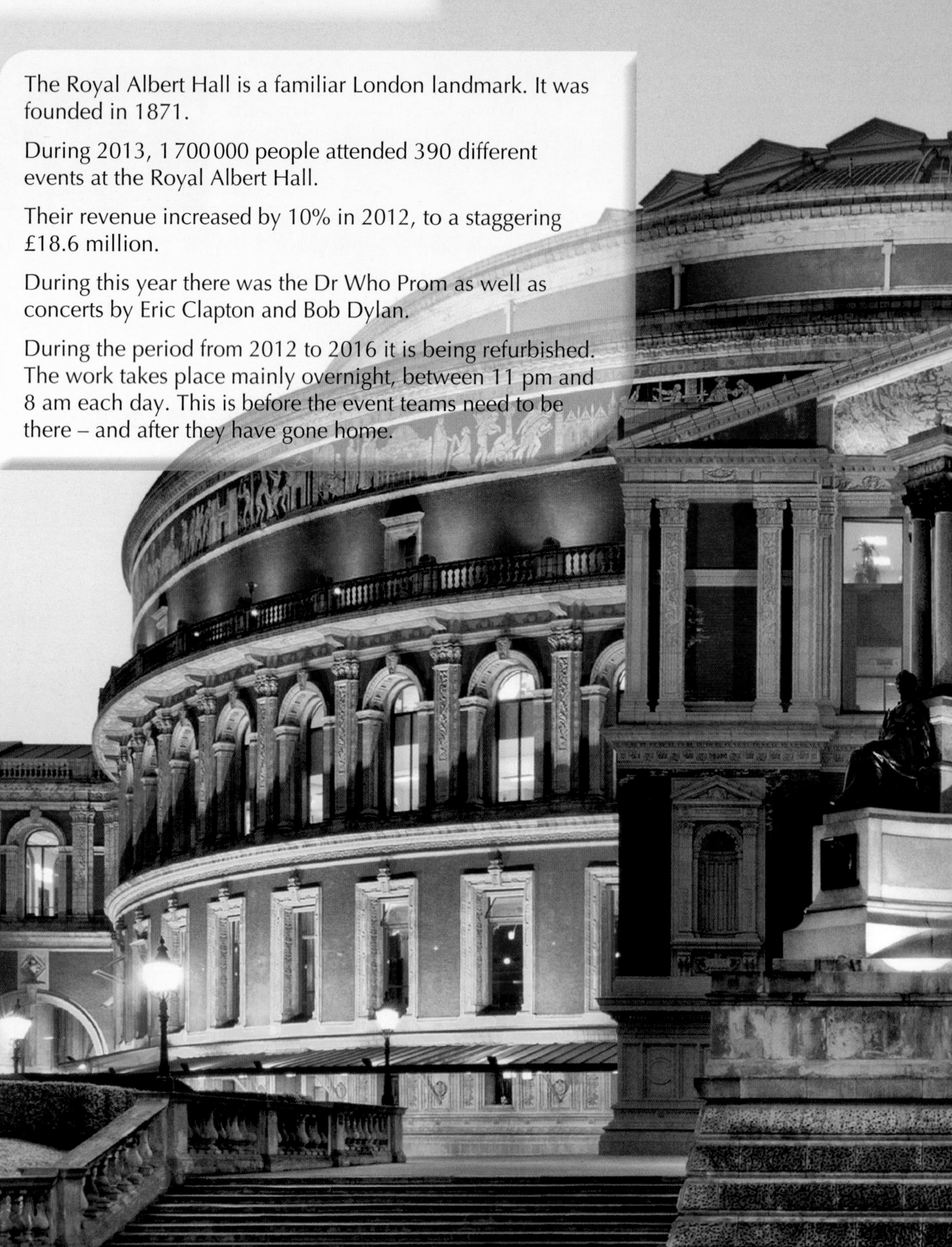

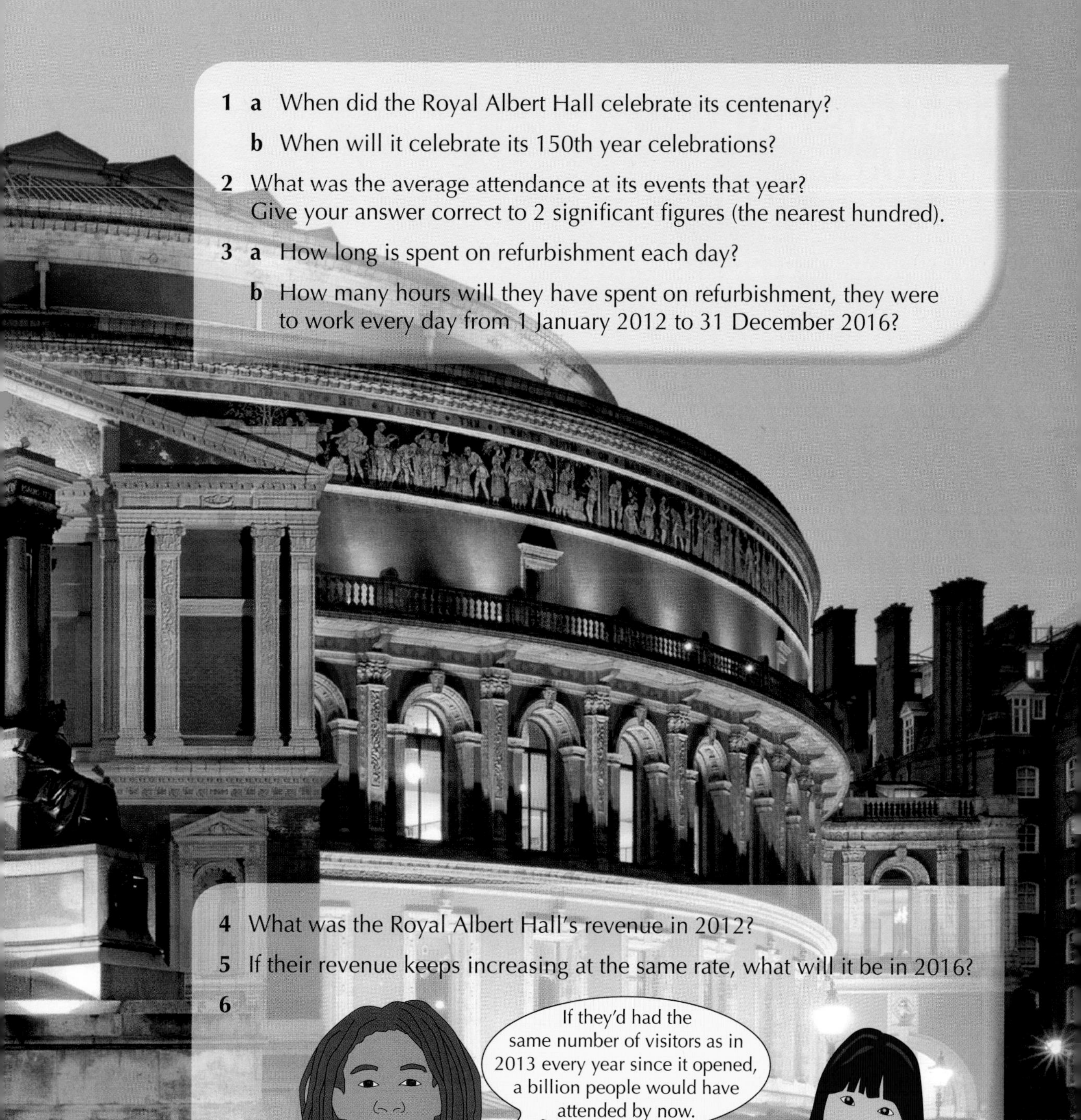

1 **a** When did the Royal Albert Hall celebrate its centenary?

b When will it celebrate its 150th year celebrations?

2 What was the average attendance at its events that year?
Give your answer correct to 2 significant figures (the nearest hundred).

3 **a** How long is spent on refurbishment each day?

b How many hours will they have spent on refurbishment, they were to work every day from 1 January 2012 to 31 December 2016?

4 What was the Royal Albert Hall's revenue in 2012?

5 If their revenue keeps increasing at the same rate, what will it be in 2016?

6

If they'd had the same number of visitors as in 2013 every year since it opened, a billion people would have attended by now.
No they wouldn't.

Who's right, James or Helen? Explain how you know.

2

Equations and formulae

This chapter is going to show you:

- how to expand brackets and factorise algebraic expressions
- how to solve equations
- how to use formulae.

You should already know:

- how to collect like terms in an expression
- how to use one or two operations to solve equations
- how to substitute values into a formula
- what a highest common factor (HCF) is.

About this chapter

During World War 2, it was vital for all sides to keep up with communications, and to try to find out what each other were doing. One very important factor, for Britain and its allies, was the work done at Bletchley Park, where mathematicians worked to break the codes used to send messages. To do this, they had to be able to use algebraic rules to solve formulae. Algebraic rules are still used by computer programmers today.

The picture shows *Colossus*, the machine developed through the work of Alan Turing, the famous code-breaker. He is acknowledged to be a pioneer of theoretical computer science and artificial intelligence. *Colossus* is now considered to be the first digital computer.

In this chapter, you will learn about some of the basic ideas that were used by Alan Turing and his colleagues.

2.1 Multiplying out brackets

Learning objective

- To multiply out brackets

Key word

expand

You already know how to multiply out a term that includes brackets. If there is a number outside the brackets you multiply each term inside the brackets by that number.

This is also called **expanding** brackets.

For example, $4(7 + 3)$ is the same as $4 \times 7 + 4 \times 3$.

$4(7 + 3) = 4 \times 10 = 40$

$4 \times 7 + 4 \times 3 = 28 + 12 = 40$

Both calculations give the same answer.

This also works if you replace some of the numbers with letters.

Example 1

Multiply out the brackets in these expressions.

a $3(x + 4)$ **b** $2(y - 4)$ **c** $2(3t + 5)$

a $3(x + 4) = 3x + 12$ $3 \times x = 3x$ and $3 \times 4 = 12$

b $2(y - 4) = 2y - 8$ $2 \times y = 2y$ and $2 \times -4 = -8$

c $2(3t + 5) = 6t + 10$ $2 \times 3t = 6t$ and $2 \times 5 = 10$

Example 2

Write down an expression for the area of the rectangle.

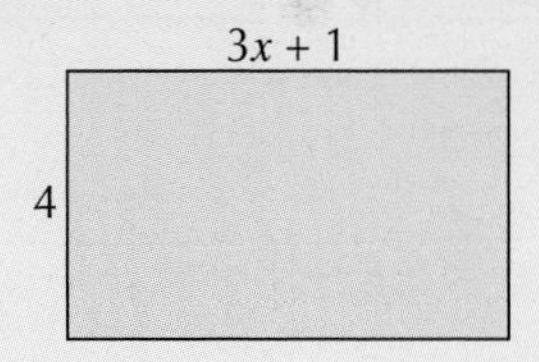

You can find the area by multiplying the two different sides.

Area $= 4 \times (3x + 1)$

Then write this more simply as $12x + 4$.

Exercise 2A

1 Simplify each of these expressions.

a $3x + 4x$ **b** $4a + 3a$ **c** $7t + t$ **d** $4y + y + 3y$

e $8m - 2m$ **f** $7k - 4k$ **g** $5n - n$ **h** $3p - 7p$

2 Simplify each expression.

a $6m + m + 3m$ **b** $2y + 4y + y$ **c** $6t + 2t + t$ **d** $5p + 2p + 4p$

e $6n + 2n + 5n$ **f** $5p + 3p + p$ **g** $4t - t + 3t$ **h** $4e - 2e + 5e$

i $7k + 2k - 3k$ **j** $6h + h - 2h$ **k** $9m - 3m - m$ **l** $5t + 3t - 2t$

3 Write down the perimeter, P, of each shape.

a
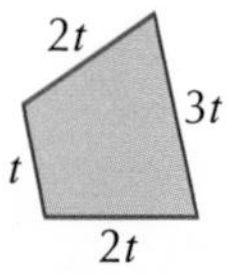

b
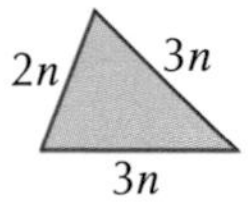

c
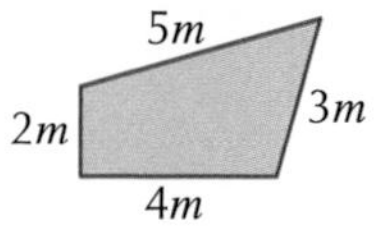

d
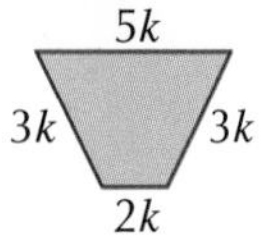

e
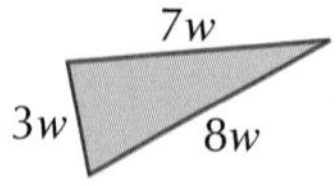

f
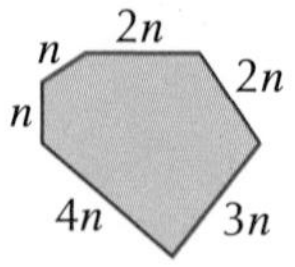

4 Simplify each expression.

a $4b + 3 + b$ **b** $5x + 6 + 2x$ **c** $q + 3 + 5q$ **d** $5k + 2k + 7$
e $4x + 5 - 2x$ **f** $7k + 3 - k$ **g** $5p + 1 - 3p$ **h** $8d + 2 - 5d$
i $6m - 2 - 4m$ **j** $5t - 3 - 3t$ **k** $5w - 7 - 2w$ **l** $6g - 5 - 2g$
m $2t + k + 5t$ **n** $4x + 3y + 5x$ **o** $3k + 2g + 4k$ **p** $5h + w + 3w$
q $7t + 3p - 4t$ **r** $8n + 3t - 6n$ **s** $p + 5q - 4q$ **t** $4n + p - 2n$

5 Write each expression as simply as possible.

a $2 \times 4x$ **b** $4 \times 3a$ **c** $5 \times 2t$ **d** $3 \times 2y$ **e** $6 \times 4k$
f $3t \times 5$ **g** $2x \times 6$ **h** $4m \times 3$ **i** $6t \times 2$ **j** $5y \times 7$

6 Multiply out the brackets.

a $3(t + 4)$ **b** $3(x + 5)$ **c** $2(m - 3)$ **d** $4(k - 2)$
e $2(3 + x)$ **f** $3(4 - k)$ **g** $4(6 - y)$ **h** $5(3 - x)$

7 Write down an expression for the area, A, of each rectangle.
The first one has been done for you.

a
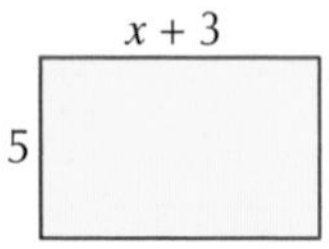

b
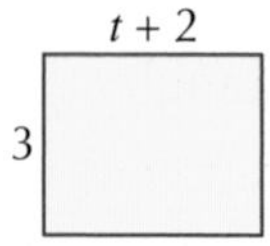

c
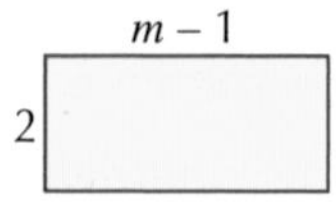

d
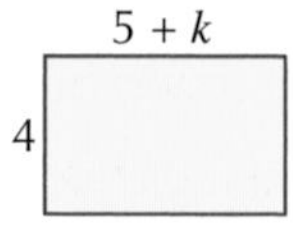

$A = 5(x + 3)$
$A = 5x + 15$

8 Multiply out the brackets.

a $2(m + 3)$ **b** $3(k - 4)$ **c** $3(a + 2)$ **d** $5(3 - p)$
e $2(3x + 4)$ **f** $5(2x + 3)$ **g** $4(2t - 1)$ **h** $5(4m + 7)$
i $3(2x + 1)$ **j** $4(3k - 2)$ **k** $2(5b + 3)$ **l** $7(2 - 4m)$
m $8(3 + p)$ **n** $5(4 - t)$ **o** $6(2 - 3g)$ **p** $8(2 + 3t)$
q $9(2k - 6)$ **r** $5(2m + 3)$ **s** $3(3t - 2)$ **t** $2(3 - 4y)$

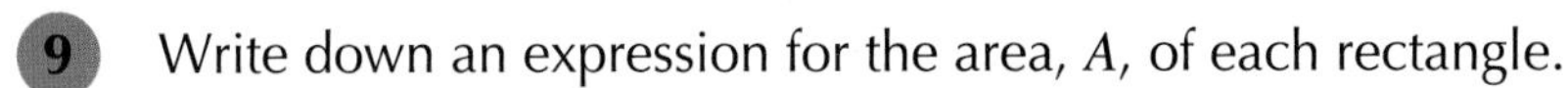

9 Write down an expression for the area, A, of each rectangle.

a

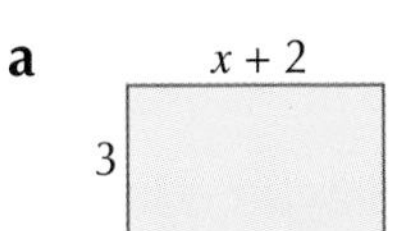

b

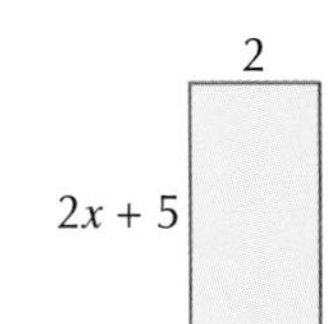

c $3m + 4$ by 5

d

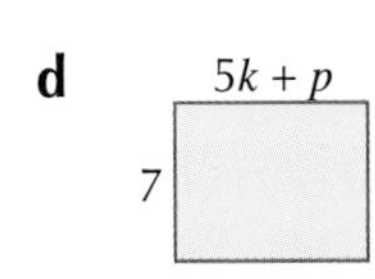

e

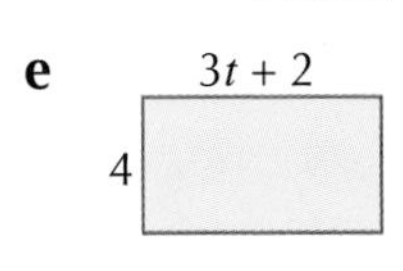

f

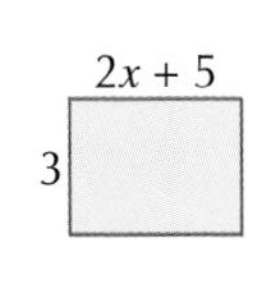

10 Show that $3(4x - 2)$ is the same as $6(2x - 1)$.

11 Leon and Aiden are talking about their homework.
Explain why Aiden is wrong.

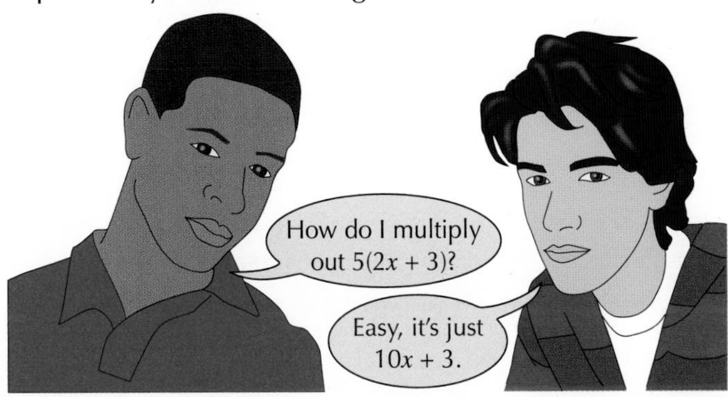

How do I multiply out 5(2x + 3)?
Easy, it's just 10x + 3.

Challenge: Code breaker

A Write down any three-digit number in which the first and last digits have a difference of more than one, for example, 472 or 513.

B Reverse the order of the digits (for the examples above, 274 and 315).

C Subtract the smaller number from the larger number.

D Reverse the digits of the answer to part **C** and add this number to the answer to part **C**.

E Multiply the answer by one million.

F Subtract 733 361 573.

G Then:

- under each 2 in your answer, write the letter P
- under each 3, write the letter L
- under each 4, write the letter R
- under each 5, write the letter O
- under each 6, write the letter F
- under each 7, write the letter A
- under each 8, write the letter I.

H Now read your letters backwards.

2.2 Factorising algebraic expressions

Learning objective

- To factorise expressions

Key word

factorise

To **factorise** an expression, you look for common factors. This is the opposite of multiplying out brackets. When you have found a factor that is common to all the terms, you take it outside the brackets.

- If you multiply out $4(x - 5)$ you get $4x - 20$.
- If you factorise $4x - 20$ you get $4(x - 5)$.

In this case, 4 is a common factor of $4x$ and 20 so you can put it outside the brackets.

Example 3

Factorise each expression. **a** $10x + 15$ **b** $8k - 12$

a The HCF of $10x$ and 15 is 5.

$10x + 15 = 5(2x + 3)$ Write the 5 outside the brackets.

The numbers inside are $10 \div 5 = 2$ and $15 \div 5 = 3$.

b The factors of $8k$ and 12 are 2 and 4. You need the HCF, which is 4.

$8a - 12 = 4(2a - 3)$ Divide 8 and 12 by 4.

Look again at part **b** of the example.

You could take out 2 as a factor and write $8a - 12 = 2(4a - 6)$.

This is not factorised completely because 4 and 6 still have a common factor of 2.

To factorise the expression completely you must take out the HCF.

Exercise 2B

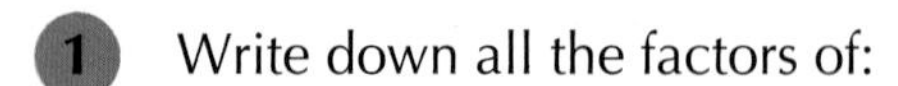

1 Write down all the factors of:

a 12 **b** 15 **c** 18 **d** 20

e 24 **f** 10 **g** 8 **h** 30.

2 Look at the numbers in the box.

Write down the numbers that are:

a factors of 12 **b** multiples of 5 **c** even numbers

d factors of 35 **e** odd numbers **f** factors of 40.

2 3 5 10
12 20 11
4 7 8 18
21 24 35

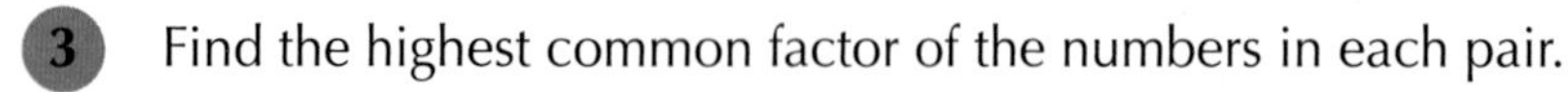

3 Find the highest common factor of the numbers in each pair.

a 4 and 6 **b** 9 and 12 **c** 12 and 18 **d** 12 and 30

4 Copy and complete each expansion.

a $\square(x + 2) = 3x + 6$ **b** $\square(t + 3) = 2t + 6$ **c** $\square(n + 2) = 4n + 8$

d $\square(q + 4) = 2q + 8$ **e** $\square(x - 3) = 3x - 9$ **f** $\square(p - 1) = 4p - 4$

g $\square(y - 2) = 5y - 10$ **h** $\square(t - 4) = 3t - 12$ **i** $\square(4 + x) = 8 + 2x$

j $\square(3 + k) = 12 + 4k$ **k** $\square(2 - t) = 12 - 6t$ **l** $\square(5 - k) = 15 - 3k$

5 Copy and complete each expansion.

a $3(\ldots) = 3t + 9$ **b** $2(\ldots) = 2m + 4$ **c** $5(\ldots) = 5p + 5$

d $4(\ldots) = 4m + 12$ **e** $6(\ldots) = 6k - 18$ **f** $3(\ldots) = 3n - 6$

g $2(\ldots) = 2x - 8$ **h** $3(\ldots) = 3q - 15$ **i** $5(\ldots) = 10 + 5x$

j $4(\ldots) = 16 + 4h$ **k** $3(\ldots) = 12 - 3t$ **l** $6(\ldots) = 18 - 6k$

6 Write down the missing length for each rectangle.

a Top: $x + 2$; Area $= 4x + 8$; Side: ?

b Top: ?; Area $= 6t + 3$; Side: 3

c Top: $2t - 3$; Area $= 4t - 6$; Side: ?

d Top: ?; Area $= 15 - 10y$; Side: 5

7 Look at Maizy's homework.

a $4x + 8 = 2(2x + 4)$ b $6t + 12 = 3(2t + 4)$

c $12 - 8p = 2(6 - 4p)$ d $20 - 16t = 2(10 - 8t)$

She has factorised each of the expressions, but not completely.

Complete the factorisations for her.

8 Copy and complete each factorisation.

a $4t + 6 = 2(\ldots)$ **b** $6x + 9 = \square(2x + 3)$ **c** $8t + 6 = 2(\ldots)$

d $9x + 6 = \square(3x + 2)$ **e** $9x - 3 = \square(3x - 1)$ **f** $10t + 5 = 5(\ldots)$

g $8x + 4 = \square(2x + 1)$ **h** $12t + 9 = 3(\ldots)$ **i** $12t + 8 = 4(\ldots)$

j $8x + 2 = \square(4x + 1)$ **k** $15t + 12 = 3(\ldots)$ **l** $9x - 6 = \square(3x - 2)$

9 Factorise each expression as much as possible.

a $16x + 10$ **b** $14x - 7$ **c** $15y + 25$ **d** $10y - 5$

e $15m - 18$ **f** $8t + 20$ **g** $12t - 8$ **h** $12 + 16k$

i $10 - 12y$ **j** $30 - 6m$ **k** $35 + 10k$ **l** $21q + 14$

  **10** Yienia has asked Bethan for help with her homework.

How can I factorise $12p + 8$?
That's easy, it's $2(6p + 4)$.

a Explain why Bethan is not quite correct.

b Explain why Bethan is also partly correct.

Investigation: Interesting numbers

A Write down any three different, whole numbers that are smaller than ten, for example, 2, 5 and 8.

B Add up these three numbers. Call this total x.

C Make all the six possible two-digit numbers from these three different numbers, for example, 25, 28, 52, 58, 82 and 85.

D Add up all six numbers. Call this total y.

E Divide y by x and write down the answer.

F **a** Repeat this for other sets of three different whole numbers that are smaller than ten.

b What do you notice?

2.3 Equations with brackets

Learning objective

- To solve equations with one or more sets of brackets

You can already solve simple equations with brackets.

Example 4

Solve the equation $3(x - 4) = 15$ by first:

a multiplying out the brackets **b** dividing by 3.

a $3(x - 4) = 15$

$3x - 12 = 15$ Multiply x and 4 by 3.

$3x = 27$ Add 12 to both sides.

$x = 9$ Divide both sides by 3.

b $3(x - 4) = 15$

$x - 4 = 5$ As 3 is outside the brackets, divide both sides by 3.

$x = 9$ Add 4 to both sides.

You can use either method to solve equations like these.

If there is more than one set of brackets, it is usually easier to multiply them both out first.

Exercise 2C

1 Solve each of these equations.

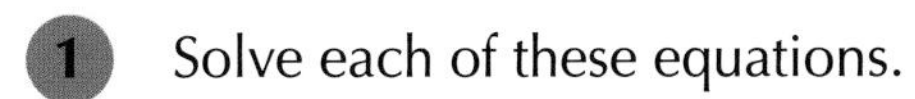

a $m + 7 = 11$ **b** $y - 2 = 10$ **c** $k + 10 = 6$ **d** $n - 3 = 12$

e $k - 6 = 2$ **f** $x + 4 = 15$ **g** $y - 5 = 3$ **h** $t - 8 = 1$

2 Solve each equation.

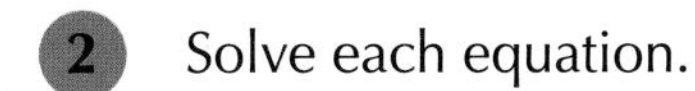

a $2(t + 3) = 12$ **b** $3(t - 5) = 15$ **c** $3(m - 5) = 12$ **d** $5(x + 3) = 15$

e $2(y + 7) = 16$ **f** $2(p - 6) = 8$ **g** $4(t - 2) = 16$ **h** $6(k - 4) = 18$

i $4(q - 3) = 20$ **j** $2(t + 3) = 8$ **k** $6(m - 7) = 24$ **l** $2(g - 5) = 10$

m $3(t - 6) = 27$ **n** $8(n - 7) = 32$ **o** $3(y + 5) = 21$ **p** $3(q - 4) = 30$

3 Solve these equations.

Hint Some of the solutions are negative.

a $m + 3 = 2$ **b** $t + 7 = 5$ **c** $n + 5 = 4$ **d** $q + 3 = 6$

e $2(t + 6) = 8$ **f** $2(k + 5) = 6$ **g** $4(p + 8) = 12$ **h** $6(t + 6) = 24$

i $4(a + 5) = 16$ **j** $2(t + 8) = 14$ **k** $6(h + 6) = 30$ **l** $2(p + 4) = 16$

m $3(d + 8) = 12$ **n** $8(x + 9) = 40$ **o** $3(t + 6) = 12$ **p** $5(m + 9) = 25$

4 The rule for the perimeter of this rectangle is:

perimeter $= 2(x + 5)$

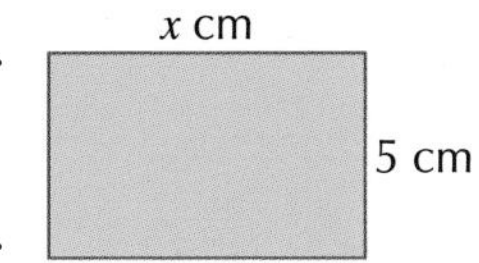

a **i** Write an equation involving x, when the perimeter is 24 cm.

ii Solve your equation.

b **i** Write an equation involving x, when the perimeter is 36 cm.

ii Solve your equation.

c Suppose the area of the rectangle is 75 cm^2.

i Write an equation involving x.

ii Solve your equation.

d What would the perimeter of the rectangle be if its area was 80 cm^2?

5 The rule for the area of this shape is:

area $= 5(8 + t)$

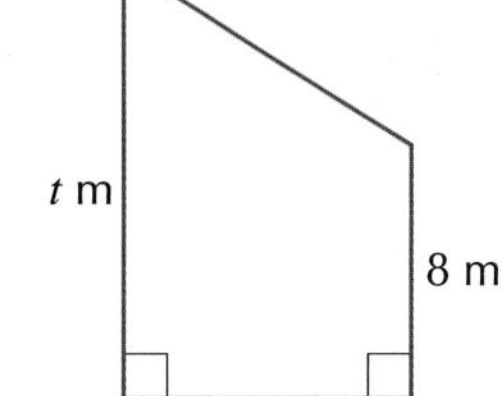

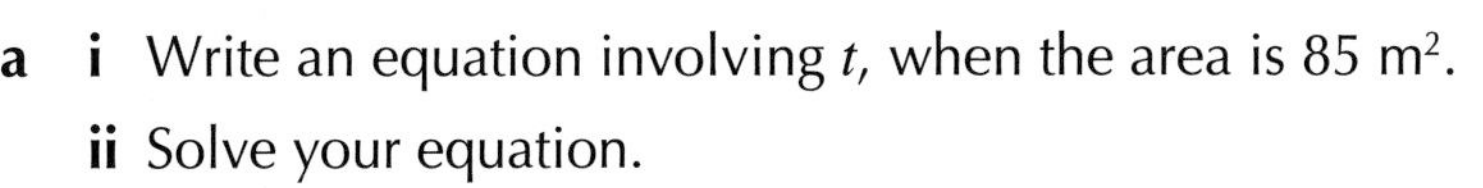

a **i** Write an equation involving t, when the area is 85 m^2.

ii Solve your equation.

b **i** Write an equation involving t, when the area is 90 m^2.

ii Solve your equation.

c Suppose the area of the trapezium is 75 m^2.

i Write an equation involving t.

ii Solve your equation.

d What would the value of t in this trapezium be, if the area was 100 m^2?

6 The rule for the sum of the interior angles of a polygon with n sides is:

angle sum = $180°(n - 2)$

a **i** Write down an equation involving n, when the angle sum of a polygon is 180°.

ii Solve your equation.

iii What is the name of a polygon with an angle sum of 180°?

b **i** Write an equation involving n, when the angle sum is 1080°.

ii Solve your equation.

iii What is the name of a polygon with an angle sum of 1080°?

c Suppose the angle sum of a polygon is 1260°.

i Write an equation involving n.

ii Solve your equation.

iii What is the name of a polygon with an angle sum of 1260°?

d How many sides has a polygon with an angle sum of 1800°?

Challenge: Primes from primes

The prime numbers, larger than 2, up to 20 are 3, 5, 7, 11, 13, 17 and 19.

A Joe said: 'Two prime numbers larger than 2 can never add together to make another prime number.' Is Joe correct? Explain your answer.

B Can the sum of three prime numbers make another prime number? Explain your answer.

2.4 Equations with fractions

Learning objective

- To solve equations involving fractions

If there is a fraction in an equation, you can remove it by multiplying the whole equation by the denominator of the fraction.

Example 5

Solve the equation $\frac{1}{2}x = 8$.

$\frac{1}{2}x = 8$

$x = 16$ Multiply both sides by 2. $8 \times 2 = 16$

Example 6

Solve the equation $\frac{x}{3} = 5$.

$\frac{x}{3} = 5$

$x = 15$ Multiply both sides by 3. $5 \times 3 = 15$

Example 7

Solve the equation $\frac{a-3}{4} = 6$.

$\frac{a-3}{4} = 6$

$a - 3 = 24$ Multiply both sides by 4. $6 \times 4 = 24$

$a = 27$ Add 3 to 24.

Exercise 2D

1 Solve each of these equations.

a $2t = 10$ **b** $3m = 12$ **c** $4y = 16$ **d** $5p = 55$

e $3x = 15$ **f** $5q = 20$ **g** $6n = 48$ **h** $7a = 21$

i $3h = 36$ **j** $7n = 42$ **k** $8x = 32$ **l** $2q = 18$

2 Solve each equation.

a $\frac{x}{2} = 10$ **b** $\frac{x}{9} = 3$ **c** $\frac{x}{3} = 4$ **d** $\frac{x}{5} = 5$

e $\frac{x}{2} = 4$ **f** $\frac{x}{4} = 2$ **g** $\frac{x}{3} = 7$ **h** $\frac{x}{4} = 5$

3 Solve these equations.

a $\frac{1}{2}(t + 3) = 5$ **b** $\frac{1}{3}(x - 2) = 2$ **c** $\frac{1}{5}(m + 5) = 3$ **d** $\frac{1}{6}(x - 1) = 1$

e $\frac{1}{4}(k + 5) = 1$ **f** $\frac{1}{3}(t - 5) = 4$ **g** $\frac{1}{6}(x + 3) = 2$ **h** $\frac{1}{8}(y - 4) = 3$

4 Solve these equations. The first one has been done for you.

a $\frac{t+3}{3} = 5$ **b** $\frac{x-2}{3} = 2$ **c** $\frac{m+5}{4} = 3$ **d** $\frac{t-1}{5} = 1$

so $t + 3 = 15$
$t = 12$

e $\frac{k+5}{8} = 1$ **f** $\frac{t-5}{10} = 4$ **g** $\frac{x+3}{4} = 2$ **h** $\frac{y-4}{5} = 3$

5 Which of these equations is the odd one out?

a $\frac{x}{2} = 3$ **b** $x - 2 = 4$ **c** $3(x + 2) = 24$ **d** $\frac{x+9}{3} = 1$

 Kathy thinks of a number.

She divides it by three.

Call Kathy's number n.

a Write down an expression for the number Kathy has worked out in her head.

Kathy says that her answer is seven.

b Write down an equation involving n and Kathy's answer.

c Solve the equation.

  Chris thinks of a number.

He adds two to the number.

He divides this answer by five.

Call Chris's number n.

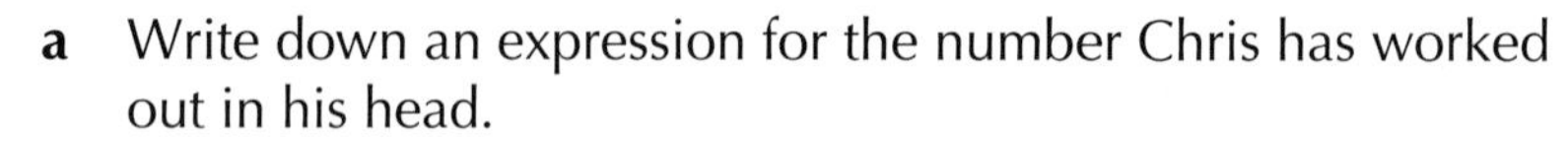

a Write down an expression for the number Chris has worked out in his head.

Chris says that his answer is three.

b Write down an equation involving n and Chris's answer.

c Solve the equation.

d If Chris had ended up with six, what number would he have started with?

  Explain why $\frac{1}{5}(x + 13) = 2$ and $\frac{x + 13}{5} = 2$ have the same solution.

Mathematical reasoning: Making equations

Lewis makes up an equation like this.

Start with the answer of 5.	$x = 5$
Add 2 to both sides.	$x + 2 = 7$
Multiply both sides by 3.	$3(x + 2) = 21$
Divide both sides by 7.	$\frac{3(x + 2)}{7} = 3$

A Solve Lewis's equation in the usual way, to check that the solution is $x = 5$.

B **a** Make up an equation by following these steps.

Start with an answer of 9.

Subtract 5 from both sides.

Multiply both sides by 2.

Divide both sides by 4.

Remember to make sure you work with integers at each stage.

b Solve your equation in part **a** to check that it gives the correct answer.

C Use the same method to make up an equation of your own.

Give it to someone else to solve.

2.5 Formulae

Learning objective

- To practise using formulae

Key words	
formula	subject
variable	

A **formula** is a rule used to work out a value from one or more **variables**.

For example, $A = ab$ is a rule, or formula, used to calculate the area, A, of a rectangle from the lengths, a and b, of two adjacent sides.

A formula also always has a **subject**, which is usually written on the left-hand side of the equals sign. For example:

$P = 2a + 2b$

Note: **This is the formula for the perimeter of a rectangle.**

The subject is P. The variables are a and b.

When a is 3 cm and b is 5 cm, the formula becomes:

$P = 2 \times 3 + 2 \times 5$

$= 16$ cm

Exercise 2E

1 You can use the formula $C = 3D$ to calculate an approximation for the circumference, C, of a circle from its diameter, D. Use the formula to calculate the approximate circumference of each of these circles.

a

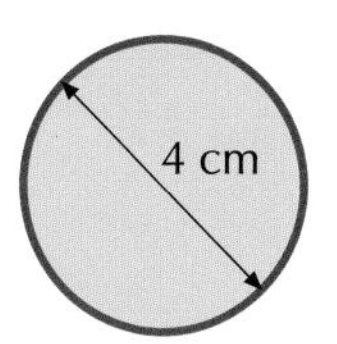

b

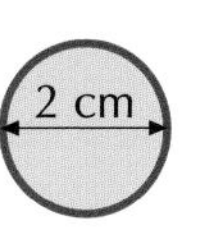

c

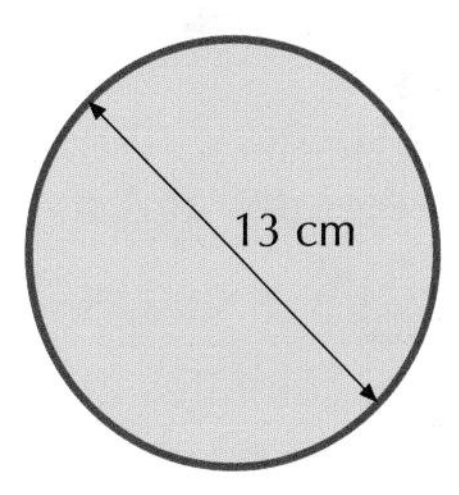

2 You can use the formula $A = 180n - 360$ to calculate the sum of the angles inside a polygon with n sides. Use the formula to calculate the sum of the angles inside each of these polygons.

a Pentagon, five sides

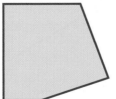

b Hexagon, six sides

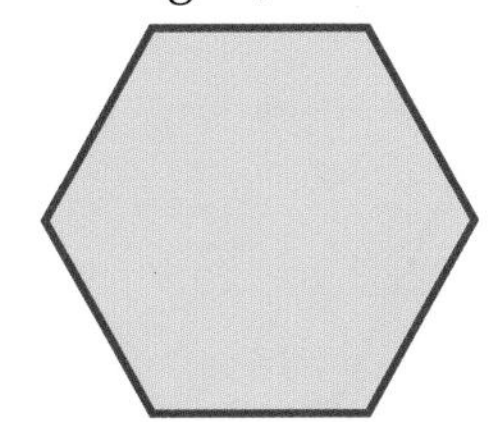

3 The cost, C, of placing an advertisement in a local newspaper is given by:

$C = £20 + £2N$

where N is the number of words used in the advertisement.

What is the cost of placing advertisements with:

a 12 words **b** 25 words?

4 Lennie, the driving instructor, used this formula to charge learner drivers for lessons:

$C = £4 + £13H$

where H is the number of hours in the driving lesson.

What is the cost of driving lessons lasting:

a 2 hours **b** from 1:00 pm to 4:00 pm?

5 The amount of money, M, that a charity can expect to collect is approximated by the formula:

$M = £(5000T + 20C)$

where T is the number of TV advertisements appearing on the day before a charity event is held, and C is the number of collectors.

Approximately, how much might each of these charities expect to collect?

a NCS had three TV advertisements and 100 collectors.

b TTU had two TV advertisements and 300 collectors.

c BCB had no TV advertisements and 500 collectors.

6 The speed, S m/s, of a rocket can be found from the formula $S = AT$, where the rocket accelerates at A m/s^2 for T seconds.

Find the speed of a rocket when the rocket:

a accelerates at 25 m/s^2 for 8 seconds

b accelerates at 55 m/s^2 for 6 seconds.

7 You can use the formula $a = \frac{1}{2}bh$ to calculate the area, a, of a triangle from its base length, b, and its height, h. Use the formula to calculate the area of each of these triangles.

a

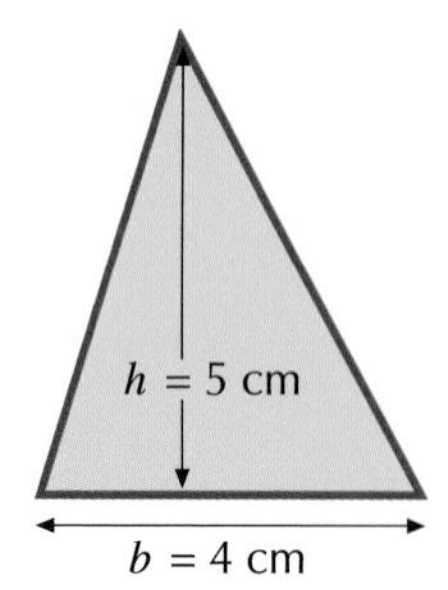

b

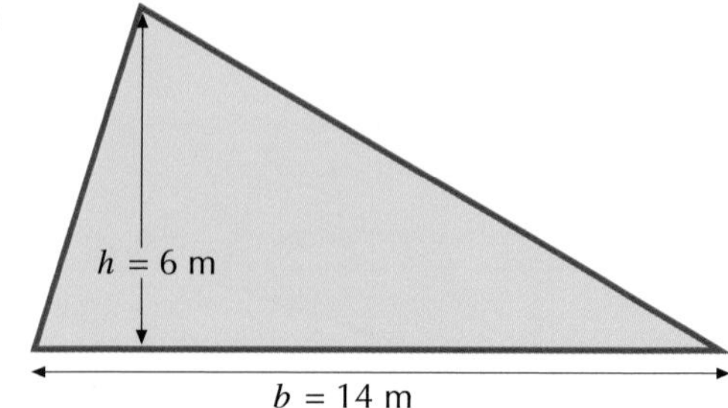

8 To calculate the cost, C, of his gigs, Disco Den uses the formula:

$$C = £55 + £3N + £5T + £10E$$

where:

N is the number of people attending the gig

T is the number of hours worked before midnight

E is the number of hours worked after midnight.

Calculate the cost of a gig for:

a 60 people from 9:00 pm to 2:00 am

b 40 people from 7 pm to 1:00 am.

Activity: Using a flowchart

Use this flowchart to work through the function $x \rightarrow \frac{3x + 5}{4}$.

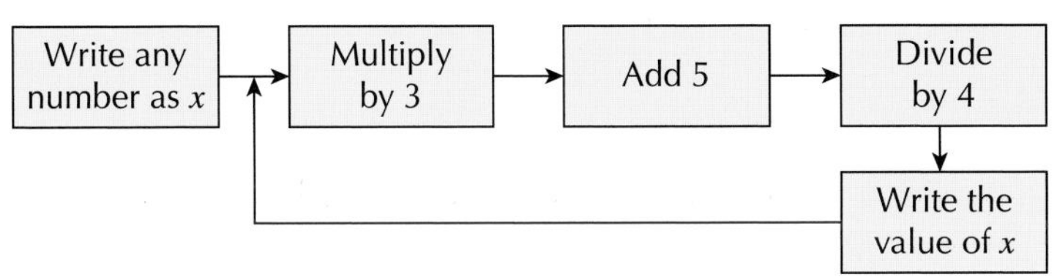

A Start with $x = 9$. Show that working through the flowchart twice gives these results.

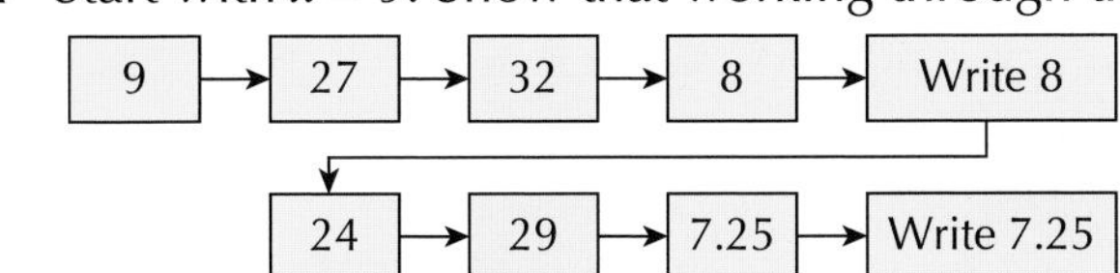

B Continue to work through the flowchart at least 8 more times.

C What do you notice?

D Does the value of the starting number make any difference?

Ready to progress?

I can expand brackets.

I can factorise simple algebraic expressions.
I can substitute into simple formulae.

I can solve equations that have brackets or fractions or both.

Review questions

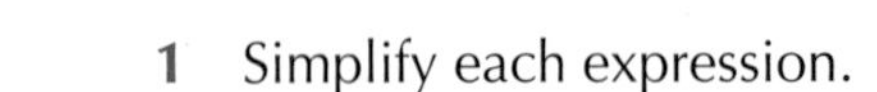

1 Simplify each expression.

a	$5p + p + 2p$	**b**	$3x + 5x + x$	**c**	$7q + 3q + q$	**d**	$6t + 3t + 2t$
e	$5n + 3n + 4n$	**f**	$3p + 5p + p$	**g**	$3m - m + 4m$	**h**	$5a - 3a + 4a$
i	$8h + 3h - 4h$	**j**	$5g + g - 3g$	**k**	$8n - 4n - n$	**l**	$6t + 4t - 5t$

2 Write down the perimeter, P, of each shape.

a

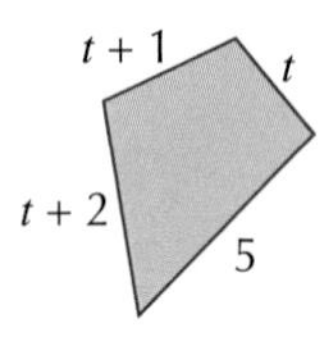

b

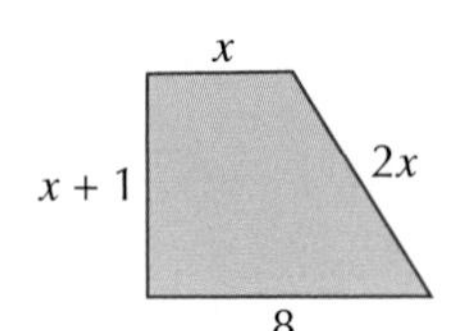

c

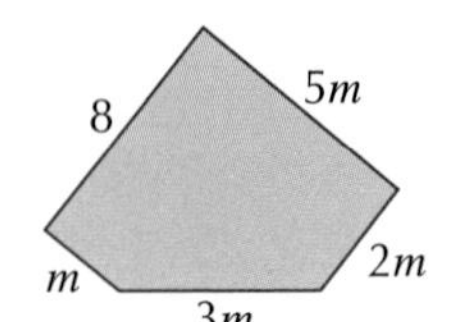

3 Look at the numbers in the box.

2	3	5	10
15	18	25	
4	7	8	9
31	36	49	

Write down numbers that are:

a factors of 15 **b** multiples of 3 **c** square numbers
d prime numbers **e** factors of 60 **f** factors of 100.

4 Solve each of these equations.

a	$t + 8 = 12$	**b**	$x - 5 = 11$	**c**	$m + 6 = 13$	**d**	$p - 5 = 8$
e	$t - 7 = 3$	**f**	$y + 6 = 14$	**g**	$n - 4 = 2$	**h**	$q - 9 = 2$

5 Solve each of these equations.

a	$2m = 12$	**b**	$3p = 18$	**c**	$4x = 20$	**d**	$5q = 45$
e	$\frac{x}{2} = 6$	**f**	$\frac{x}{4} = 5$	**g**	$\frac{x}{3} = 9$	**h**	$\frac{x}{4} = 6$

6 Multiply out the brackets.

a $3(m + 5)$ **b** $3(t + 7)$ **c** $2(x - 5)$ **d** $4(t - 3)$

e $2(7 + y)$ **f** $4(3 - h)$ **g** $4(5 - t)$ **h** $5(2 - t)$

7 Copy and complete each expansion.

a $6t + 4 = 2(\ldots)$ **b** $9x + 12 = \square(3x + 4)$ **c** $4t + 8 = 4(\ldots)$

d $12x + 8 = \square(3x + 2)$ **e** $9x - 6 = \square(3x - 2)$ **f** $15t + 5 = 5(\ldots)$

g $2x + 6 = \square(x + 3)$ **h** $14t + 7 = 7(\ldots)$ **i** $18t + 12 = 6(\ldots)$

j $10x + 6 = \square(5x + 3)$ **k** $16t + 12 = 4(\ldots)$ **l** $6x - 9 = \square(2x - 3)$

8 Write down the missing length for each rectangle.

a Top: $2x - 1$; Area $= 6x - 3$; side: ?

b Top: ?; Area $= 8t + 6$; side: 2

c Top: ?; Area $= 10 - 15t$; side: 5

9 Solve these equations.

a $\frac{1}{2}(m + 5) = 3$ **b** $\frac{1}{3}(t - 3) = 4$ **c** $\frac{1}{5}(x + 6) = 2$ **d** $\frac{1}{6}(y - 3) = 4$

e $\frac{n + 6}{8} = 1$ **f** $\frac{k - 3}{10} = 2$ **g** $\frac{x + 5}{4} = 1$ **h** $\frac{t - 3}{5} = 4$

10 Raghib thinks of a number.

He adds three to the number.

He divides this answer by four.

Call Raghib's number n.

a Write down an expression for the number Raghib now has in his head.

Raghib says that his answer is two.

b Write down an equation involving n and Raghib 's answer.

c Solve the equation.

d If Raghib had ended up with seven, what number would he have started with?

11 Jane takes in ironing for other people. To work out how much to charge, she uses the formula:

$C = £(7 + 9H)$

where H is the number of hours she spends ironing.

What is the charge of ironing:

a for 2 hours **b** from 9:30 am to 1:00 pm.

Financial skills

Wedding day

The cost of a wedding can include many hidden extras.

Let's look at the possible costs of Joy and Chris's wedding.

Church

The fees included £60 for the organist, £75 for the priest and the hire of the church.

This was calculated as:

cost = £100 + £90 × number of hours open

Cars

The charge for each car was calculated as:

cost = £150 + £8 × number of miles driven

Bedsock Priory

The charge for the venue was £12 000 which included flowers and drinks at the reception.

The wedding breakfast cost £21 per person, up to a maximum of 120 guests.

The evening buffet cost £13 per person, up to a maximum of 250 guests.

The evening disco costs were calculated as:

cost = £140 + number of hours × £60

Clothes

Chris decided to hire morning suits for himself, his best man, the bride's father and the four groomsmen. It cost £85 to hire each suit.

Joy was resplendent in a dress costing £1400. The dresses for each of her four bridesmaids cost £350 each.

1 The church service started at 11:00 am. The service was due to last an hour, so they booked the church for an hour before and an hour after the start of the service.

How much would the church, priest and organist cost altogether?

2 They needed two cars.
The car for the bridegroom and the best man had to make journeys of 27 miles.
The car for the bride had to take her to the church, and then to the wedding reception, a total of 31 miles.

What is the total cost of the cars?

3 They booked the reception at Bedsock Priory.
They invited, 110 guests to the wedding breakfast and 250 people to the evening reception.
They also booked the disco from 7:30 pm to midnight.

What will be their total cost for the reception at Bedsock Priory?

4 What is the total cost of the suits, the bride's dress and the bridesmaids' dresses?

5 Joy's father was expected to pay for all of the things listed above. How much did his daughter's wedding cost him?

3 Polygons

This chapter is going to show you:

- the names of different polygons
- the difference between an irregular polygon and a regular polygon
- how to work out the sum of the interior angles of a polygon
- how to work out the size of each interior angle in regular polygons.

You should already know:

- the different names for triangles and quadrilaterals
- that the sum of the interior angles in a triangle is 180°
- that the sum of the interior angles in a quadrilateral is 360°.

About this chapter

Flanked by the wild North Atlantic Ocean on one side and a landscape of dramatic cliffs on the other, for centuries the Giant's Causeway has inspired artists, stirred scientific debate and captured the imagination of all who see it. Storytellers have their own explanation for this captivating stretch of coast, and many tales endure to the present day. The most famous legend associated with the Giant's Causeway is that of an Irish giant, Finn McCool. The causeway was believed to be the remains of the bridge that Finn built, to link Ireland to Scotland. The landscape became so imbued with the spirit of this legend that it gave rise to the name – the Giant's Causeway.

The formation consists of about 40 000 interlocking basalt columns, most of which are hexagonal. The columns form huge stepping stones, some as high as 39 feet (about 12 metres), which slope down to the sea. Some of the columns have four, seven or eight sides. Weathering of the rock formation has also created circular structures, which the locals call 'giant's eyes'.

The Giant's Causeway was actually formed by intense volcanic activity, about 50 million years ago. As the lava rapidly cooled, it contracted into the distinctive shapes.

3.1 Polygons

Learning objectives

- To know the names of polygons
- To know the difference between an irregular polygon and a regular polygon

Key words	
concave polygon	convex polygon
decagon	heptagon
hexagon	irregular polygon
nonagon	octagon
pentagon	polygon
regular polygon	

A **polygon** is a 2D shape that has straight sides.

The names of the most common polygons are listed in this table.

Number of sides	Name of polygon
3	triangle
4	quadrilateral
5	**pentagon**
6	**hexagon**
7	**heptagon**
8	**octagon**
9	**nonagon**
10	**decagon**

These are examples of some of the different polygons.

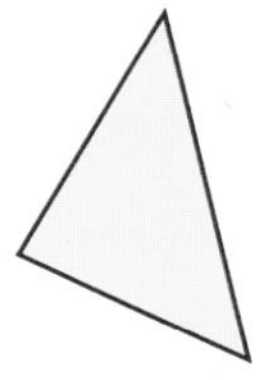

Triangle
3 sides

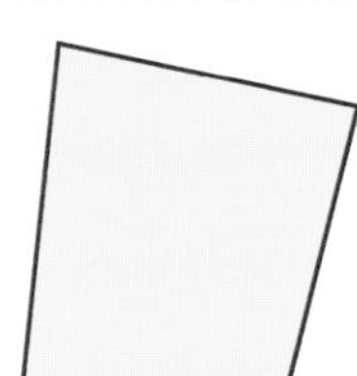

Quadrilateral
4 sides

Pentagon
5 sides

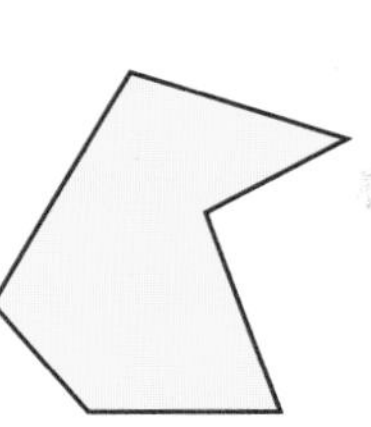

Hexagon
6 sides

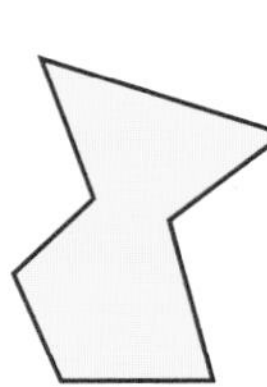

Heptagon
7 sides

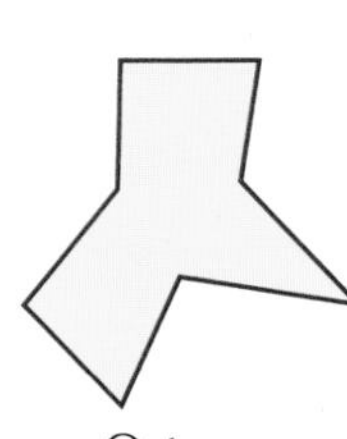

Octagon
8 sides

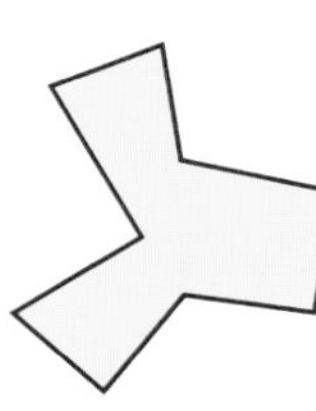

Nonagon
9 sides

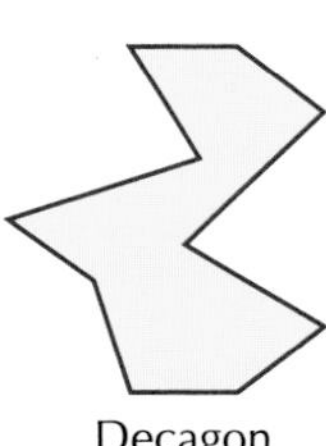

Decagon
10 sides

These are all **irregular polygons**. The angles are not all equal and the sides are not the same length.

In a **regular polygon**, all the angles are equal and all the sides are the same length.

Here are some examples of regular polygons.

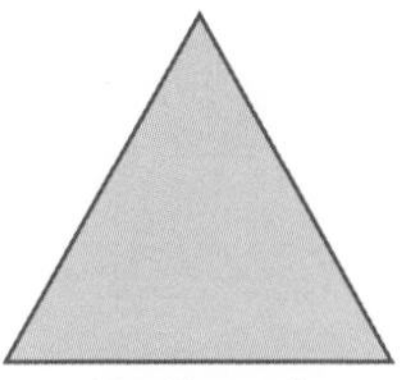

Equilateral triangle

Square

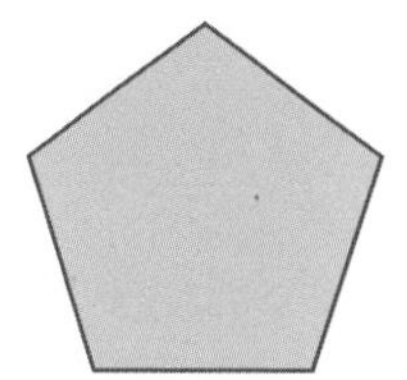

Regular pentagon

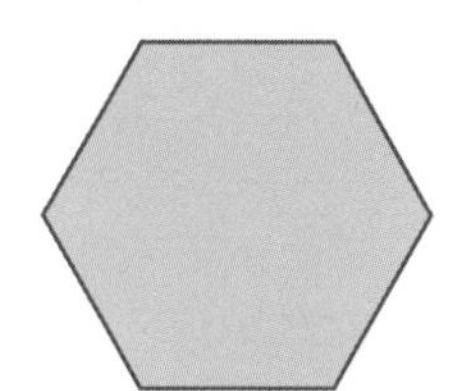

Regular hexagon

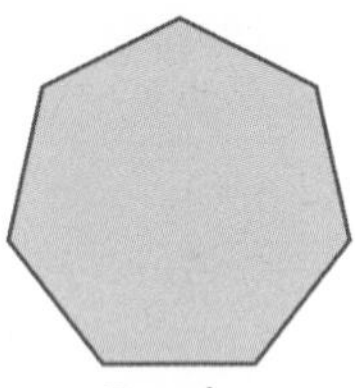

Regular heptagon

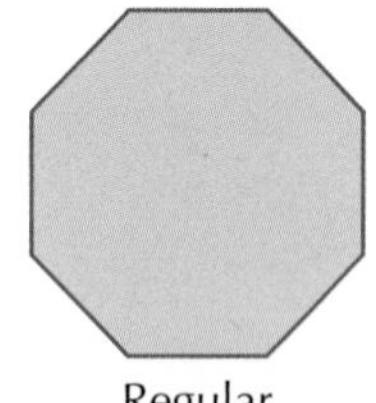

Regular octagon

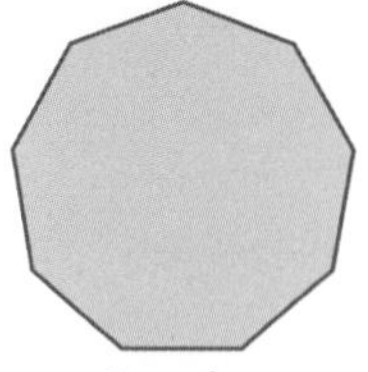

Regular nonagon

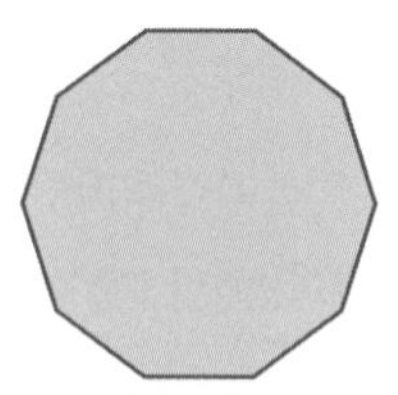

Regular decagon

A **convex polygon** has all of its diagonals inside the polygon.

A **concave polygon** is one with at least one diagonal outside the polygon.

Remember that a diagonal is a line that joins two non-adjacent corners of a 2D shape.

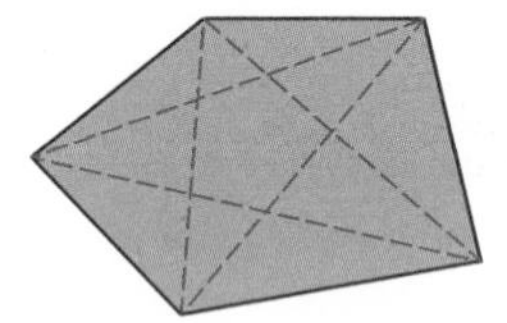

A convex pentagon (all diagonals inside)

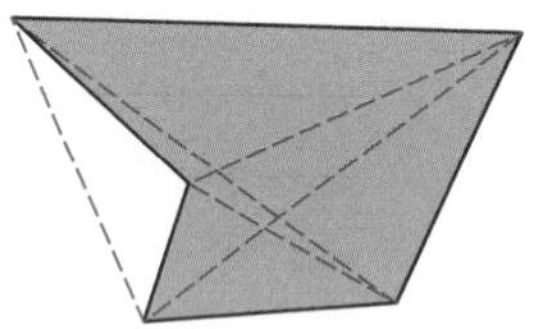

A concave pentagon (one diagonal outside)

Exercise 3A

1 Write down the name of each irregular polygon.

a

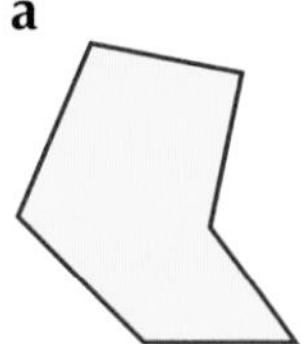

b

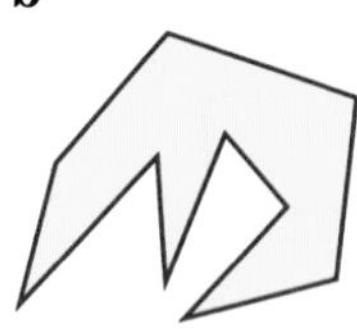

c

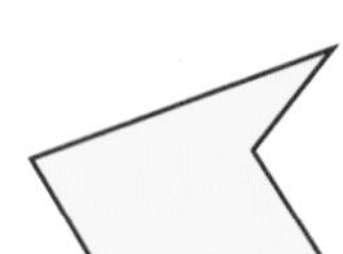

d

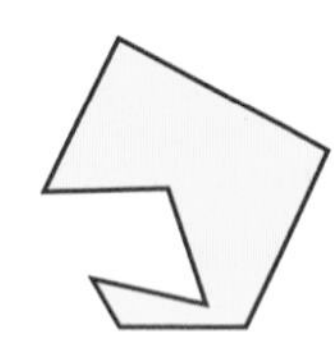

e

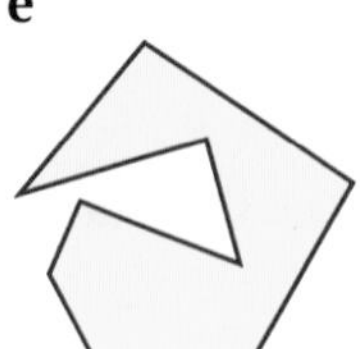

f

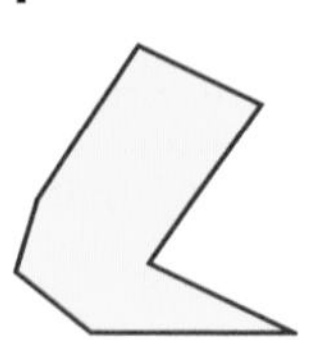

2 Which shapes below are polygons? If they are polygons, write down their names.

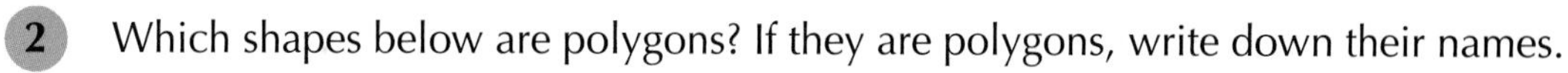

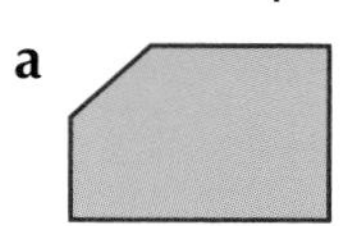

a **b** **c** **d** **e**

3 Which shapes below are regular polygons?

a **b** **c** **d** **e**

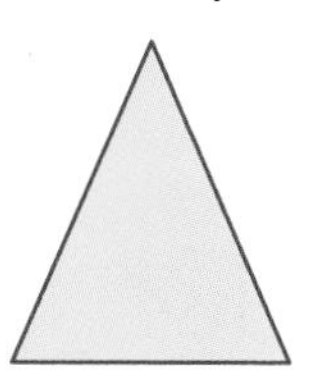

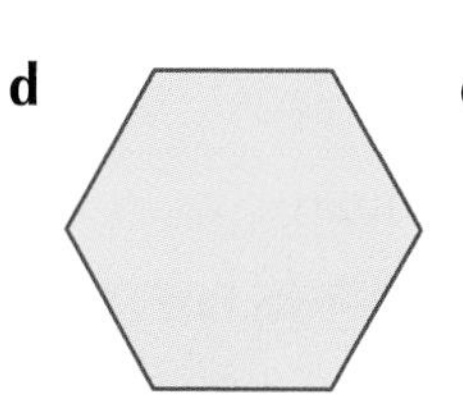

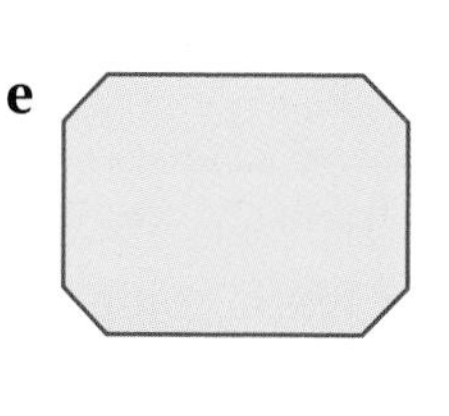

4 State whether each of these shapes is a convex polygon or a concave polygon.

a **b** **c** **d** **e**

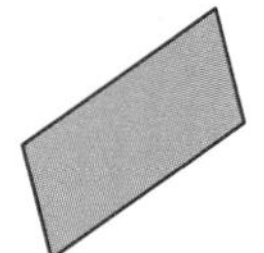

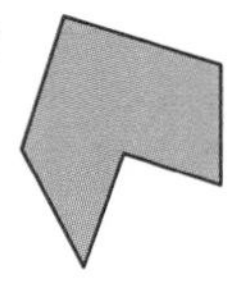

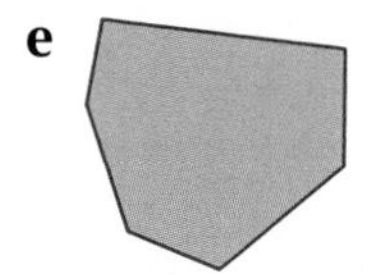

MR **5** This pentagon has one right angle.

a Draw a pentagon that has two right angles.

b Draw a pentagon that has three right angles.

c Explain why you cannot draw a pentagon that has four right angles.

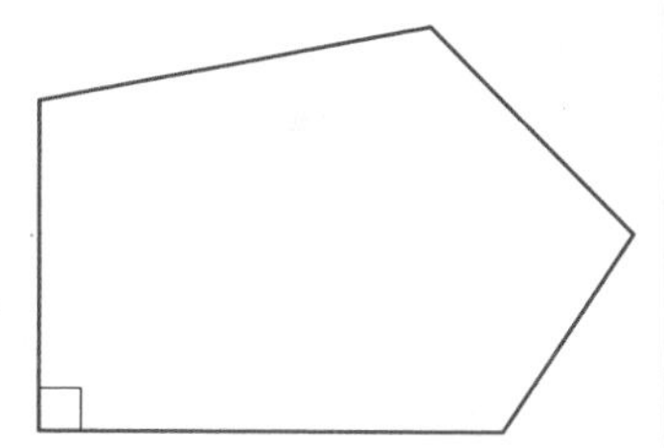

6 Here is a square ABCD.

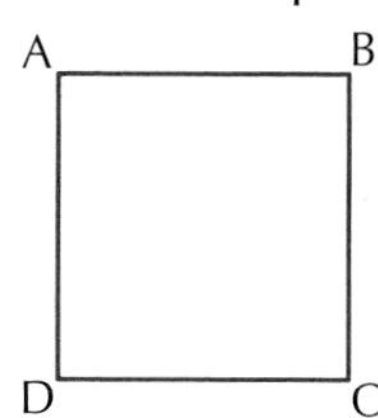

AB is parallel to DC and AD is parallel to BC.

a Here is a regular hexagon ABCDEF.
Copy and complete this sentence.

AB is parallel to… and AF is parallel to… and BC is parallel to….

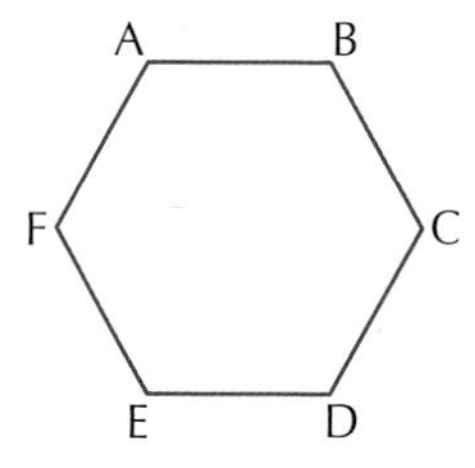

b Here is a regular octagon ABCDEFGH.
Copy and complete this sentence.

AB is parallel to… and BC is parallel to… and CD is parallel to… and DE is parallel to….

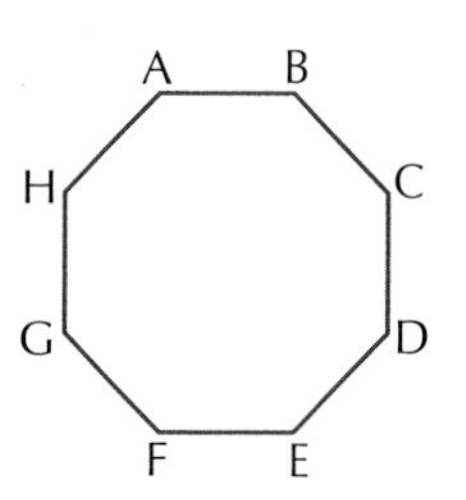

Investigation: Overlapping squares

Here are two squares of different sizes.

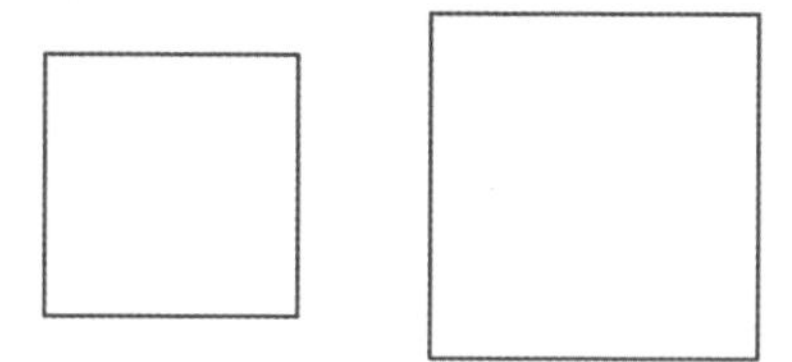

Overlap the two squares, as in the diagram below.

Overlapping the squares has produced four polygons A, B, C and D.

Polygon A is a triangle, polygon B is a pentagon, polygon C is a triangle and polygon D is a heptagon.

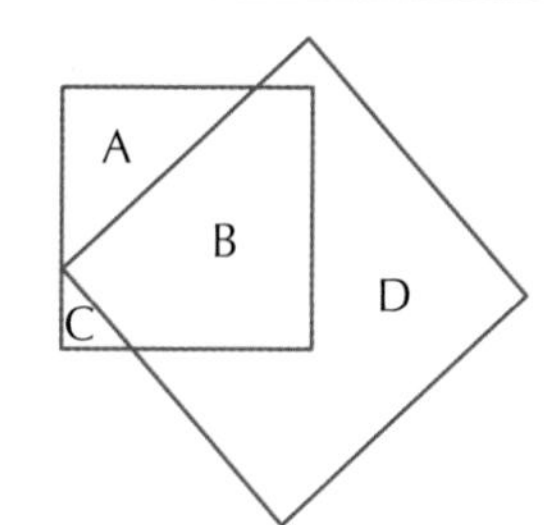

A Now overlap the squares in this way.

Write down the names of the polygons A, B and C.

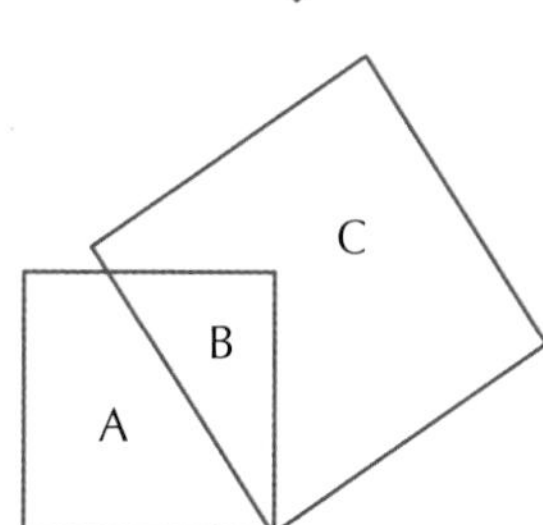

B Now overlap the squares in this way.

Write down the names of the polygons A, B and C.

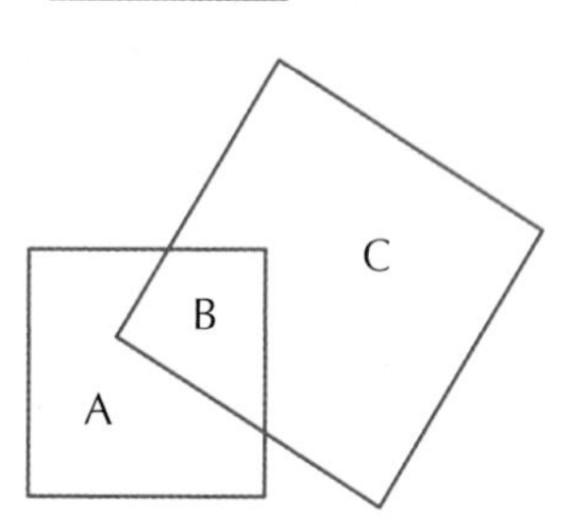

C What polygons can you make by overlapping three squares?

3.2 Angles in polygons

Learning objective

- To work out the sum of the interior angles of a polygon

Key words

interior angle

The angles inside a polygon are called **interior angles**.

You already know that:

- for any triangle, the sum of the interior angles, a, b and c, is 180°.

  $a + b + c = 180°$

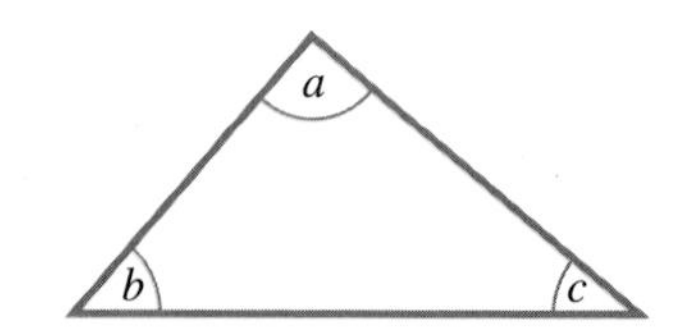

- for any quadrilateral, the sum of the interior angles, a, b, c and d, is 360°.

$a + b + c + d = 360°$

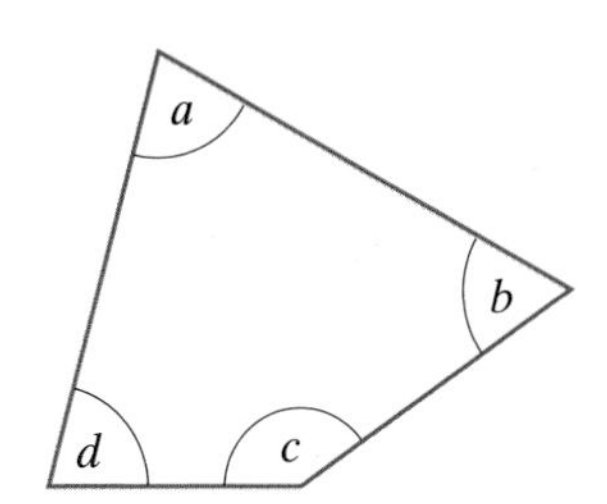

This is a pentagon.

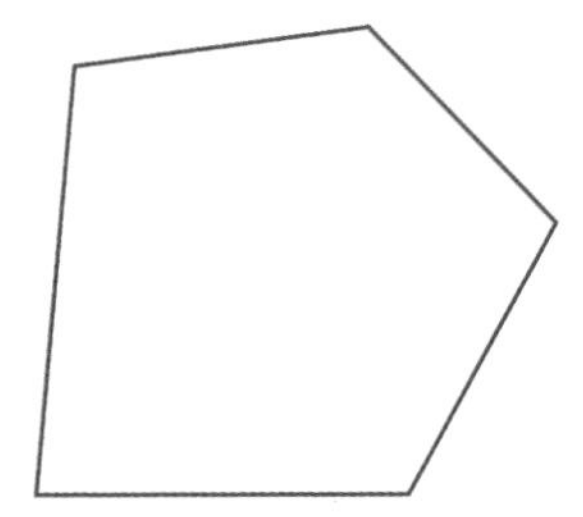

Starting at any vertex, you can draw in diagonals to split it into triangles.

This shows that a pentagon can be split into three triangles.
So the sum of the interior angles of a pentagon is $3 \times 180° = 540°$.

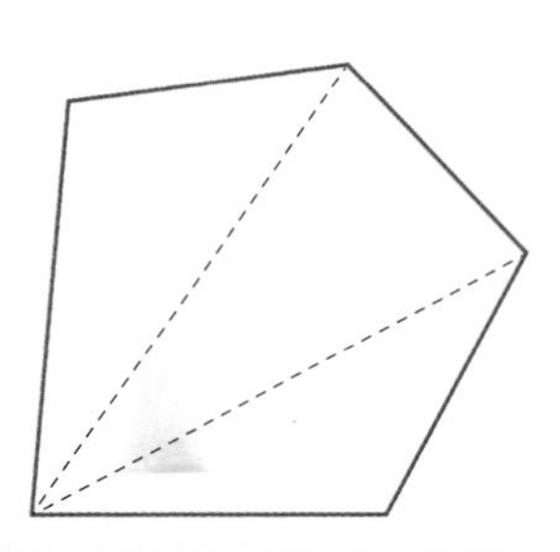

Example 1

Work out the size of the angle labelled a in this pentagon.

The sum of the interior angles of a pentagon is 540°.

So $a = 540° - 150° - 60° - 130° - 90°$

$= 110°$.

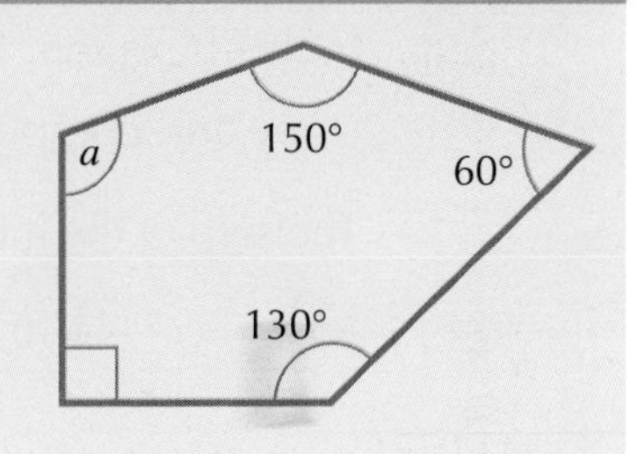

Exercise 3B

1 This hexagon has been split into triangles.

Copy and complete these sentences.

A hexagon can be split into… triangles.

So the sum of the interior angles of a hexagon is given by:

$\ldots \times 180° = \ldots°$

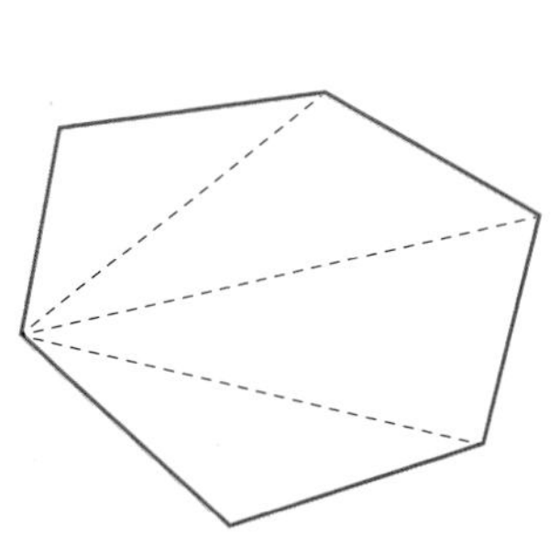

2 This heptagon has been split into triangles.

Copy and complete these sentences.

A heptagon can be split into… triangles.

So the sum of the interior angles of a heptagon is given by:

$\ldots \times 180° = \ldots°$

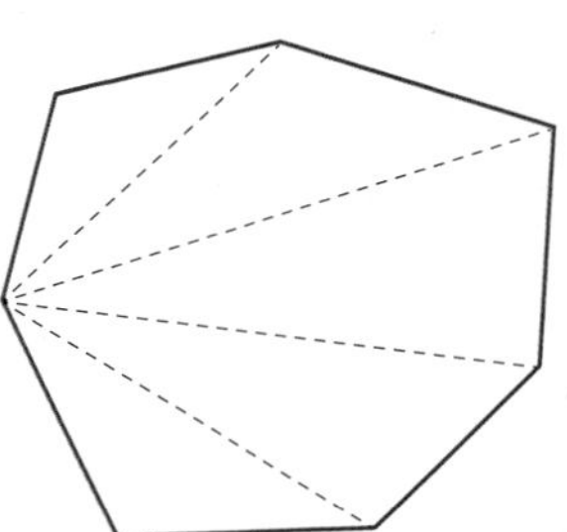

3 This octagon has been split into triangles.

Copy and complete these sentences.

An octagon can be split into … triangles.

So the sum of the interior angles of an octagon is given by:

$\ldots \times 180° = \ldots°$

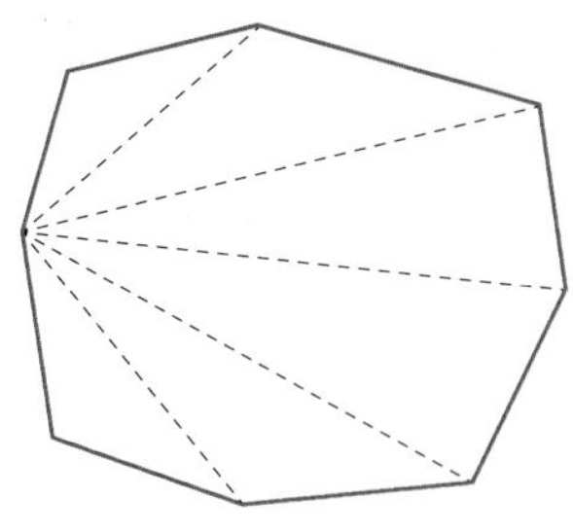

4 Copy and complete this table. The first three rows have been done for you.

Do not draw the polygons.

Hint Look for patterns in the table.

Name of polygon	Number of sides	Number of triangles inside polygon	Sum of interior angles
Triangle	3	1	180°
Quadrilateral	4	2	360°
Pentagon	5	3	540°
Hexagon	6		
Heptagon	7		
Octagon	8		
Nonagon	9		
Decagon	10		

5 In this hexagon, there is an unknown angle labelled a.

Copy and complete these sentences.

The sum of the interior angles of a hexagon is …°.

So $a = \ldots° - 150° - 70° - 140° - 130° - 120°$

$= \ldots°$.

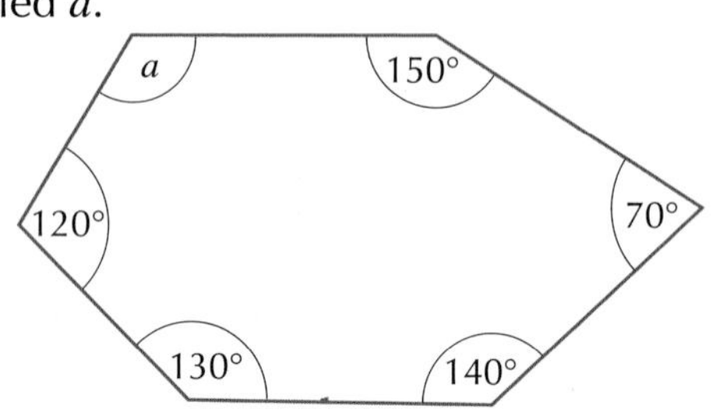

6 Work out the unknown angle in each polygon.

a

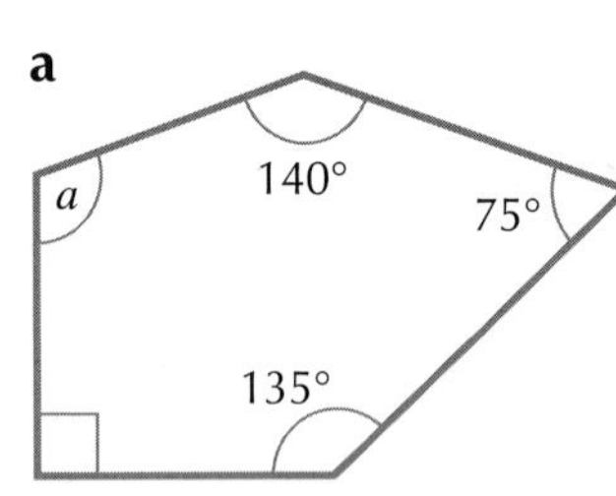

b

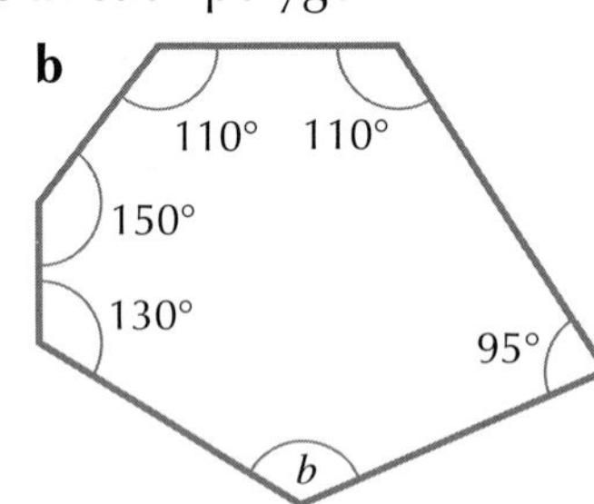

c

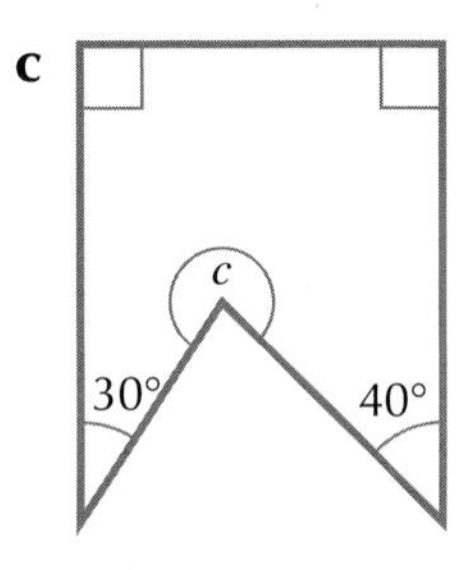

PS **7** Work out the value of a in this pentagon.

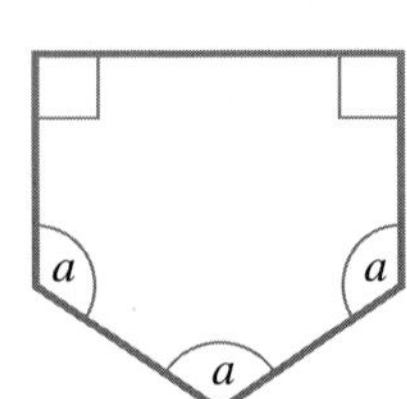

Problem solving: Polygons and diagonals

A triangle has no diagonals, a quadrilateral has two diagonals and a pentagon has five diagonals

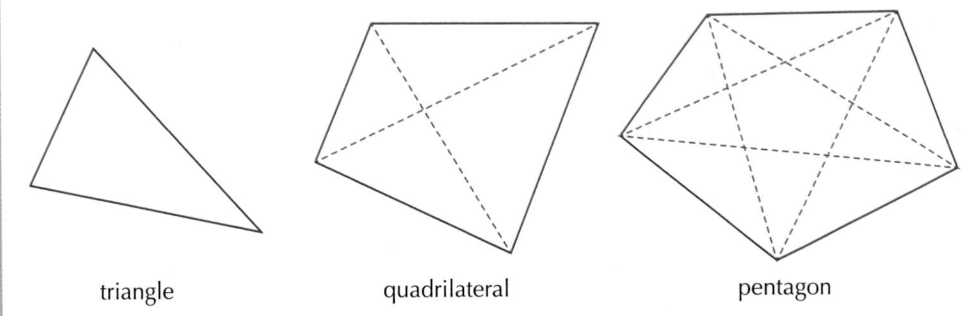

A Copy this hexagon and show that it has nine diagonals.

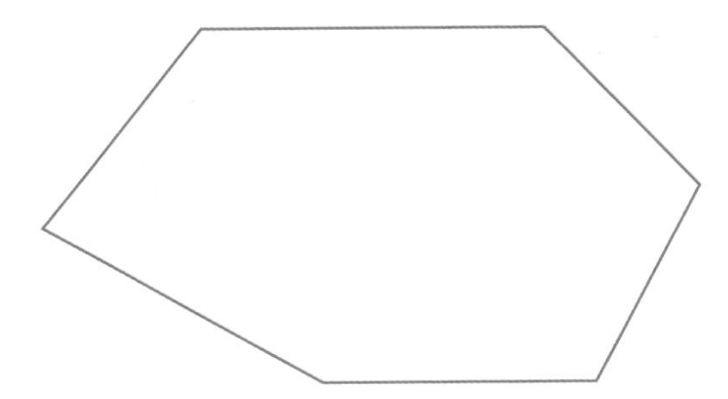

B This is a formula that connects the number of diagonals, d, to the number of sides, s, for any polygon.

$$d = \frac{s(s-3)}{2}$$

For the hexagon, $s = 6$, and so the number of diagonals is:

$$d = \frac{6(6-3)}{2} = \frac{6 \times 3}{2} = \frac{18}{2} = 9$$

a Use the formula to show that the number of diagonals in a heptagon is 14.

b Use the formula to show that the number of diagonals in an octagon is 20.

C Copy and complete the table for the numbers of diagonals in different polygons.

Number of sides	Name of polygon	Number of diagonals
3	triangle	0
4	quadrilateral	2
5	pentagon	5
6	hexagon	9
7	heptagon	14
8	octagon	20
9	nonagon	
10	decagon	

3.3 Interior angles of regular polygons

Learning objective

- To work out the sizes of the interior angles in regular polygons

You have just learnt how to work out the sum of the angles in a polygon. Now you will find out how to work out the size of each angle in any regular polygon.

Example 2

This is a regular pentagon. Work out the size of its interior angles.

The sum of the interior angles is 540°.

Because it is a regular polygon, all the angles are equal.

So the size of each interior angle is $540° \div 5 = 108°$.

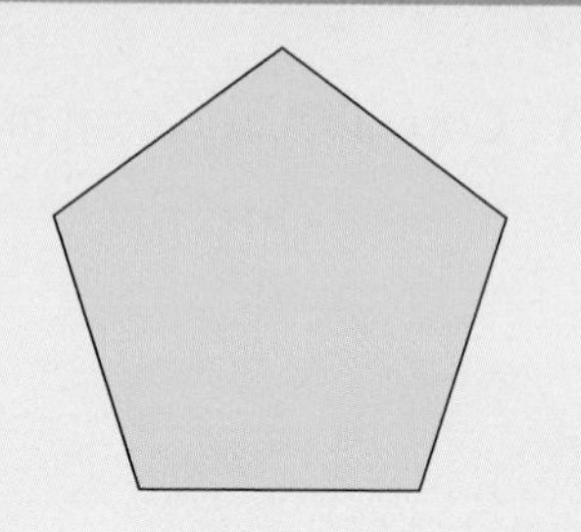

Exercise 3C

1 This is a regular hexagon.

a Write down the sum of the interior angles.

b Work out the size of each interior angle.

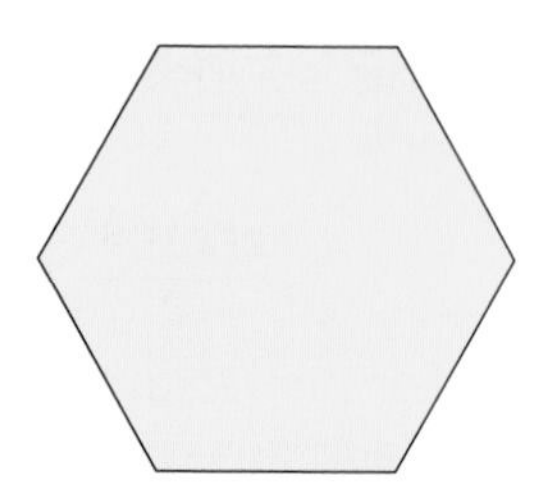

2 This is a regular octagon.

a Write down the sum of the interior angles.

b Work out the size of each interior angle.

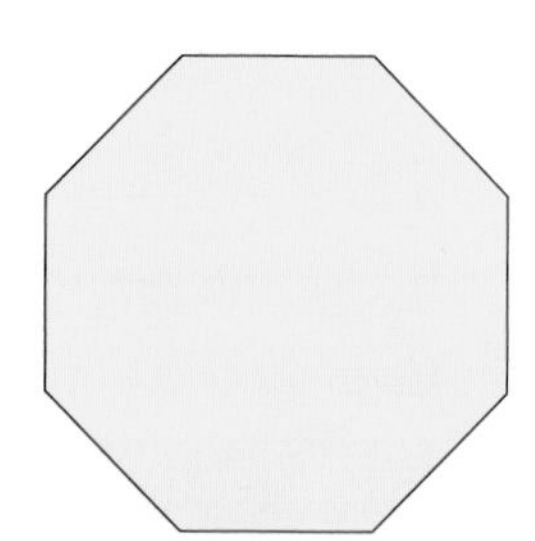

3 Copy and complete the table. The first three lines have been done for you.

Name of polygon	Number of sides	Sum of interior angles	Size of each interior angle
Triangle	3	180°	60°
Quadrilateral	4	360°	90°
Pentagon	5	540°	108°
Hexagon	6		
Octagon	8		
Nonagon	9		
Decagon	10		

4 ABCDE is a regular pentagon.

Copy and complete these sentences.

a Angle AED = ...°

b Triangle AED is an ... triangle.

c Angle EAD = ...°

d Angle BAC = ...°

e Angle DAC = ...°

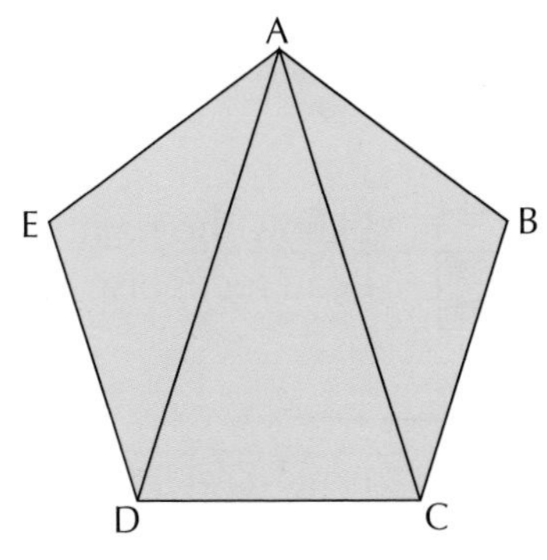

5 ABCDEF is a regular hexagon.

Work out the size of angle FAE, marked x on the diagram.

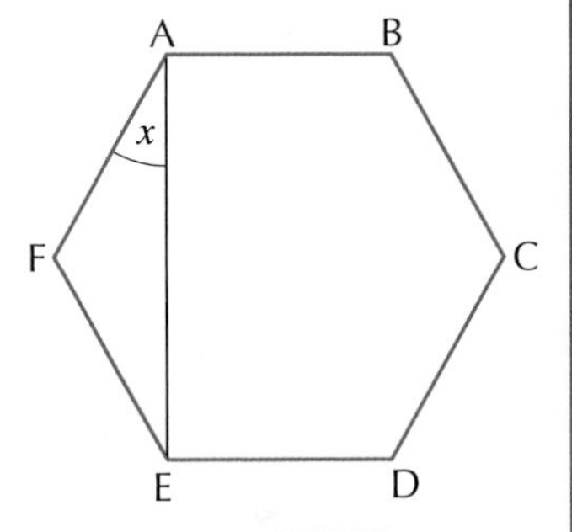

6 This is a regular dodecagon.

All 12 interior angles are equal and all 12 sides are the same length.

Copy and complete these sentences.

a A regular dodecagon can be split into ... triangles.

b The sum of the interior angles of a regular dodecagon is given by:

... × 180° = ...°

c The size of each interior angle of a regular dodecagon is given by:

...° ÷ 12 = ...°

Challenge: Interior angles in a regular heptagon

The interior angles of some regular polygons are not whole numbers of degrees.

Work out the size of each interior angle of a regular heptagon.

Give your answer as a mixed number.

Use the $a\,{}^b/_c$ key or the key on your calculator, to work out fractions.

Ready to progress?

I know the names of different polygons.
I can recognise regular, irregular, convex and concave polygons.

I can work out the sum of the interior angles for different polygons.
I can work out the size of each interior angle for different regular polygons.

Review questions

1 a Write down the name of each irregular polygon.

b Write down whether it is a convex or a concave polygon.

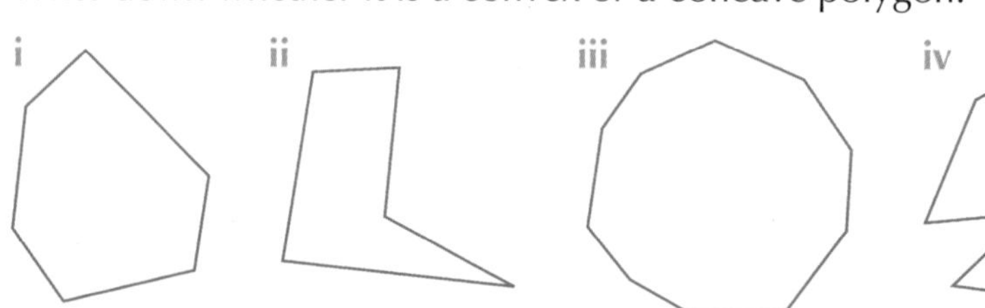

2 Which hexagon is the odd one out?

Give a reason for your answer.

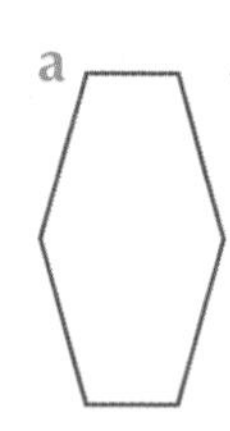

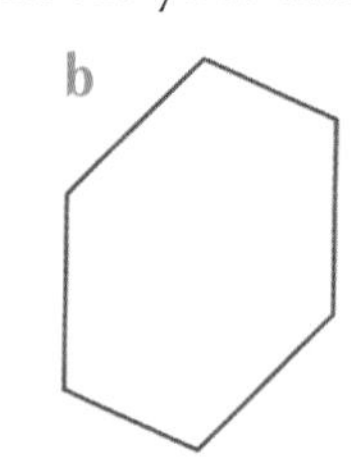

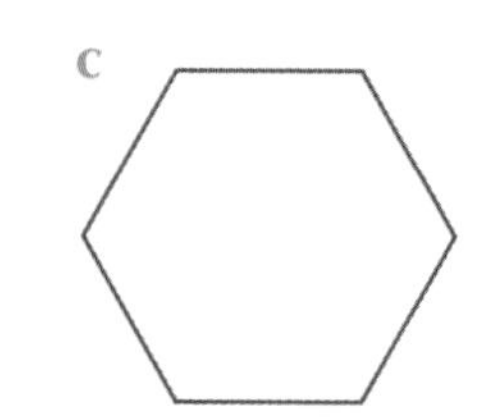

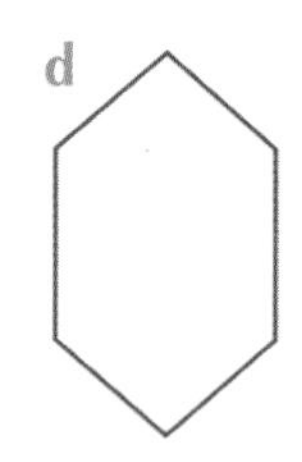

3 Draw a diagram to show that the sum of the interior angles of a pentagon is 540°.

4 a Copy and complete these sentences.

i The sum of the interior angles of a pentagon is 540°.

ii The sum of the interior angles of a hexagon is ...°.

iii The sum of the interior angles of a heptagon is . . . °.

iv The sum of the interior angles of an octagon is . . . °.

b Use this information to work out the unknown angle in each polygon.

i

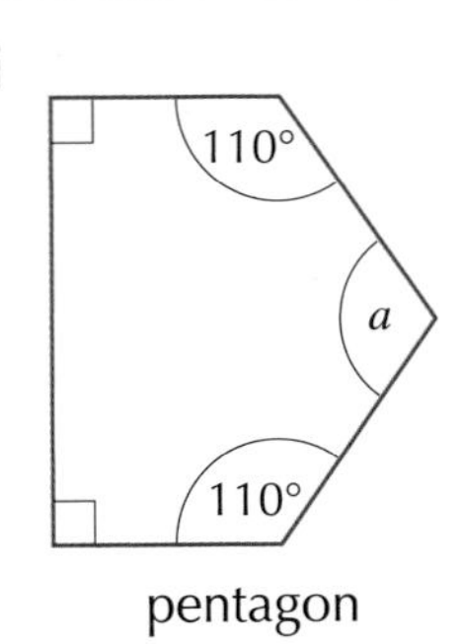

pentagon

ii

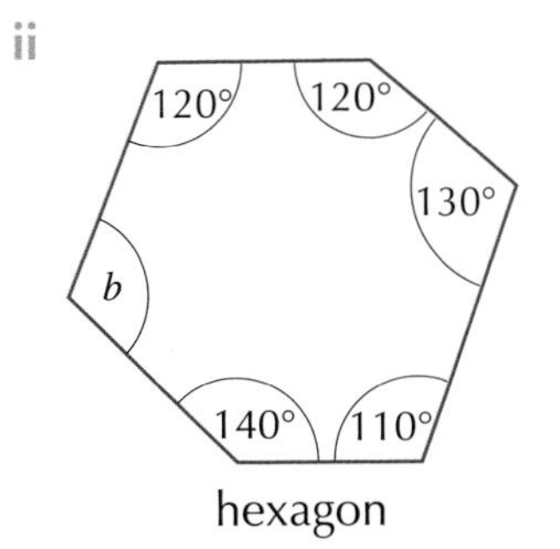

hexagon

iii

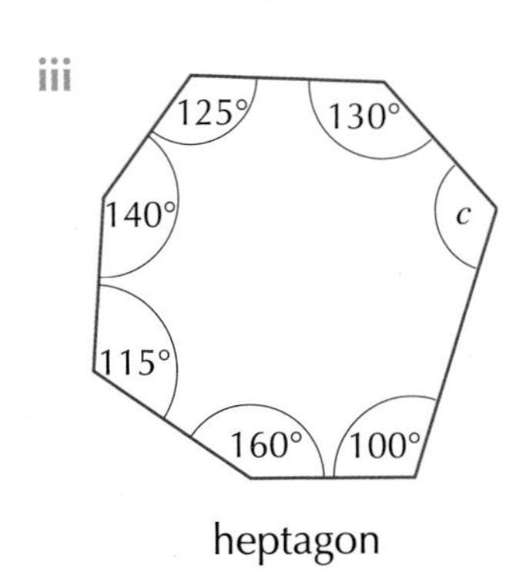

heptagon

iv

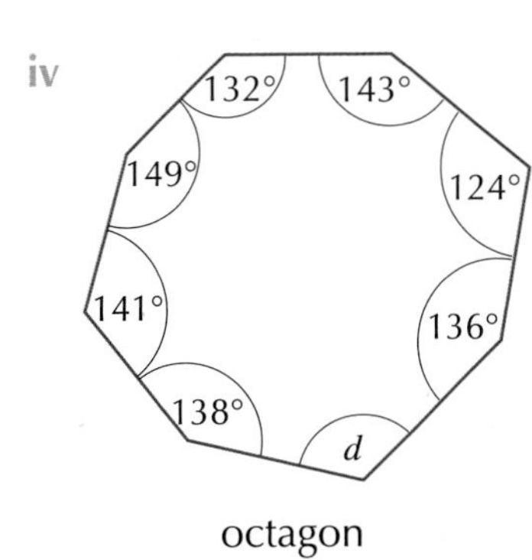

octagon

5 ABCDEFGH is a regular octagon.

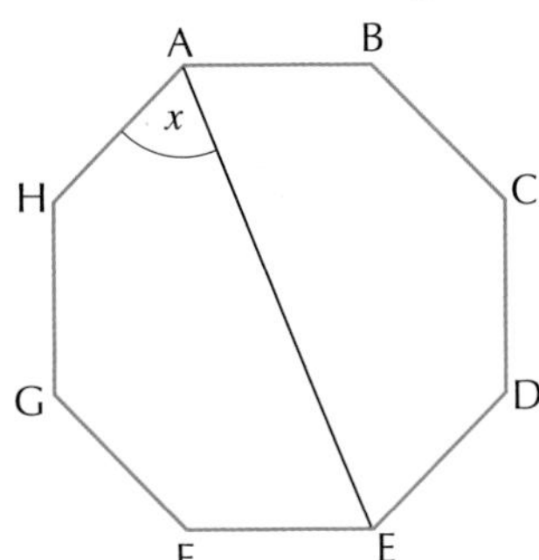

Work out the size of angle HAE, marked x on the diagram.

6 The five interior angles in a pentagon are x, $2x$, $3x$, $4x$ and $5x$.

Work out the size of each of the five angles.

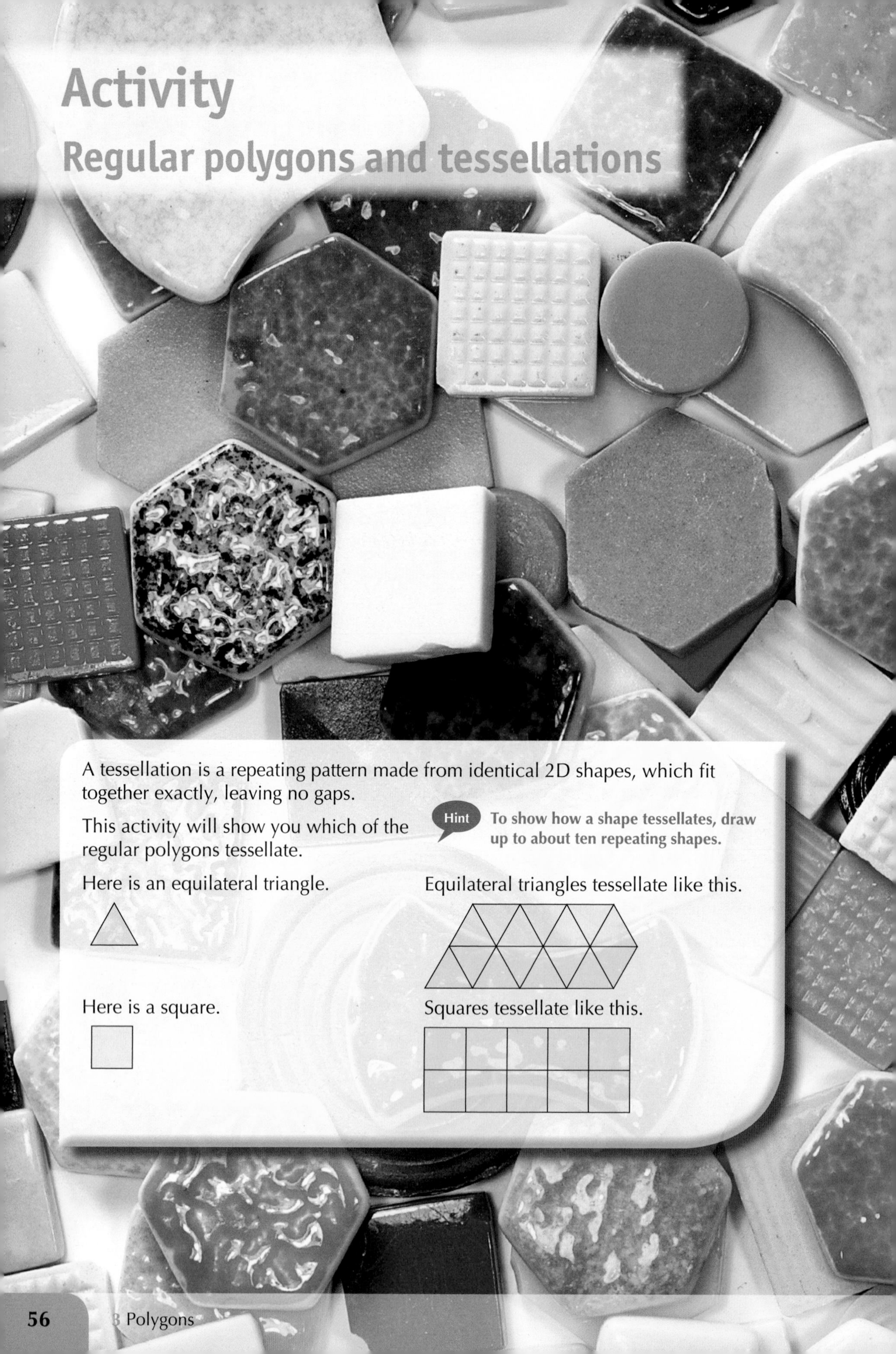

Activity

Regular polygons and tessellations

A tessellation is a repeating pattern made from identical 2D shapes, which fit together exactly, leaving no gaps.

This activity will show you which of the regular polygons tessellate.

Hint To show how a shape tessellates, draw up to about ten repeating shapes.

Here is an equilateral triangle.

Equilateral triangles tessellate like this.

Here is a square.

Squares tessellate like this.

1 Trace this regular pentagon onto card and cut it out to make a template. Use your template to see if a regular pentagon tessellates.

2 Trace this regular hexagon onto card and cut it out to make a template. Use your template to see if a regular hexagon tessellates.

3 Trace this regular octagon onto card and cut it out to make a template. Use your template to see if a regular octagon tessellates.

4 Copy and complete this table.
The first two rows have been completed for you.

Regular polygon	**Does it tessellate?**
Equilateral triangle	Yes
Square	Yes
Regular pentagon	
Regular hexagon	
Regular octagon	

4 Using data

This chapter is going to show you:

- how to recognise correlation from scatter graphs
- how to construct and interpret two-way tables
- how to compare two sets of data from statistical diagrams
- how to plan a statistical investigation.

You should already know:

- how to calculate averages
- how to use a suitable method to collect data
- how to draw and interpret graphs for discrete data
- how to use mode, median, mean and range to compare two sets of data.

About this chapter

When the weather is hot, many people go to the beach. Sales of sunscreen, ice cream, swimwear and sunshades are high. What happens when the weather is cold?

It is easy to see that some things are related. People like to keep cool and protect themselves from the hot sunshine.

However, it is not always clear whether other information is related in the same way. Do tall people have large handspans? Do cars go faster the bigger their wheels are?

In this chapter, you will find out how to compare two sets of data, to find out if they are related to each other.

4.1 Scatter graphs and correlation

Learning objective

- To infer a correlation from two related scatter graphs

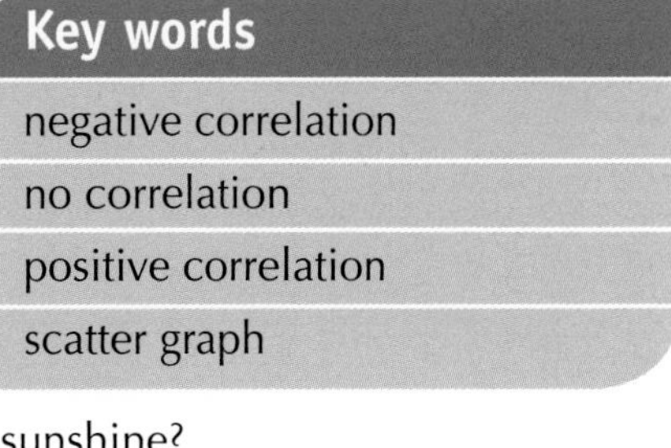

Key words
negative correlation
no correlation
positive correlation
scatter graph

An amateur weather forecaster recorded the maximum temperature, rainfall and hours of sunshine each day in a town on the south coast of England.

She plotted these two **scatter graphs** from her data. Is it possible to work out the relationship between rainfall and hours of sunshine?

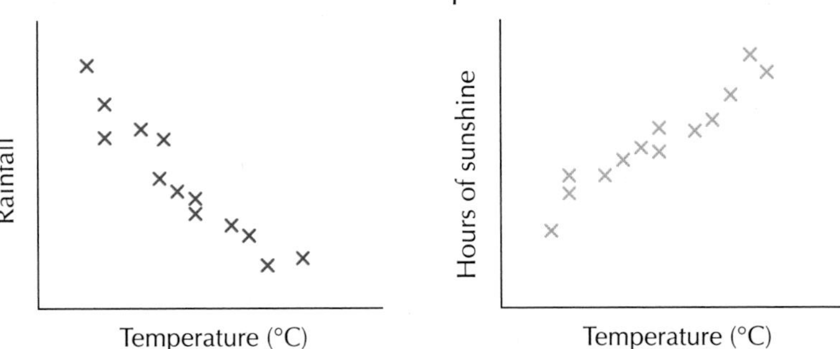

- The graph on the left shows **negative correlation**. In this case, it means that the higher the temperature, the less rainfall there is.
- The graph on the right shows **positive correlation**. In this case, it means that the higher the temperature, the more hours of sunshine there are.

Now look at this graph. It shows **no correlation** between the temperature and the number of fish caught daily off Rhyl – as you might expect.

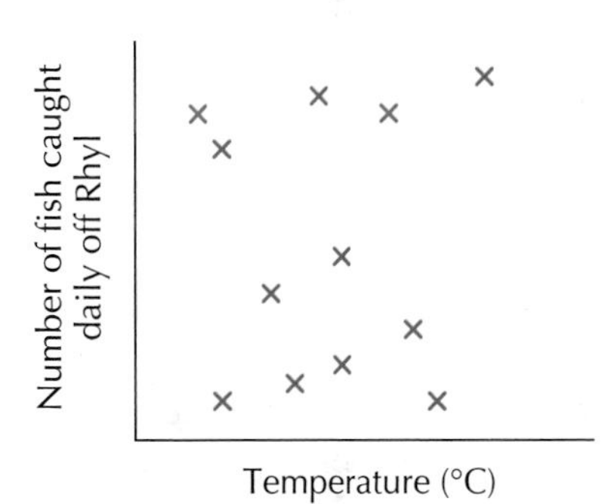

Exercise 4A

1 In a competition there are three different games.

These scatter diagrams were drawn from the scores of everyone who played all three games.

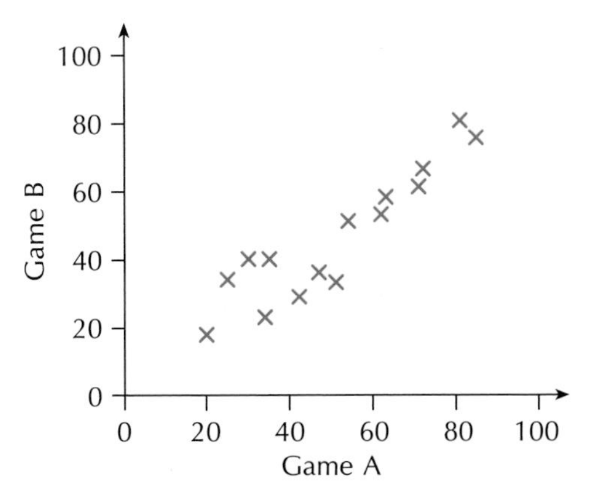

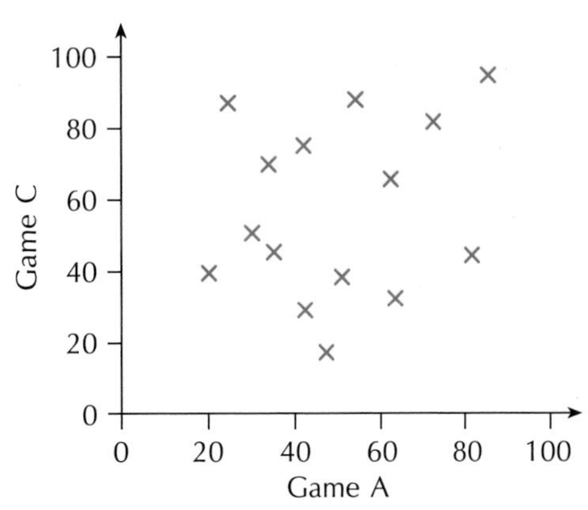

Look at the scatter diagrams and describe any correlation between:

a Game A and Game B **b** Game A and Game C.

2 A company compared how they were charged for posting parcels.

The results are shown on these scatter graphs.

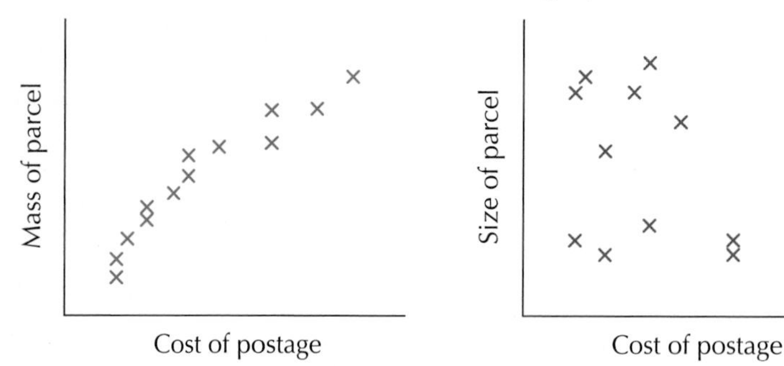

a Describe the type of correlation between the mass of parcels and the cost of postage.

b Describe the type of correlation between the size of parcels and the cost of postage.

3 A fitness club made a comparison between the ages of a group of men with the length of their hair and also with their masses. The results are shown on the two scatter graphs.

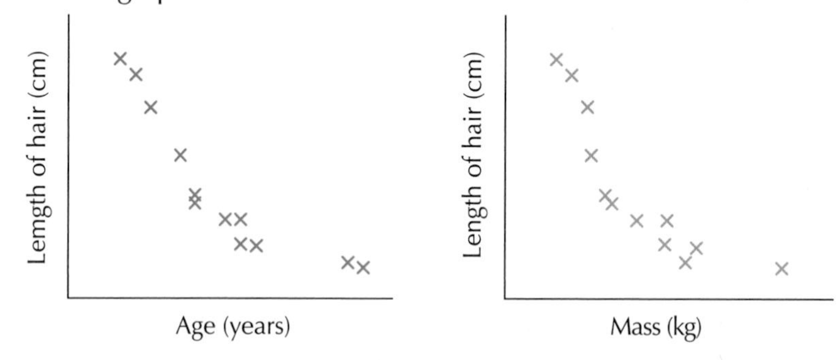

a Describe the type of correlation between their age and the length of their hair.

b Describe the type of correlation between their mass and the length of their hair.

4 Josh lived in Scarborough. In the summer he carried out a survey on the midday temperature and the number of seagulls he could see at that time.

His results are in this table.

Day	1	2	3	4	5	6	7	8	9	10	11	12	13	14	15	16	17	18
Temperature (°C)	13	18	21	15	19	21	18	19	21	22	19	18	17	18	16	23	22	20
Number of seagulls	36	25	20	31	23	19	26	24	19	19	25	24	26	25	28	17	18	21

a Draw a scatter diagram to represent the data in the table.

Label the horizontal axis as the temperature, from 10 °C to 25 °C.

Label the vertical axis as the number of seagulls, from 10 to 40.

b What correlation can you see from your scatter diagram?

c If one midday Josh could see 22 seagulls, explain how he could use the scatter diagram to estimate the temperature.

5 A magazine suggested some correlations between various subjects often studied in schools. Their results showed:

- a positive correlation between mathematics and music
- a positive correlation between mathematics and physics
- a negative correlation between English and art
- no correlation between mathematics and art.

Draw simple scatter diagrams to illustrate these relationships.

6 **a** The price of oil and the cost of petrol show a positive correlation.
Draw a scatter diagram to illustrate this relationship.

b The cost of petrol and the price of food have a positive correlation.
Draw a scatter diagram to illustrate this relationship.

7 The number of heavy coats sold has a negative correlation with the average daily temperature.
Draw a scatter diagram to illustrate this relationship.

8 The number of newspapers sold has no correlation with the hours of sunshine.
Draw a scatter diagram to illustrate this relationship.

Investigation: Comparing marks

A teacher collected test marks for ten pupils in five different subjects.

These are shown below.

	Pupil 1	Pupil 2	Pupil 3	Pupil 4	Pupil 5	Pupil 6	Pupil 7	Pupil 8	Pupil 9	Pupil 10
Mathematics	35	48	72	23	59	27	65	53	39	46
Science	21	37	58	18	43	16	5	44	32	28
Art	74	54	31	85	38	68	41	43	60	51
English	43	46	52	58	85	55	79	75	69	65
History	32	37	43	49	74	44	68	64	58	54

Investigate whether there is any correlation between any of the subjects.

Comment on anything that you notice.

4.2 Interpreting graphs and diagrams

Learning objective

- To use and interpret a variety of graphs and diagrams

In this section you will learn how to interpret a variety of graphs and diagrams as well as how to evaluate and criticise statements made about the data they contain.

Example 1

The diagram shows how some pupils spend their time during one normal week.

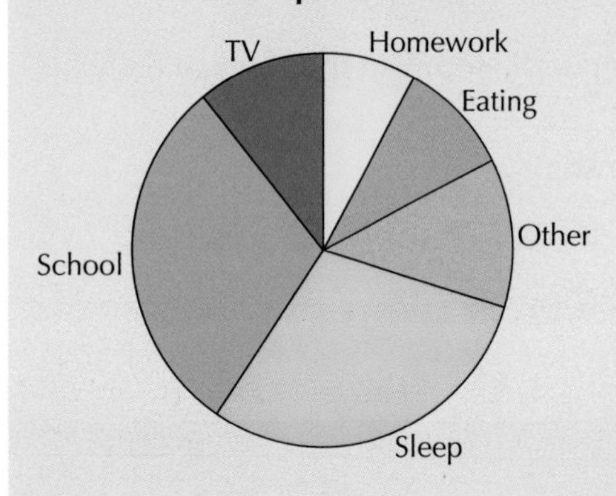

Do you agree with the statements made by Aron and Stefania?

Give your reasons.

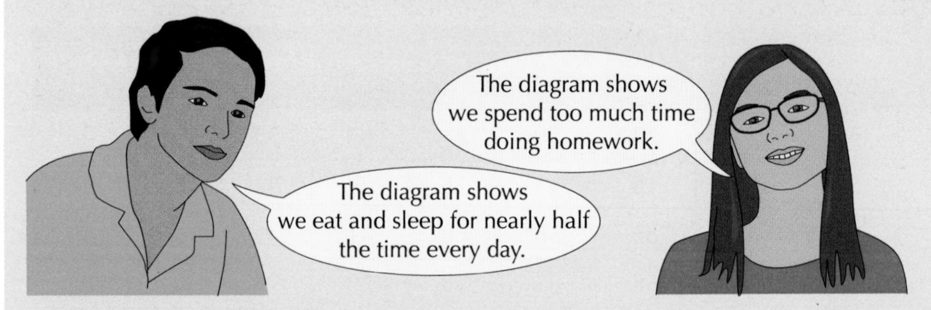

Stefania's comment:

The pie chart shows that more time is spent watching TV than doing homework, so this may not be a fair criticism.

Aron's comment:

This is not true, because the section on 'other' is between them. If this is not included, then sleeping and eating take up only about one third of the time.

Exercise 4B

1 The distance–time graph shows Amy's journey when she visited a friend. There are four stages and three stops.

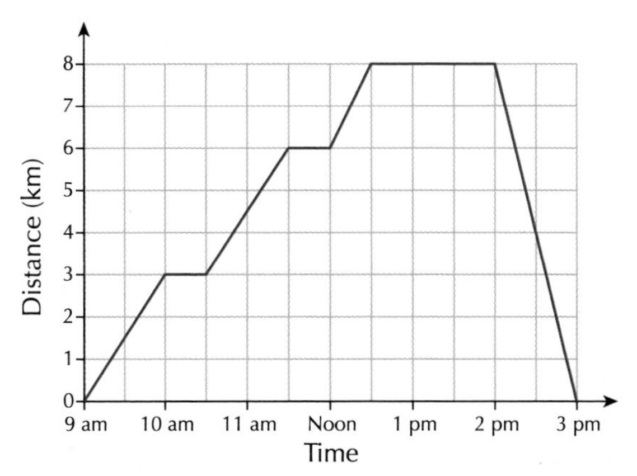

a How far was the first stage of Amy's journey?

b How long was the first stop?

c Kath said: 'I can tell that Amy travelled faster on the third stage than on the first two.'

Explain how Kath could tell that this was so.

d Pete said: 'Amy only travelled 8 km.'

Explain how you can tell that Pete is wrong.

2 The bar chart shows the results of a welly-throwing competition.

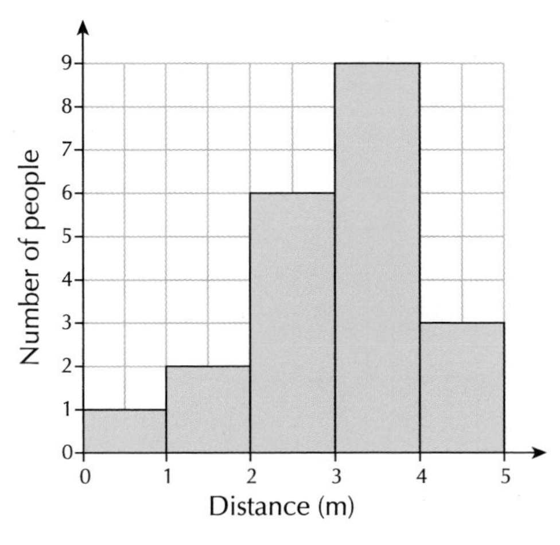

a How many people threw their wellies between 2 and 3 metres?

b How many people threw their wellies between 3 and 5 metres?

c Jo said: 'The shortest throw was only 70 cm.'
Could he be correct? Explain your answer.

d Andreis said: 'There were over 20 people in the competition.'
Is he correct? Explain how you know.

MR **3** The pie chart shows how many different spring flowers were seen in a school garden.

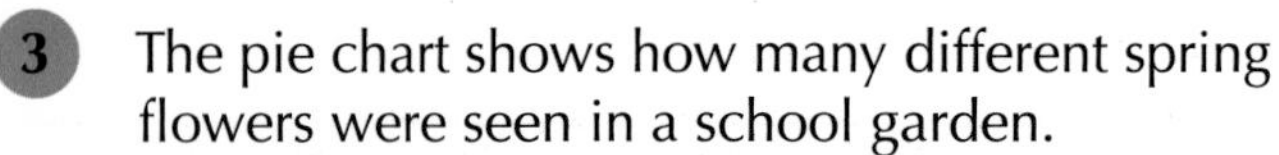

a Anna said: 'More than half the flowers in the garden are daffodils.'

Is Anna correct? Explain how you know.

b Helena said: 'There are ten times as many anemones as there are crocuses.'

How would you find out if she was correct?

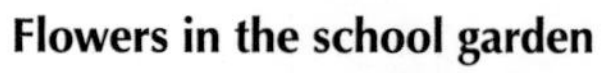

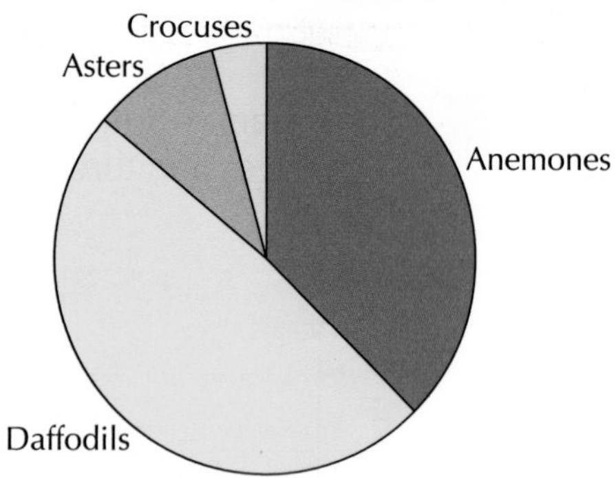

MR **4** These bar charts are extracts from a report into how succesful some managers of England's cricket team were.

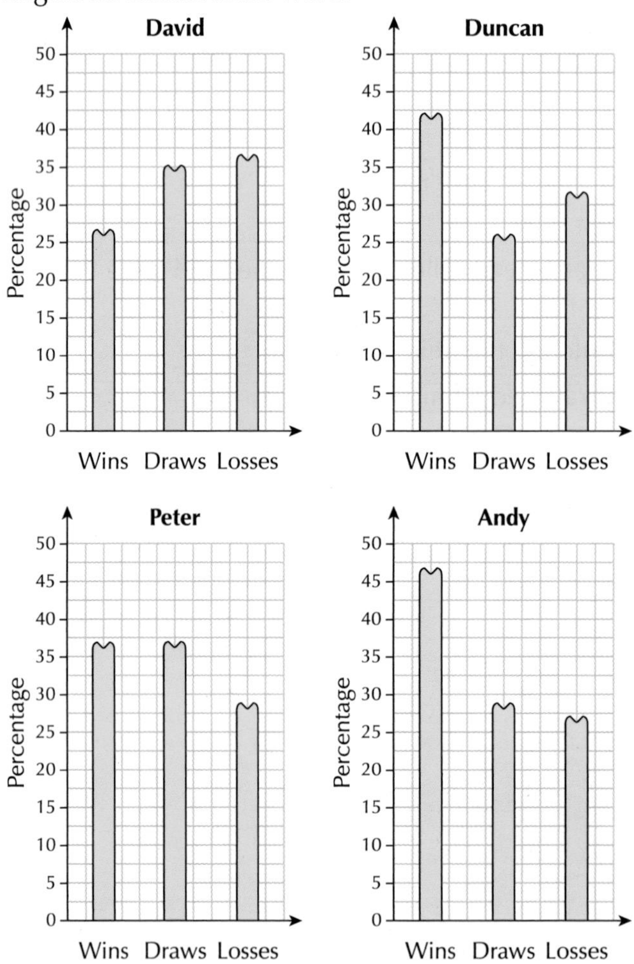

a Which manager won the highest percentage of his games?

b Which manager drew the least percentage of his games?

c Michael said: 'It's clear that Andy won more games than any of the other managers.'

Explain why Michael is incorrect.

d Which of these managers seems to be the least successful?

Give reasons to support your choice

e Why might these graphs not provide a fair comparison?

5 The table gives information about four dairy farms.

Farm	Number of cows in herd (2009)	Number of cows in herd (2014)
Dale Farm	230	199
Daisy Down Farm	74	77
Bannerdale	264	518
Carter Knowle	62	38
Total	630	832

a Which dairy farms had increased their herds from 2009 to 2014?

b Which dairy farm had shown the greatest decrease in their herd size?

c A TV headline was: 'Dairy farms are increasing in size!'
Is this statement true? Give reasons to support your answer.

Challenge: Off their trolley

The table shows the 2013 performance of the leading four supermarket chains.

Store	Number of stores	Sales (£ billion)	Profit (£ million)	Market share (%)
Tesco	3146	65	3500	30
Sainsbury	1106	23	788	18
Asda	565	23	484	18
Morrison's	500	18	879	12

A From the data in the table, draw a suitable diagram that clearly shows which chain had most sales.

B Now draw a diagram that makes it more difficult to see which of the four had the lowest:

a number of stores **b** market share.

4.3 Two-way tables

Learning objective

- To interpret a variety of two-way tables

Key words

two-way table

As you study mathematics, you will see many different types of table. **Two-way tables** are used for many types of data, from timetables to data-analysis tables. In this section you will practise interpreting a variety of them.

Example 2

Ollie and Eve go to the school car park and take note of the different cars that are there one day. They created this two-way table to show their results.

		Colour of cars				
		Silver	Red	Black	Blue	Other
Make of car	Peugeot	6	1	2	3	1
	Ford	8	2	1	1	3
	Honda	5	3	0	1	2
	BMW	2	1	4	0	0
	Other	3	2	1	2	3

a How many silver Peugeots were there?

b How many BMWs were there?

c How many blue cars were there?

d How many more Fords than Hondas were there?

a There were 6 silver Peugeots.

b There were $2 + 1 + 4 + 0 + 0 = 7$ BMWs.

c There were $3 + 1 + 1 + 0 + 2 = 7$ blue cars.

d There were 15 Fords and 11 Hondas, so there were 4 more Fords than Hondas.

Exercise 4C

1 Five friends, Neil, Helen, Robert, Jenny and Paul, belong to a club.
The table shows the activities they do one week.

	Monday	Tuesday	Wednesday	Thursday	Friday
Gymnastics	Neil Jenny		Jenny	Jenny Paul	Neil
Running	Robert	Helen Neil	Paul Robert Helen		
Badminton		Robert Jenny		Neil	Jenny Paul

a How many of the friends go to gymnastics?

b How many of the friends do not go to badminton?

c Who goes to all three activities?

PS **2** Six cousins took part in a family competition.
The table shows how many games each person won, drew or lost.

	Won	Drew	Lost
Philip	0	1	3
Pete	1	1	3
Brian	6	0	0
David	2	2	1
Malcolm	2	2	2
Kevin	0	2	2

a Which two were knocked out of the competition after the first round?

Four of them went into the semi-finals, but only two could then go into the final.

b Which two played in the final and who won?

c Which two lost in the semi-final?

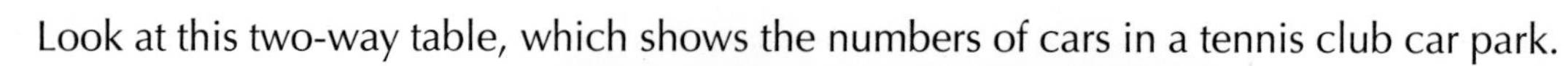

3 Look at this two-way table, which shows the numbers of cars in a tennis club car park.

		Colour of car				
		Red	White	Blue	Black	Other
Make of car	Volkswagen	9	2	5	2	3
	Kia	12	3	5	3	5
	Lotus	6	5	1	1	1
	Nissan	2	3	3	1	2
	Other	5	2	2	0	1

a How many blue Nissans are there?

b How many Kias are there?

c How many white cars are there?

d How many more Volkswagens than Lotuses are there?

MR PS **4** Reikie, Jana, Milo and Boris had a games competition.

They played two games, noughts and crosses, and boxes.

Each played each other at both games.

Jana recorded how many games each person won.

Reikie	///
Jana	///
Milo	//
Boris	////

Reikie recorded who won each game.

Noughts and crosses	Jana, Milo, Jana, Boris, Reikie, Jana
Boxes	Boris, Boris, Reikie, Boris, Milo

a Reikie has missed one name from her table.
Use Jana's table to say which name is missing.

b Who won the most games of noughts and crosses?

c Give a reason why Jana's table is a good way of recording the results.

d Give a reason why Reikie's table is a good way of recording the results.

e Create a two-way table showing as much of this information as possible.

5 The table shows the numbers of pupils who have school lunch in Y7, Y8 and Y9.

	Have school lunch	Do not have school lunch
Y7	120	64
Y8	97	87
Y9	80	104

a How do the numbers of pupils who have school lunch change as the pupils get older?

b Between which two years is the greatest change?
Explain your answer.

c Looking at the changes in the table, approximately how many pupils would you expect to have school lunch in Y10? Explain how you got your answer.

6 The table shows the percentages of boys and girls, by age group, who have android phones.

		Boys (%)	Girls (%)
Age	10	29	25
	11	32	29
	12	53	50
	13	64	67
	14	67	70
	15	73	75

a Work out the differences in the percentages for boys and girls at ages 10 to 15.

b Write down what you notice about the differences in the percentages for boys and girls.

Activity: A tall story

The table shows the heights of 70 Year 9 pupils.

		Boys	Girls
Height (cm)	130–139	4	2
	140–149	3	3
	150–159	11	11
	160–169	15	10
	170–179	7	4

A Put this information into a dual bar chart to show clearly the differences between the heights of the boys and the girls.

B Use the chart to decide if boys are taller than girls in Y9. Explain your answer.

4.4 Comparing two or more sets of data

Learning objective

- To compare two sets of data from statistical tables and diagrams

Data can be shown in all sorts of forms. You will often want to compare two sets of data, but where do you start? Follow through these examples to see two different ways data may be presented, allowing you to compare and comment on them.

Example 3

Two cars, X and Y, each cost £10 000 when they were new.

The bar charts show how the values of the cars fell over the next eight years.

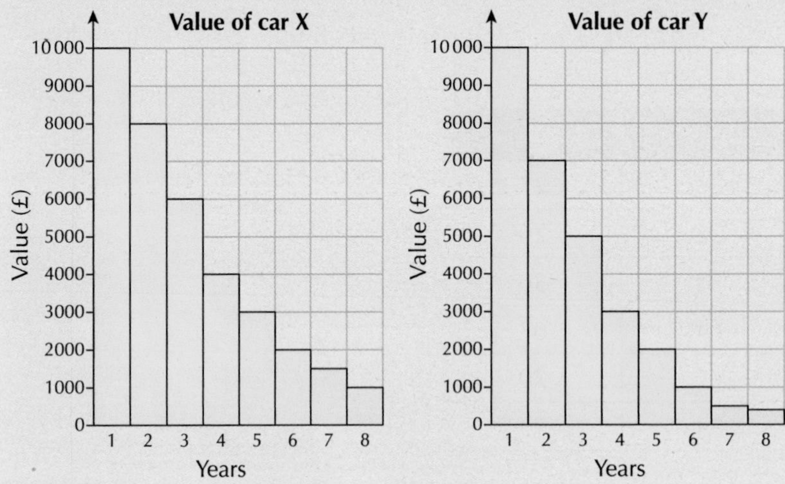

Comment on the differences between the values of the cars over the 8 years.

Looking at the two bar charts, you can make several observations.

- The value of car Y fell more quickly than the value of car X.
- Each year after the first, the value of car Y was less than that of car X.
- After 8 years, the value of car Y was less than half the value of car X.

Example 4

A teacher is comparing the reasons for the absence of pupils who have had time off school.

The charts show the reasons for absence of two different year groups.

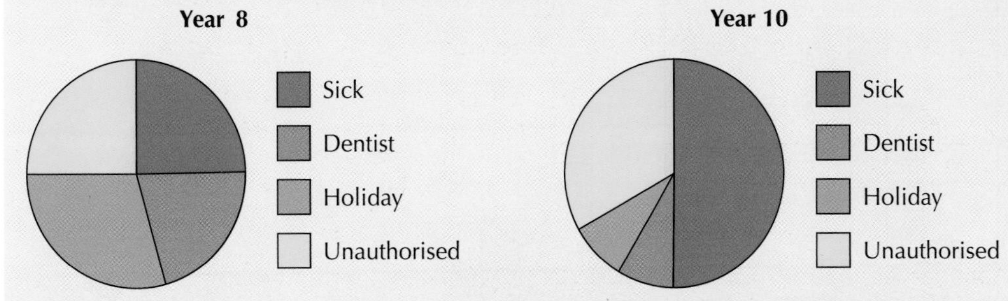

One hundred pupils in Year 8 and 40 pupils in Year 10 had time off school.

The teacher says: 'The charts show that more pupils in Year 10 than in Year 8 were absent because they were sick.'

Explain why the charts do not show this.

The number of Year 8 pupils absent because they were sick was a quarter of 100, which is 25.

The number of Year 10 pupils absent because they were sick was half of 40, which is 20.

So fewer Year 10 pupils than Year 8 pupils were absent because they were sick.

When comparing data sets, you must take care to ensure that you also take note of any large differences in sample sizes, as this could change your first impression of the comparison.

Exercise 4D

  A survey was taken in a class to find out girls' and boys' favourite seasons.

These charts show the results.

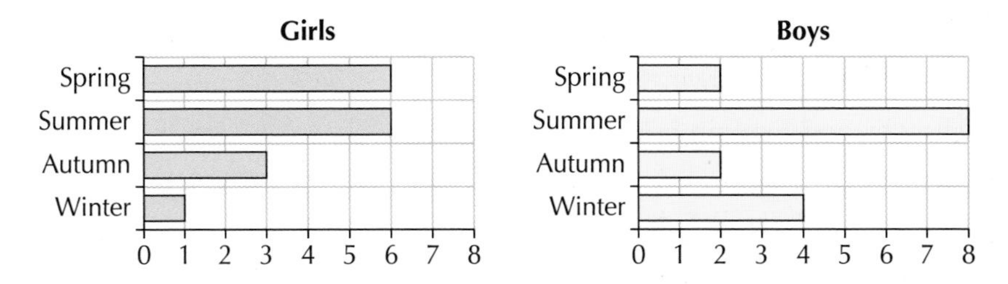

a How many chose summer as their favourite season?

b Which was the least favourite season among the girls?

c Ava said: 'Three times as many girls as boys like spring best.'

Is she correct? Explain how you know.

2 A tomato grower experimented by growing two similar tomato plants, one under blue light and one under normal light. At the end of the experiment he counted how many tomatoes there were on each truss on the plant. The charts show his results.

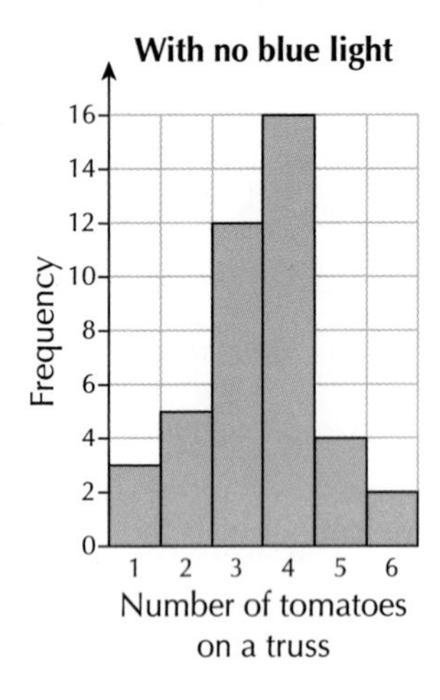

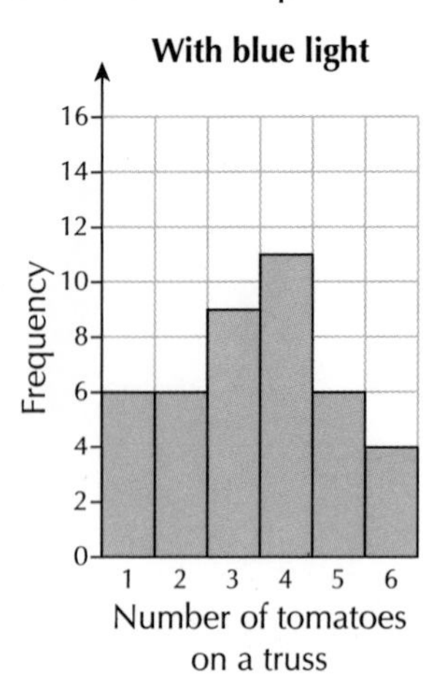

a Which tomato plant had most tomatoes on the plant?

b Comment on the effectiveness of growing the tomatoes under the blue light.

3 A survey was taken in two different schools about how pupils travel to school.

These are the results.

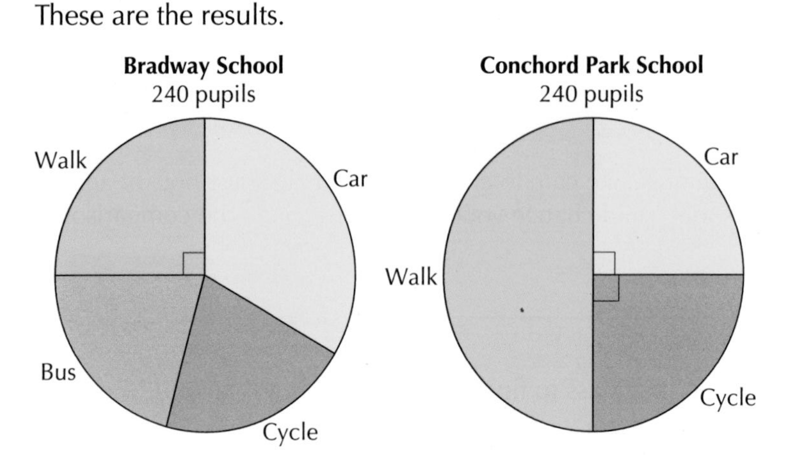

a Lily says: 'The number of pupils who cycle to school is higher for Conchord Park School than for Bradway School.'

Is she correct? Explain how you know.

b Explain how you know for sure that more Conchord Park School pupils walk to school than Bradway School pupils.

c Suggest two different reasons why no pupil from Conchord Park School uses a bus to get to school.

4 The graph shows the attendance at two park events, a rock band and a brass band. Comment on the proportions of children attending each event.

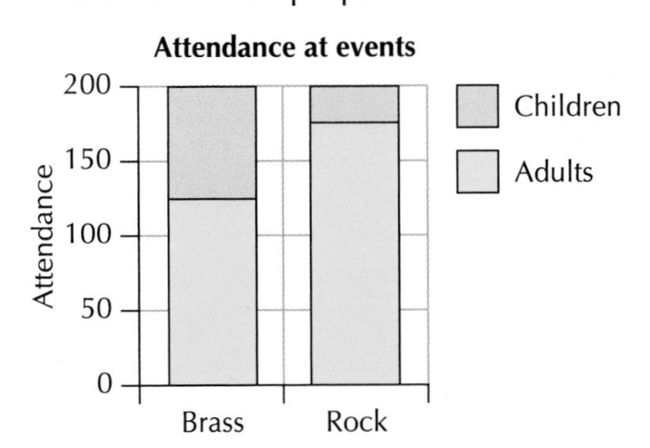

5 Eighty pupils took two English tests: a reading test and a written test.

The results are shown on the graph.

Which test did the pupils find more difficult?

Explain your answer.

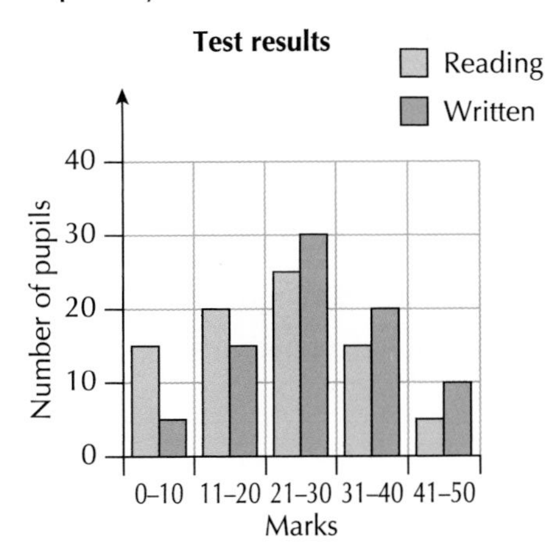

6 The chart shows the percentages of buses that were on time and late during one day in a UK city.

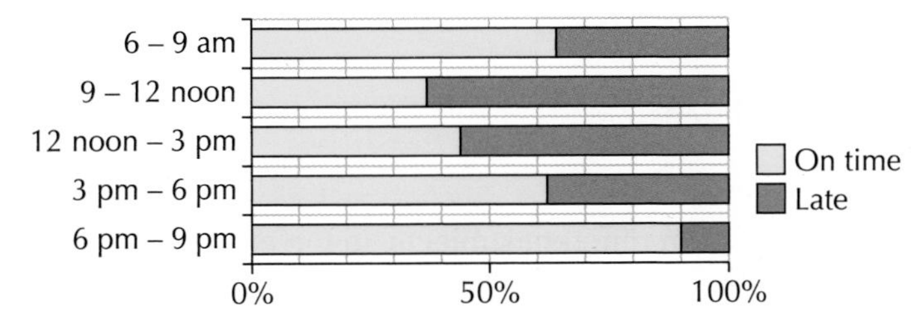

a Compare the lateness for different parts of the day.

b Comment on what you would expect to happen between 9 pm and 12 midnight.

  7 The table shows the circulation figures for some UK comics over a four-year period.

	2010	2012	2013
The Beano	46 700	36 000	32 000
Peppa Pig	83 000	86 900	98 900
Barbie	55 000	50 000	42 900
Toxic	40 200	47 000	53 000

a Which comic had the greatest increase of readers?

b Which comic had the greatest loss of readers?

c Draw a diagram illustrating this changing pattern of comic sales.

Activity: How many?

The table shows the population forecasts for the UK and Afghanistan.

Year	2015	2020	2025	2030	2035	2040	2045	2050
UK (millions)	62	63	64	64	65	65	64	63
Afghanistan (millions)	37	40	44	48	53	58	61	66

A Construct a graph for each country. Use the same scale on both graphs.

B Estimate the year when the populations of the two countries will be about equal.

C Estimate the populations of both countries in 2055. Give reasons for your answers.

4.5 Statistical investigations

Learning objective

- To plan a statistical investigation

Investigating a problem will involve several steps.

Look at this example of an overall plan.

You will apply the ideas to three examples, from different subjects, in the exercise.

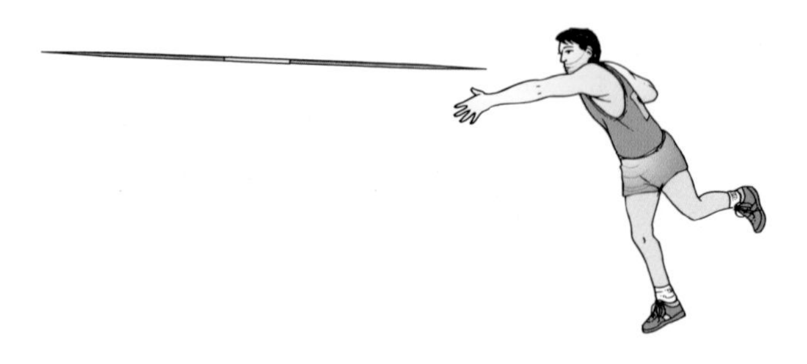

Step	Plan
1 Decide which general topic to study.	For this investigation, I am going to find out how to improve pupils' performance in sport.
2 Specify in more detail.	In particular, I am going to investigate the ability of pupils to throw a cricket ball.
3 Guess what you think could happen. (This is called 'Stating your hypotheses'.)	I will consider whether a run-up improves performance. I will also compare pupils of similar heights, as it is possible that height would also affect performance.
4 Conjectures	I think that using a run-up will improve the distance thrown – but if the run-up is too long it might then fail to improve performance. I think that Year 11 pupils of the same height may be physically stronger and would therefore throw further.
5 Sources of information required	I will carry out a survey of the distance thrown with different lengths of run-up.
6 Relevant data	I am going to choose pupils from Year 9 and Year 11, arranged in three groups according to height: short, medium height and tall. I will use 5 boys and 5 girls in each group. I will try to use pupils of different sporting abilities. Each pupil will have 3 throws, one with no run-up, one with a 5 metre run-up and one with a 10 metre run-up.
7 Possible problems	I will allow each pupil the same length of time, 5 minutes, to warm up. I will organise the event so that the throws are always taken in the same order. For example, the first throw for every pupil has no run-up. This should produce more reliable results.
8 Possible problems	I will put each pupil into a category according to their height and year group. I will then record the distance for each throw.
9 Decide on appropriate level of accuracy.	I will round all measurements to the nearest 10 cm.
10 Determine sample size.	In order to collect all this information effectively, I will ask a group of friends to help me.
11 Construct tables for large sets of raw data in order to make work manageable.	I will create a two-way table to record my results for each group.
12 Decide which statistic is most suitable.	I will calculate the mean for each group of results and then compare its value with my predictions.

This is an example of a recording sheet for Year 9 pupils of medium height.

Year 9 Medium height	Pupil 1	Pupil 2	Pupil 3	Pupil 4	Pupil 5
No run-up					
5 m run-up					
10 m run-up					

Exercise 4E

List in order the missing words from each plan given below.

The missing words for the science plan are:

car bias not books petrol nearest investigate engine

The missing words for the geography plan are:

mean incomes information compare sample average internet housing

Step	Q1 Science plan	Q2 Geography plan
1 Decide which general topic to study.	I am going to... the effect of engine size on a car's acceleration.	I am going to... life expectancy against the cost of housing.
2 Specify in more detail.	I will begin by studying only one make of....	I will compare house prices in Yorkshire with those in the south-east.
3 Guess what you think could happen. (State your hypotheses.)	I am going to try to find out if a bigger... always means that a car can accelerate faster.	I am going to investigate whether people in expensive... tend to live longer.
4 Conjectures	It may be that more powerful engines tend to be in heavier cars and therefore the acceleration is... affected. I am sure that larger engines in the same model of car will improve acceleration.	As people in expensive housing have greater..., they may also have a longer life expectancy.
5 Sources of information required	I am going to use car magazines and... to find information on engine sizes and the acceleration times for 0–60 mph.	I am going to use the library and search the... for census data for each area.
6 Relevant data	I am using 0–60 mph because the government requires car manufacturers to publish the time taken to accelerate from 0 to 60 mph.	I will record the... cost of housing for each area and also the life expectancy for each area.

7 Possible problems	I will keep a record of the make of car, the engine size and the acceleration time. I will only compare petrol engines with other… engines and not with diesel engines to avoid… in my results.	
8 Possible problems	I will also find out and record the mass of each car, as this is part of my guess at what will affect the results.	When I find the… that I need, I will make a note of where it came from.
9 Decide on appropriate level of accuracy.	I will round engine sizes to the… 100 cm³. For example, a car with an engine capacity of 1905 cc (this is the same as cm³ but is what the motor trade use) will be recorded as 1900 cc.	
10 Determine sample size.		
11 Construct tables for large sets of raw data in order to make work manageable.		I will group the data about the population in age groups of five-year intervals.
12 Decide which statistic is most suitable.		I will make sure that I look at at least 30 pieces of data for each area so that my… is large enough to calculate the… and have a reliable answer.

Investigation: On your bike

Think of a problem related to pupils who cycle to school. Collect as much data as you can, and write up your plan. Use the planning steps in this section as your guides.

Ready to progress?

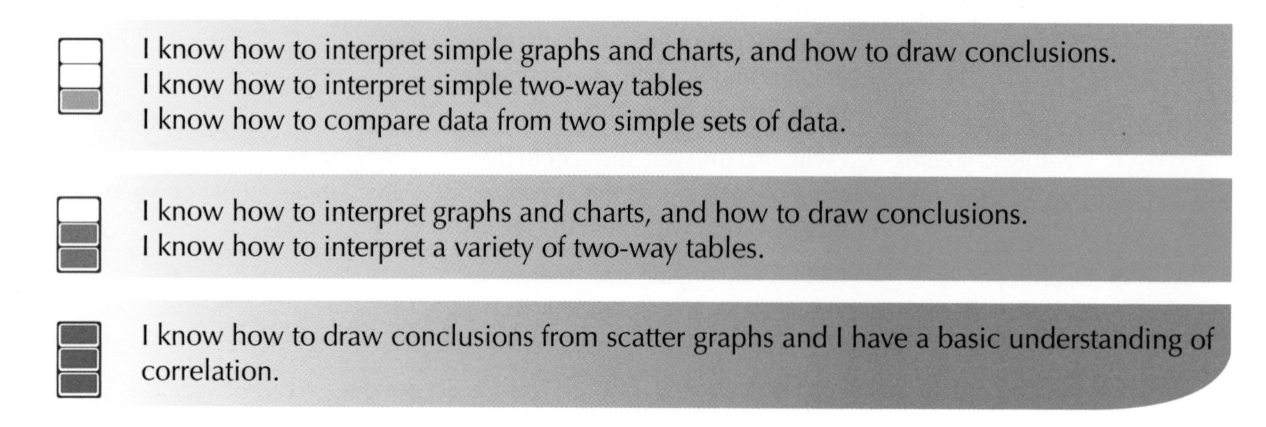

I know how to interpret simple graphs and charts, and how to draw conclusions.
I know how to interpret simple two-way tables
I know how to compare data from two simple sets of data.

I know how to interpret graphs and charts, and how to draw conclusions.
I know how to interpret a variety of two-way tables.

I know how to draw conclusions from scatter graphs and I have a basic understanding of correlation.

Review questions

MR **1** Iain asked 30 pupils: 'Do you like school dinners?' He offered three choices: 'yes', 'no' or 'don't know'.

The bar chart shows his results.

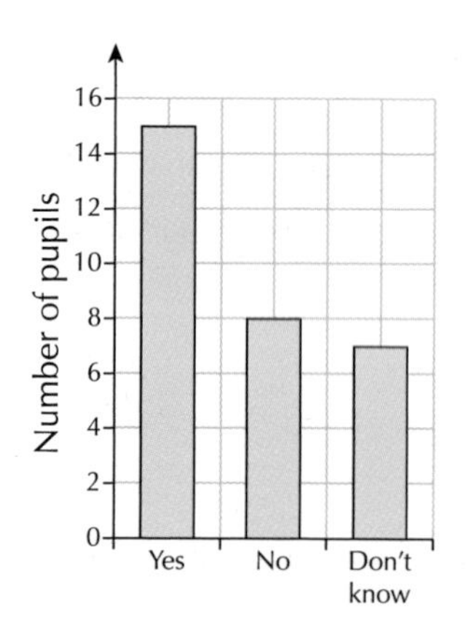

Duncan also asked 30 people the same question.

The pictogram shows his results.

Yes	○ ○ ○ ○ ○ ○
No	○ ○ ○ ○ ○ ○ ○
Don't know	○ ○

a Comment on the differences between the two boys' results.

b Combine all the data from both charts into one suitable chart.

MR **2** At the Taj Mahal in India, they keep a record of the number of Indians visiting. as well as the number of tourists.

The pie chart illustrates the numbers for one week.

Visitors to the Taj Mahal

Indian men
Indian women
Female tourists
Male tourists

a Harri said: 'There are twice as many Indian men as Indian women.'

Is Harri correct? Explain how you know.

b Riva said: 'There are twice as many men as there are women.'

Is Riva correct? Explain how you know.

MR **3** Paul grew potatoes. His friend suggested growing them in straw rather than soil.

He decided to try this and set the same number of seed potatoes in soil as he did in straw. He waited for them to grow.

He made two charts showing how many potatoes each seed had produced.

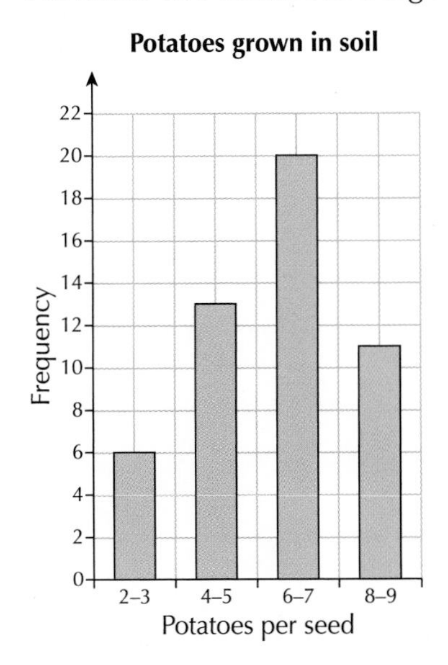

Potatoes grown in straw

Frequency

0 2 4 6 8 10 12 14 16 18 20 22

2–3 4–5 6–7 8–9

Potatoes per seed

a How many seed potatoes did Paul set to grow?

b Make comments about the effectiveness of growing the potatoes in straw.

4 During the summer cricket season, Abass thought that it was most likely that the warmer the day, the more people would come to local cricket matches.

He decided to keep a record one season of all the games and the midday temperature.

His results are shown in the table.

Match	1	2	3	4	5	6	7	8	9	10	11	12
Temperature (°C)	17	23	19	20	16	19	21	22	21	24	22	20
Attendance	90	120	115	115	70	110	125	125	120	130	130	150

a Draw a scatter diagram to display the data in the table.

Label the horizontal axis as temperature, from 15 °C to 25 °C.

Label the vertical axis as attendance, from 60 to 160.

b What correlation can you see from your scatter diagram?

c If one midday the temperature was 18 °C, estimate what you think the attendance would have been.

d One point looks out of place. Which game is this and can you think of a reason why it should be so?

Challenge

Rainforest deforestation

Since 1970, over 600 000 km^2 of Amazon rainforest have been destroyed. This is an area larger than Spain.
Between the years 2000 and 2005, Brazil lost over 132 000 km^2 of forest – an area about the same size as England.
Between the years 2005 and 2013, Brazil lost over 90 000 km^2 of forest – an area about the same size as Scotland.
The table below shows how much of the rainforests in Brazil have been lost each year since 1988.

Deforestation figure	
Year	**Deforestation (sq km)**
1988	21 000
1989	18 000
1990	14 000
1991	11 000
1992	14 000
1993	15 000
1994	15 000
1995	29 000
1996	18 000
1997	13 000
1998	17 000
1999	17 000
2000	18 000
2001	18 000
2002	21 000
2003	25 000
2004	27 000
2005	19 000
2006	14 000
2007	10 000
2008	9000
2009	7000
2010	6000
2011	5000
2012	6000

Use the information on the opposite page to answer these questions.

1 Draw a bar chart showing the deforestation of Brazil over the years from 1988 to 2012.

2 From 1988 to 1991, Brazil had an economic slowdown.
What was happening to the rate of deforestation during that time?

3 From 1992 to 1995, Brazil had economic growth.
What was happening to the rate of deforestation during that time?

4 What do think was happening to Brazil's economy:

a from 1998 to 2004

b from 2005 to 2011?

5 What does the chart and the information given in questions 1 and 2 suggest about the link between deforestation in Brazil and the economy?

The pie chart below shows the three main reasons for deforestation in the Amazon from 2000 to 2014.

Deforestation in the Amazon, 2000–2014

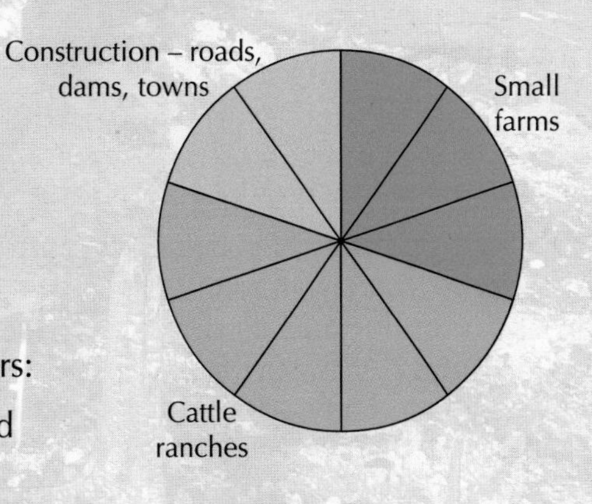

6 What appears to be the main reason for the deforestation in the Amazon between the years 2000 and 2014?

7 What percentage of the deforestation was caused by construction?

8 It was suggested that over the next two years:

- the same amount of deforestation would take place.
- the amount of construction work would increase by a half
- the number of small farms would halve
- the number of cattle ranches would increase slightly.

Draw a new pie chart reflecting the reasons for deforestation suggested for 2016.

5

Circles

This chapter is going to show you:

- how to use π
- how to use π to calculate the circumference of a circle
- how to use π to calculate the area of a circle.

You should already know:

- that the formula to work out the approximate length of the circumference of a circle is $C = 3d$.

About this chapter

Circles are all around us.

The circle is probably the most important shape in the universe. It is also the most mysterious.

You can only measure the circumference (perimeter) of a circle in terms of a number called π. You say this as 'pi'. This number is also mysterious. It cannot be written exactly as a number and its decimal places go on for ever. Its decimal digits never settle into a permanent repeating pattern. They appear to be randomly distributed, although no one has proved this yet.

In the 20th and 21st centuries, mathematicians and computer scientists, helped by increasing computational power, extended the decimal representation of π to over 10 trillion digits. Memorising π is now a sport, with records set by people who can remember more than 67 000 digits.

5.1 The formula for the circumference of a circle

Learning objective

- To calculate the circumference of a circle

Key word

π

In Book 2.1, you found that the circumference, C, of a circle with diameter d, is given approximately by the formula $C = 3d$.

In fact, for a more accurate value for the circumference, you need to multiply the diameter by a number that is slightly larger than 3.

This special number is represented by the Greek letter π (pronounced pi). It is impossible to write down the value of π exactly, as a fraction or as a decimal, so you will use approximate values.

The most common of these are:

- $\pi = 3.14$ (as a decimal rounded to two decimal places)
- $\pi = 3.142$ (as a decimal rounded to three decimal places)
- $\pi = 3.141\,592\,654$ (on a scientific calculator)

Mathematicians have used computers to calculate π to trillions of decimal places. So far, no repeating pattern has ever been found.

Look for the **π** key on your calculator.

On some calculators, you may need to key in **SHIFT** **×10^x** to use π.

So, the formula for calculating the circumference, C, of a circle with diameter d is:

$$C = \pi d$$

Example 1

Calculate the circumference of this circle.

Give your answer correct to one decimal place (1 dp).

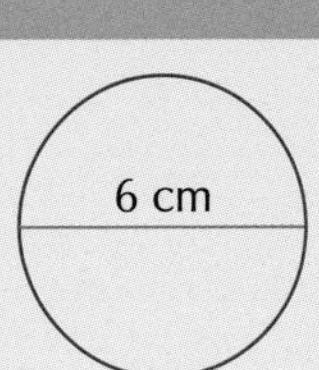

You are given the diameter and the units are centimetres.

For π, use 3.14 or use the **π** key on your calculator.

The diameter $d = 6$ cm, which gives:

$C = \pi d$

$= \pi \times 6$

$= 3.14 \times 6$

$= 18.8$ cm (to 1 dp)

or

$C = \pi d =$ **π** $\times 6$

$= 18.849\,555\,92$

$= 18.8$ cm (to 1 dp)

Example 2

Calculate the circumference of this circle.

Give your answer correct to one decimal place (1 dp).

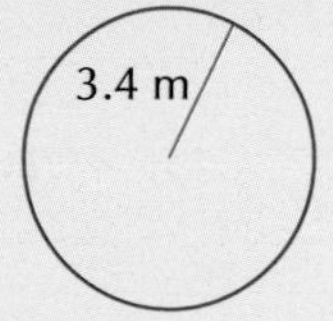

You are given the radius and the units are metres.

$r = 3.4$ m and so $d = 3.4 \times 2 = 6.8$ m

This gives:

$C = \pi d$

$= \pi \times 6.8$

$= 3.14 \times 6.8$

$= 21.4$ m (to 1dp)

or

$C = \pi d =$ **π** $\times 6.8$

$= 21.362\ 830\ 04$

$= 21.4$ m (to 1dp)

Exercise 5A

In this exercise, take $\pi = 3.14$ or use the π key on your calculator.

1 Calculate the circumference of each circle.

Give each answer correct to one decimal place.

a

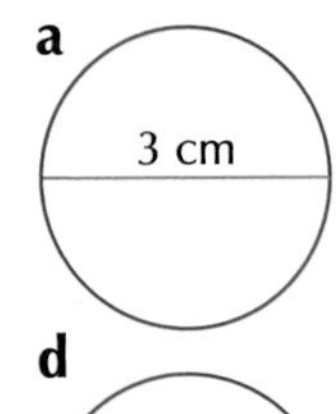

b

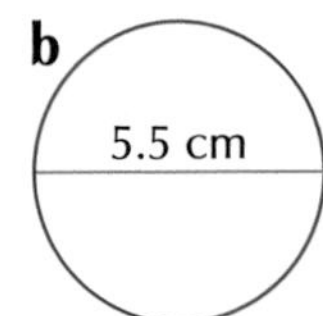

c

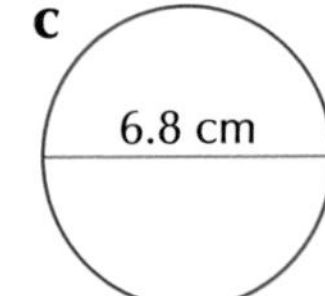

d

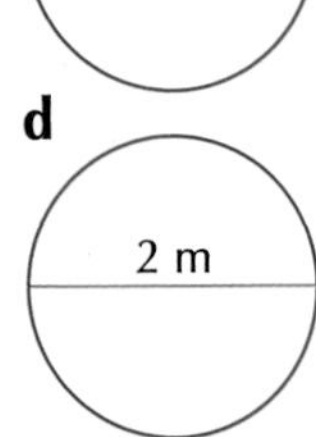

e

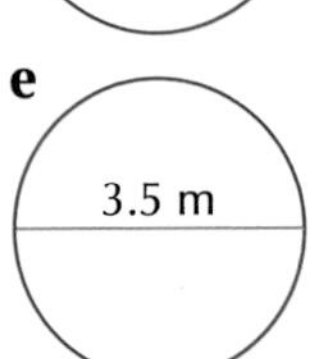

f

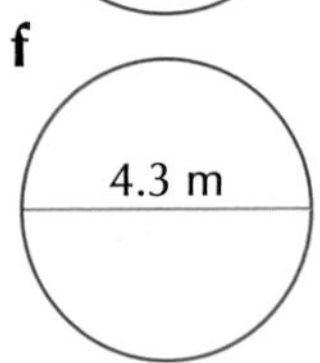

2 Calculate the circumference of each circle.

Give each answer correct to one decimal place.

a

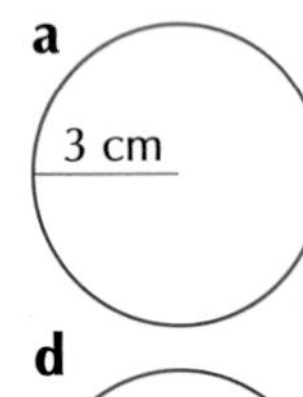

b

c

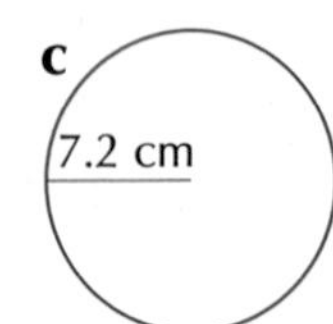

d

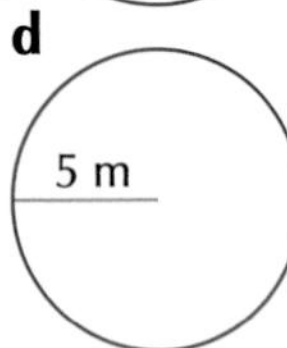

e

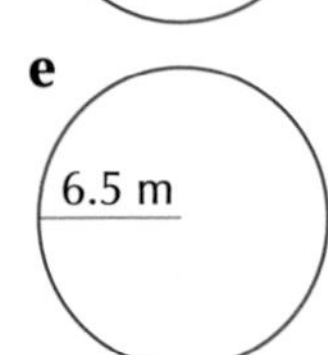

f

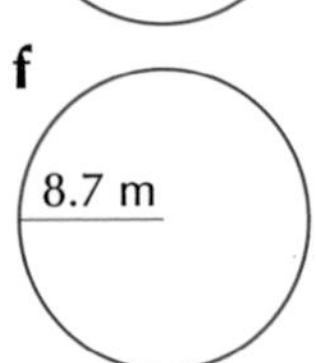

3 The table shows the diameters of British coins.

Copy the table and complete the 'circumference' column.

Give each answer correct to the nearest millimetre.

Coin	Diameter (mm)	Circumference
1p	20.3	
2p	25.9	
5p	18	
10p	24.5	
£1	22.5	
£2	28.4	

4 The diagram shows the diameter of a circular running track at a sports centre.

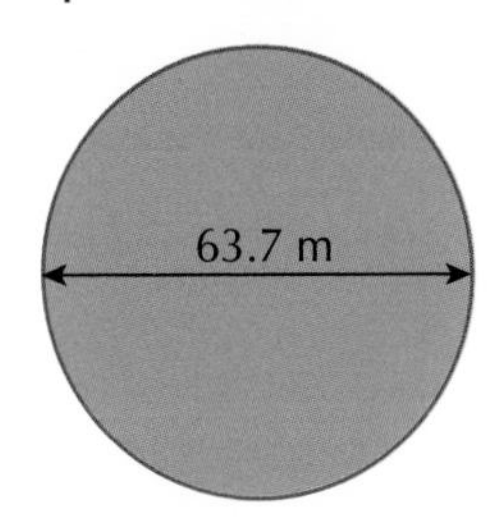

Calculate the distance round the track.

Give your answer correct to the nearest metre.

5 The length of each side of this square is 5 cm.

The radius of the circle is 3 cm.

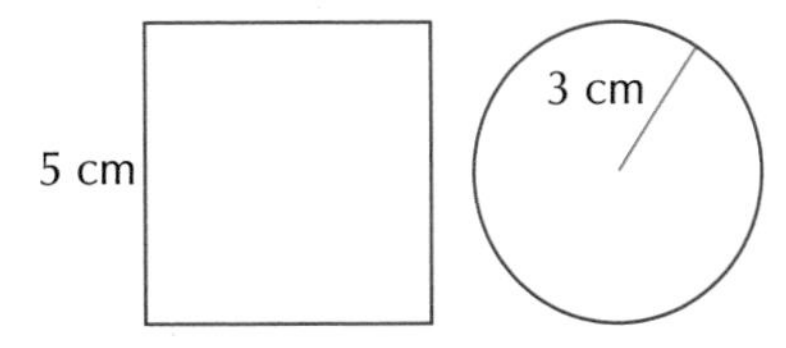

Which shape has the greater perimeter?

Explain your answer.

6 The diameter of a wheel on a BMX bike is 50.8 cm.

The diameter of a wheel on a mountain bike is 66 cm.

Each wheel turns 500 times.

How much further does the mountain bike travel?

Give your answer correct to the nearest metre.

Problem solving: To calculate the perimeter of a semicircle

Calculate the perimeter, P, of this semicircle.

Give your answer correct to one decimal place.

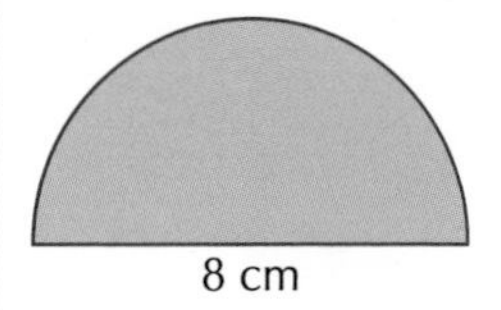

First, calculate the circumference of a circle with $d = 8$ cm.

$$C = \pi d$$
$$= \pi \times 8$$
$$= 25.13 \text{ cm (to 2 dp)}$$

To work out the length of the curved part of the semicircle, divide the circumference by 2.

$$25.13 \div 2 = 12.57$$

Now add the diameter, to work out the total perimeter.

The diameter is 8 cm.

So $P = 12.57 + 8$

$$= 20.57$$
$$= 20.6 \text{ cm (to 1 dp)}$$

Calculate the perimeter of these semicircles.

Give your answers correct to one decimal place.

A

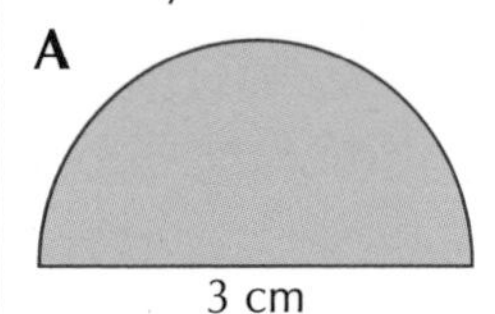

B

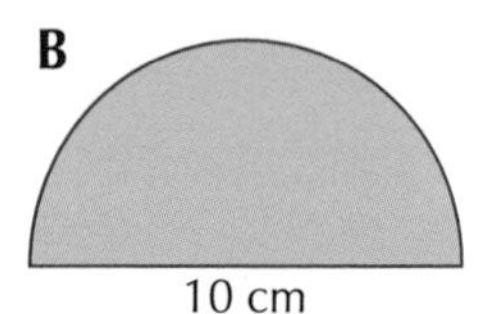

5.2 The formula for the area of a circle

Learning objective

- To calculate the area of a circle

This circle has been split into 16 equal sectors. These sectors have been placed together to form a shape that is roughly a parallelogram.

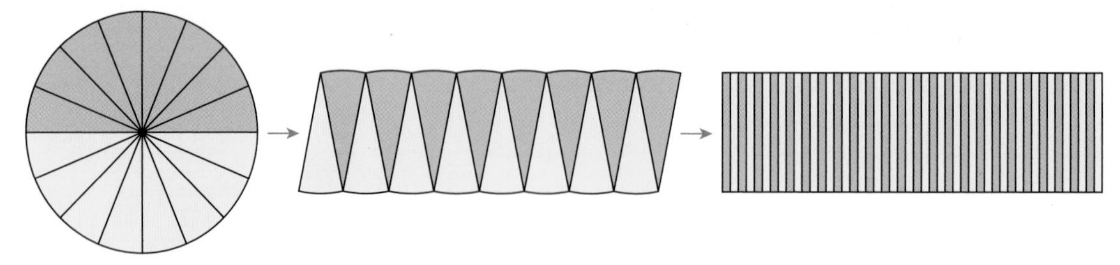

As the circle is split into more and thinner sectors, and then these are placed together, the resulting shape eventually becomes a rectangle.

The area of this rectangle will be the same as the area of the circle.

The length of the rectangle is half the circumference, C, of the circle and its width is the radius, r, of the circle.

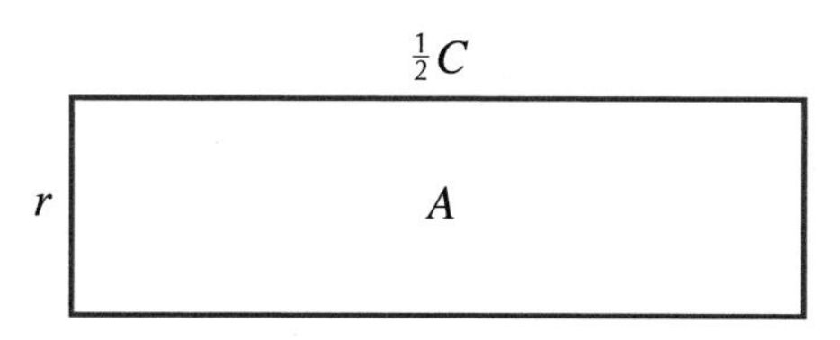

So, the area, A, of the rectangle is given by:

$$A = \tfrac{1}{2}C \times r$$

The formula for the circumference of a circle is:

$$\begin{aligned} C &= \pi d \\ &= \pi \times 2 \times r \\ &= 2\pi r \end{aligned}$$

So $\tfrac{1}{2}C = \pi r$

$$\begin{aligned} A &= \pi r \times r \\ &= \pi r^2 \end{aligned}$$

Hence, the formula for the area, A, of a circle of radius r is:

$$A = \pi r^2$$

Example 3

Calculate the area of this circle.

Give your answer correct to one decimal place.

3 cm

You are given the radius and the units are centimetres.

The radius, $r = 3$ cm, which gives:

$$\begin{aligned} A &= \pi r^2 \\ &= \pi \times 3^2 \\ &= 3.14 \times 9 \\ &= 28.3 \text{ cm}^2 \text{ (to 1 dp)} \end{aligned}$$

When you are using a calculator, you can use the 'square' key x^2.

Simply key in: π $\times$ 3 x^2 = 28.274 333 88 = 28.3 cm^2 (to 1 dp).

Example 4

Calculate the area of this circle.

Give your answer correct to one decimal place.

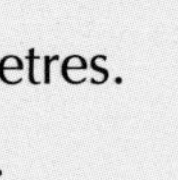

You are given the diameter and the units are metres.

The diameter $d = 3.4$ m, so $r = 3.4 \div 2 = 1.7$ m.

This gives:

$$A = \pi r^2$$
$$= \pi \times 1.7^2$$
$$= 3.14 \times 2.89$$
$$= 9.1 \text{ m}^2 \text{ (to 1 dp)}$$

When using a calculator, you can use the 'square' key x^2.

Simply key in: π × 1 · 7 x^2 = 9.079 202 769 = 9.1 m² (to 1 dp).

Exercise 5B

In this exercise, take $\pi = 3.14$ or use the π key on your calculator.

1 Calculate the area of each circle.

Give your answers correct to one decimal place.

a 2 cm **b** 5 cm **c** 7.2 cm

d 1 m **e** 4 m **f** 5.4 m

2 Calculate the area of each circle.

Give your answers correct to one decimal place.

a 5 cm **b** 7 cm **c** 8.4 cm

d 6 m **e** 11 m **f** 12 m

3 The diameter of a 5p coin is 18 mm.

Calculate the area of one face of the coin.

Give your answer correct to the nearest square millimetre.

4 Jenny is working out the area of this circle.

4 cm

This is her working.

Area $= \pi \times d$

$= \pi \times 8$

$= 8\pi$ cm

Explain why Jenny's working is wrong.

Write down the correct answer to the problem.

5 Finlay has cut out a circle of radius 3 cm and Jackson has cut out a circle of radius 6 cm.

Read what Jackson says to Finlay.

My circle is twice the size of yours, so the area of my circle should be twice the area of yours.

Calculate the areas of both circles, giving your answers correct to two decimal places.

Is Jackson correct?

Give a reason for your answer.

6 Calculate the area of a circular tablemat with a diameter of 21 cm.

Give your answer correct to the nearest square centimetre.

7 Four circles are cut out from a square sheet of card with edge length 12 cm.

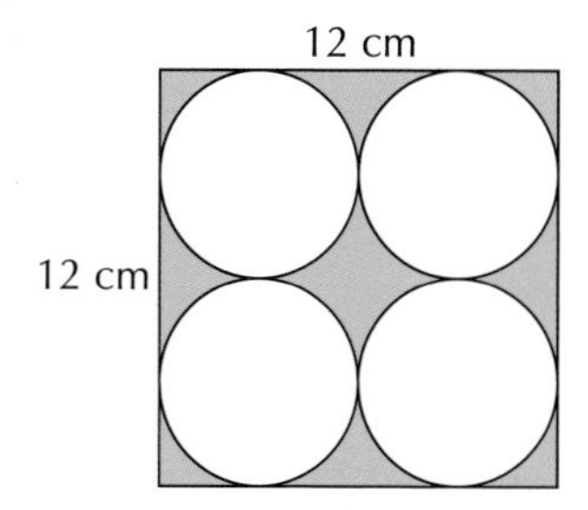

Work out the radius of one of the circles first.

Work out the area of the remaining card after the circles have been cut out. Give your answer correct to the nearest square centimetre.

Problem solving: Area of a semicircle

Calculate the area of this semicircle.

Give your answer correct to one decimal place.

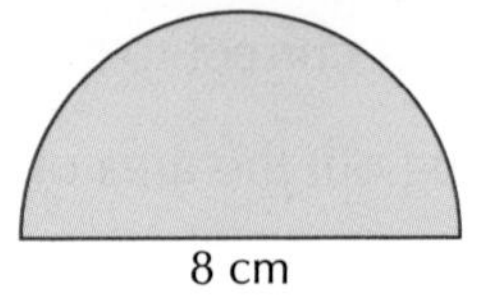

First, calculate the area of a circle.

$d = 8$ cm, so $r = 4$ cm.

$A = \pi r^2$

$= \pi \times 16$

$= 50.27$ cm^2 (to 2 dp)

To work out the area of the semicircle, divide this by 2.

$A = 50.27 \div 2$

$= 25.1$ cm^2 (to 1 dp)

Calculate the area of each of these semicircles.

Give your answers correct to one decimal place.

A

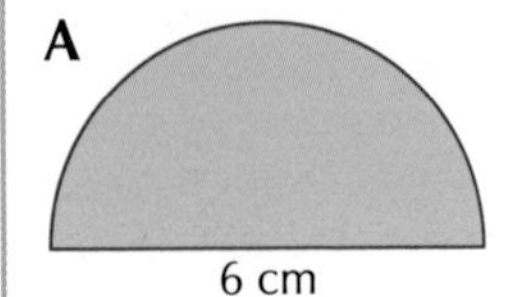

B

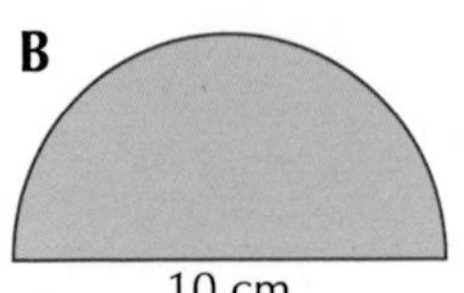

5.3 Mixed problems

Learning objective

- To solve problems involving the circumference and area of a circle

Exercise 5C

In this exercise, take $\pi = 3.14$ or use the π key on your calculator.

1 A circular paddling pool has a diameter of 6 m.

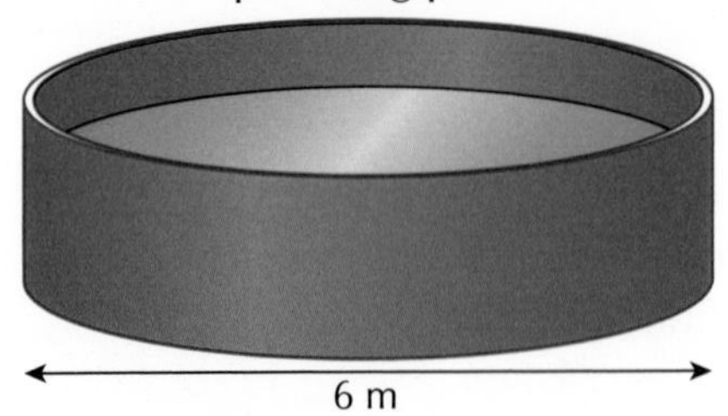

Work out the distance around the pool.

Give your answer correct to one decimal place.

2 A circular carpet has a diameter of 2.4 m.

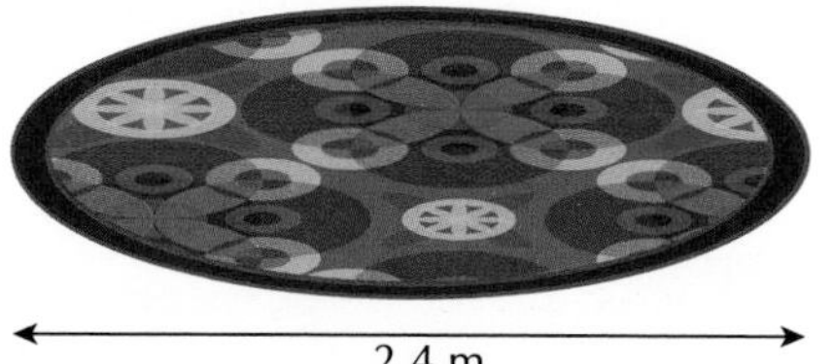

Work out the area of the carpet.

Give your answer correct to one decimal place.

3 The radius of the Earth is approximately 6400 km.

Work out the approximate distance around the equator.

Give your answer correct to the nearest 100 kilometres.

4 The radius of a circular lawn is 5 m.

The contents of a bottle of weedkiller are enough to treat 50 m^2 of the lawn.

How many bottles are needed to treat the whole lawn?

5 The radius of a circular table top is 60 cm.

a Work out the area of the table top.

Give your answer correct to the nearest square centimetre.

b Work out the area of the table top, in square metres.

Give your answer correct to two decimal places.

6 The diagram represents a running track.

The two semicircles make a complete circle.

The bends are semicircles.

Work out the total distance around the track.

Give your answer correct to the nearest metre.

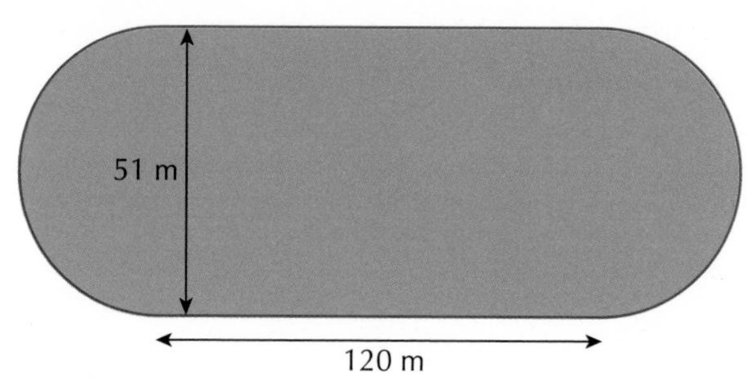

Ready to progress?

I can calculate the circumference of a circle.
I can calculate the area of a circle.

Review questions

In this exercise, take $\pi = 3.14$ or use the π key on your calculator.

1 Work out:

i the circumference

ii the area of each circle.

Give you answers correct to one decimal place.

a 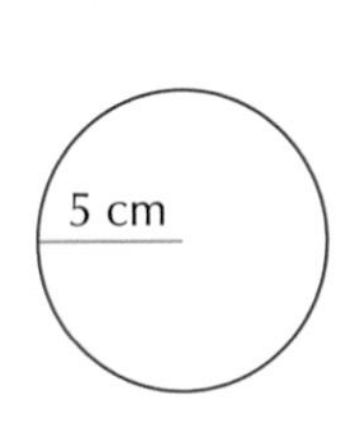

b 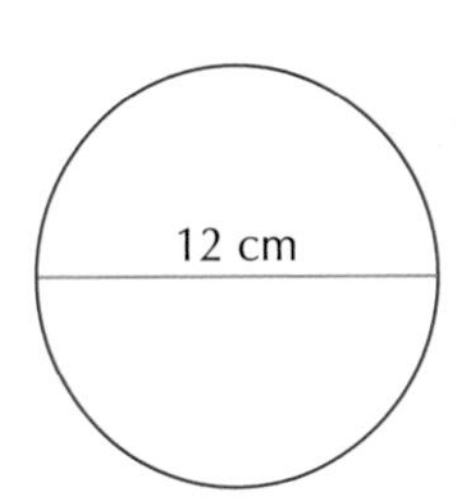

c 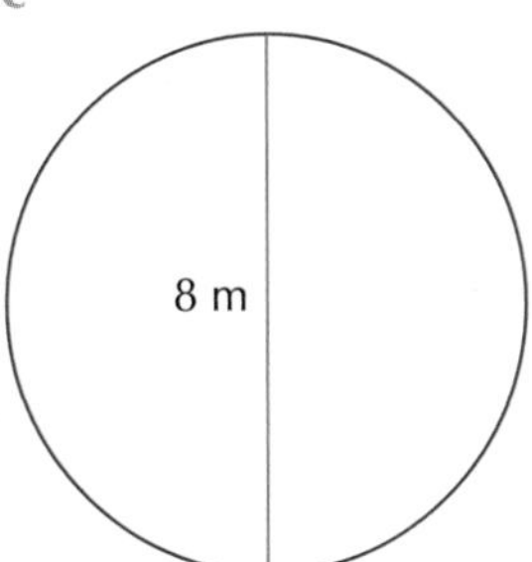

2 A circle has a radius of 2 cm. Work out:

a the circumference

b the area of the circle.

Give you answers correct to one decimal place.

Write down anything you notice about your answers.

3 Becky runs around a circular lake five times.

The lake has a diameter of 80 m.

How far does she run?

Give your answer to the nearest 10 metres.

MR 4 The dimensions of a rectangle and a circle are given in the diagram.

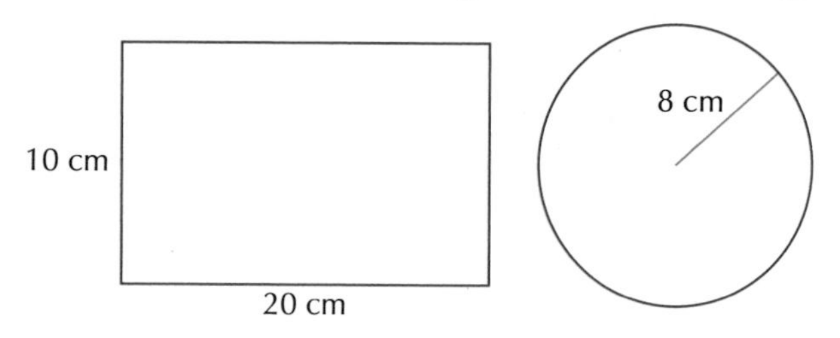

Which shape has the greater area?

Explain your answer.

5 The diagram shows a cotton reel.

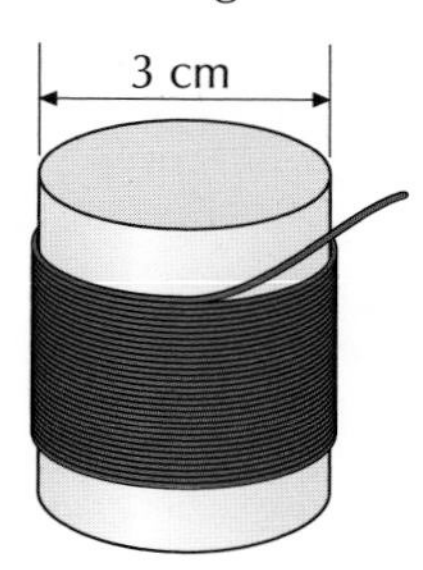

The diameter of the circular end is 3 cm.

a Work out the circumference of the circle.

Give your answer correct to one decimal place.

b The reel holds 50 m of cotton thread.

About how many times is the cotton wrapped around the reel?

Round your answer to the nearest ten.

6 A circle has a circumference of 50 cm.

Work out the diameter of the circle.

Give your answer correct to one decimal place.

Financial skills

Athletics stadium

An athletics stadium is being redeveloped.

1 The shot-put circle is going to be resurfaced.

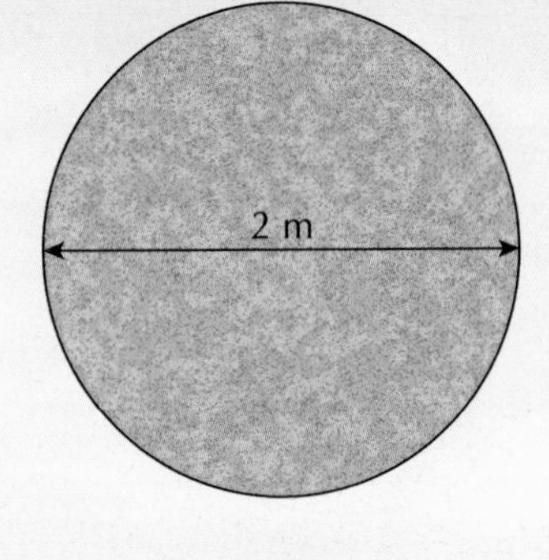

a What is the circumference of the circle? Give your answer correct to one decimal place.

b What is the area of the circle? Give your answer correct to one decimal place.

c The cost for resurfacing is £32 per square metre. How much will it cost to resurface the shot-put circle? Give your answer to the nearest pound.

2 The high-jump zone is also going to be resurfaced.

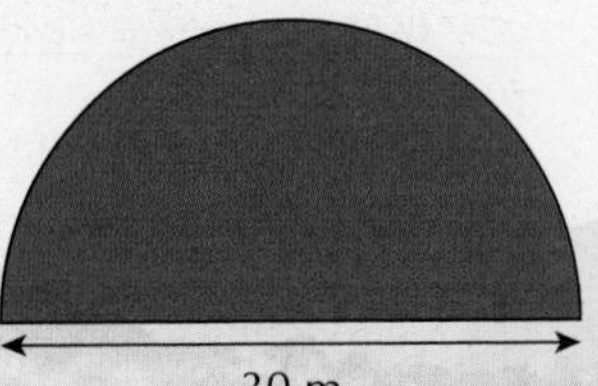

a What is the area of the high-jump zone? Give your answer correct to the nearest square metre.

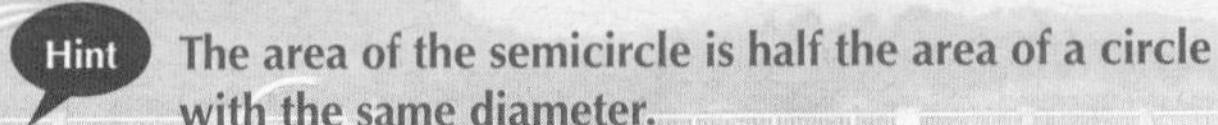

Hint The area of the semicircle is half the area of a circle with the same diameter.

b Resurfacing costs £60 per square metre. How much will it cost to resurface the high-jump zone? Give your answer correct to the nearest £1000.

3 The area inside the circular running track is going to be returfed. This diagram shows the diameter of the running track.

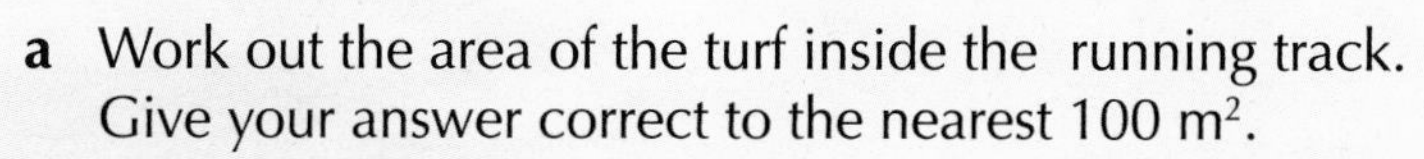

64 m

a Work out the area of the turf inside the running track. Give your answer correct to the nearest 100 m^2.

b A quote for returfing is £3.60 per square metre (including labour costs). How much will it cost to returf the area? Give your answer correct to the nearest £100.

4 This diagram shows the dimensions of the sandpit for the long jump.

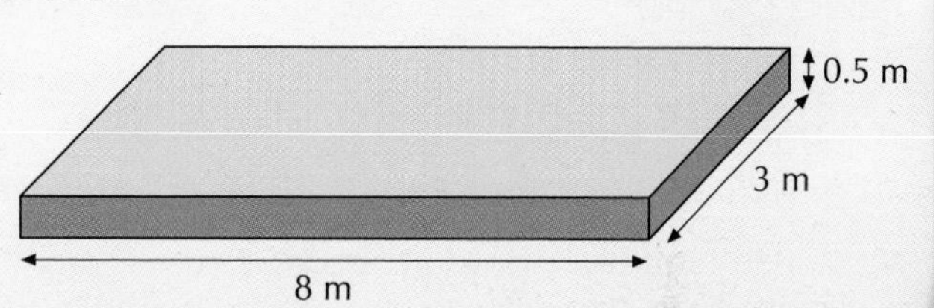

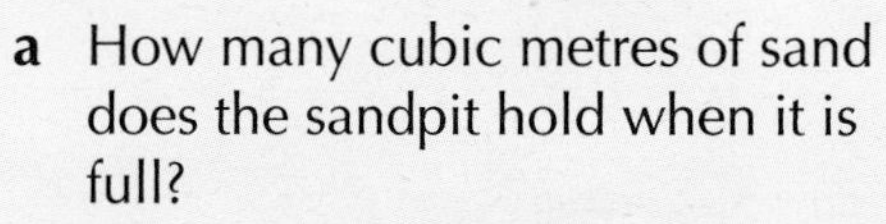

a How many cubic metres of sand does the sandpit hold when it is full?

b The cost of one cubic metre of sand is £45. How much will it cost to fill the sandpit?

5 A ramp for wheelchair access into the stadium is going to be built. This diagram shows the dimensions of the ramp.

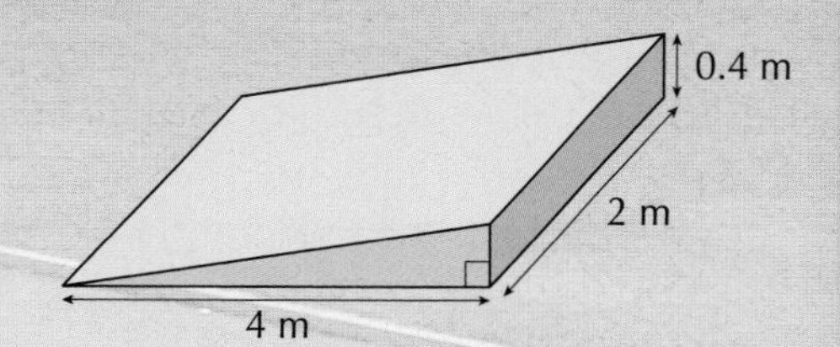

a What volume of concrete will be needed to make the ramp?

The volume of the ramp is half the volume of a cuboid with the same measurements.

b Concrete costs £50 per cubic metre. How much will it cost for the concrete?

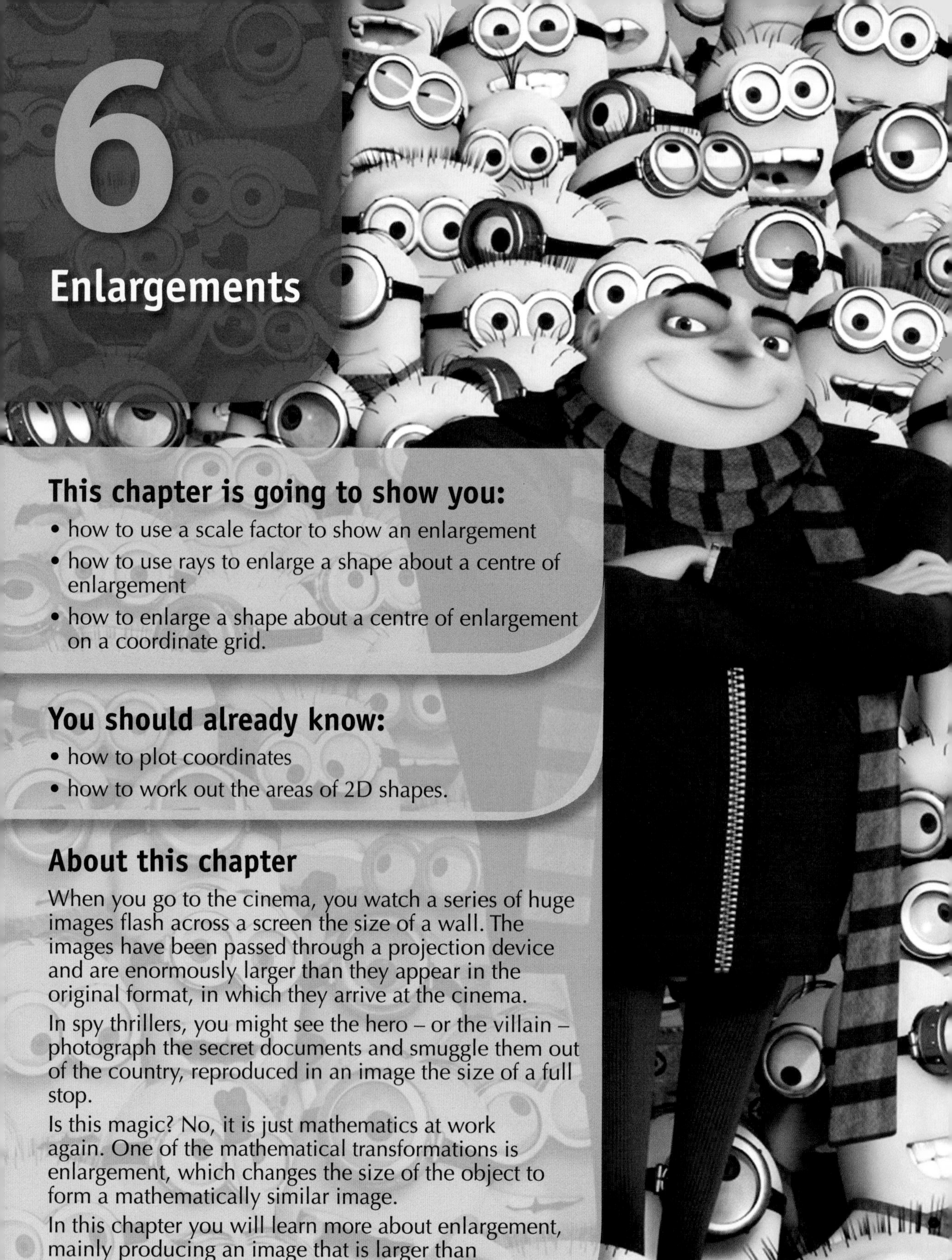

6 Enlargements

This chapter is going to show you:

- how to use a scale factor to show an enlargement
- how to use rays to enlarge a shape about a centre of enlargement
- how to enlarge a shape about a centre of enlargement on a coordinate grid.

You should already know:

- how to plot coordinates
- how to work out the areas of 2D shapes.

About this chapter

When you go to the cinema, you watch a series of huge images flash across a screen the size of a wall. The images have been passed through a projection device and are enormously larger than they appear in the original format, in which they arrive at the cinema.

In spy thrillers, you might see the hero – or the villain – photograph the secret documents and smuggle them out of the country, reproduced in an image the size of a full stop.

Is this magic? No, it is just mathematics at work again. One of the mathematical transformations is enlargement, which changes the size of the object to form a mathematically similar image.

In this chapter you will learn more about enlargement, mainly producing an image that is larger than the object. In later work, you will discover that 'enlargements' can also produce images that are smaller than the object, but that is mathematics for you!

6.1 Scale factors and enlargements

Learning objective

- To use a scale factor to show an enlargement

Key words	
enlarge	enlargement
scale factor	similar

Rectangles B, C and D are all **enlargements** of rectangle A.

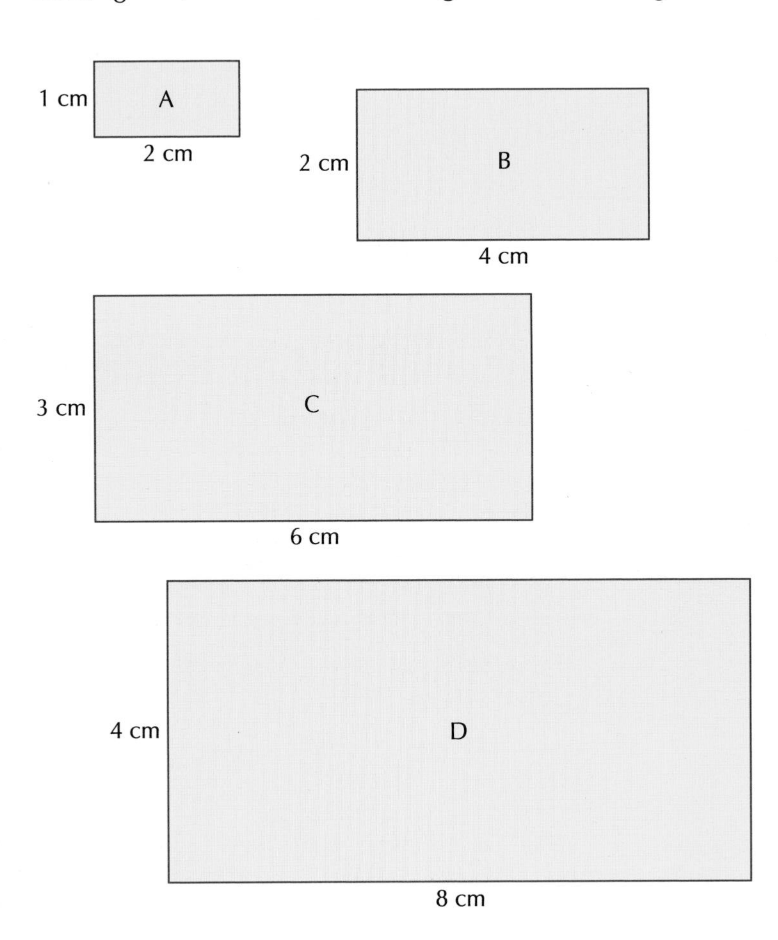

The **scale factor** describes the size of the enlargement.

Rectangle A has been **enlarged** by a scale factor of 2 to produce rectangle B. This means that the lengths of the sides of rectangle B are twice those of rectangle A.

Rectangle C is an enlargement of rectangle A with a scale factor of 3.

Rectangle D is an enlargement of rectangle A with a scale factor of 4.

When a shape is enlarged to make another shape, the two shapes are **similar** to each other.

Example 1

Rectangle Y is an enlargement of rectangle X.

Work out the scale factor for the enlargement.

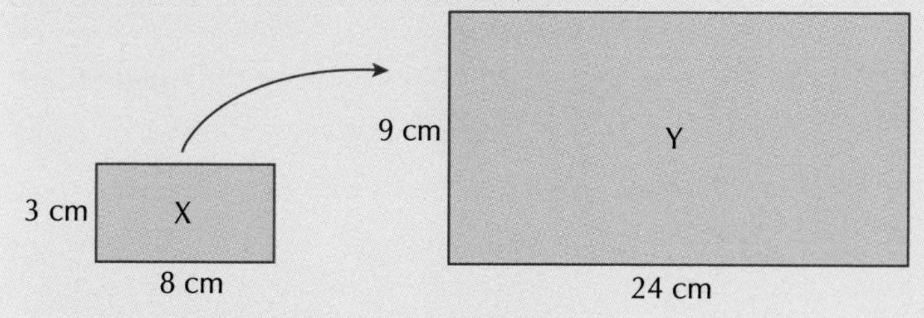

Rectangle X measures 3 cm by 8 cm.

The rectangle is enlarged and rectangle Y is 9 cm by 24 cm.

3 cm × 3 = 9 cm and 8 cm × 3 = 24 cm

So the scale factor is 3.

Example 2

The right-angled triangle ABC is enlarged by a scale factor of 4 to give the right-angled triangle DEF.

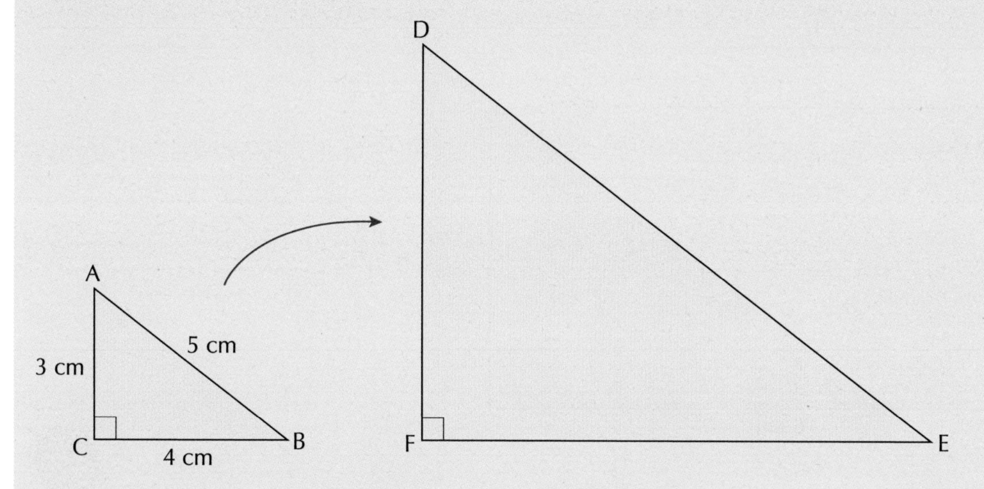

Write down the lengths of the sides of triangle DEF.

The scale factor is 4, so multiply all the lengths in triangle ABC by 4 to get the lengths in triangle DEF.

DF: 3 cm × 4 = 12 cm

FE: 4 cm × 4 = 16 cm

DE: 5 cm × 4 = 20 cm

Exercise 6A

1 Write down the scale factor for each enlargement.

a

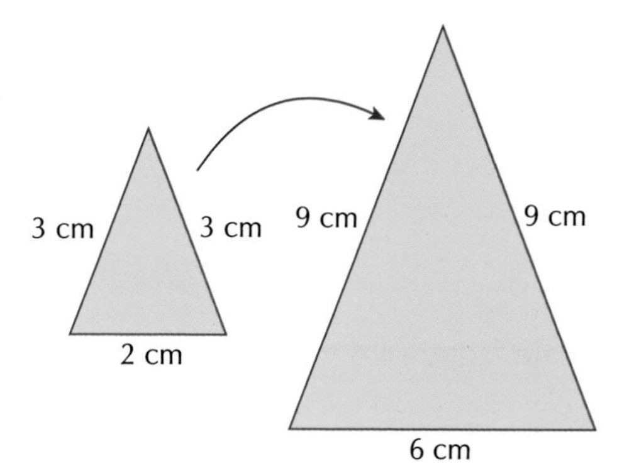

b

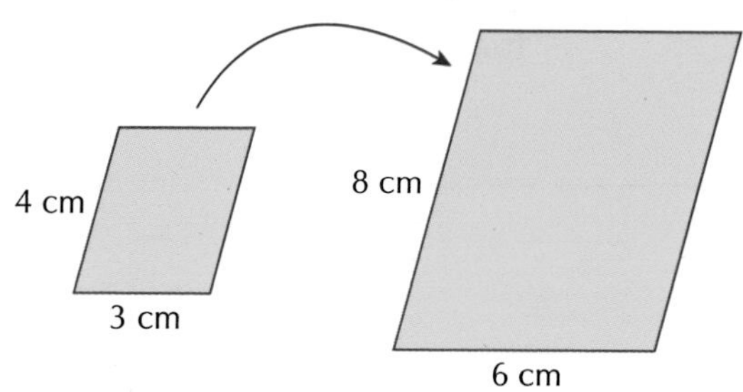

c

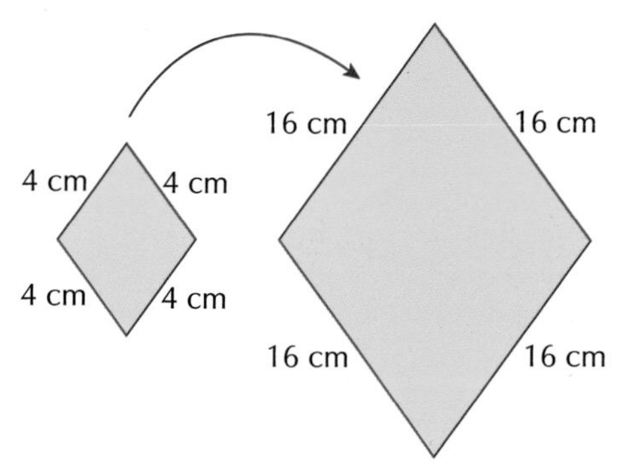

d

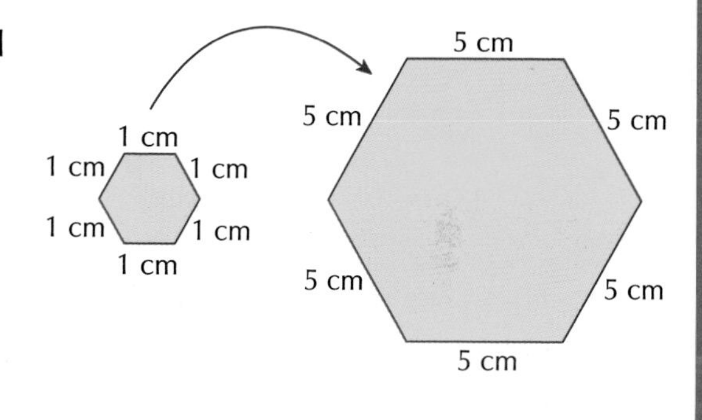

MR **2** The rectangles in one of these pairs are similar.
Which are they?
Give reasons for your answer.

a

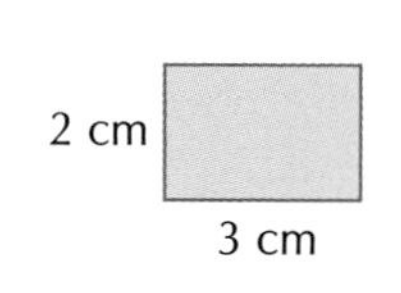

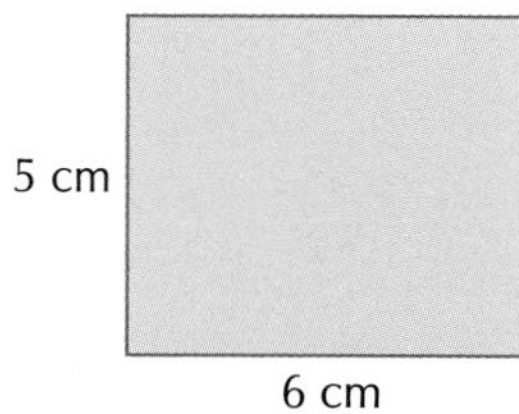

b

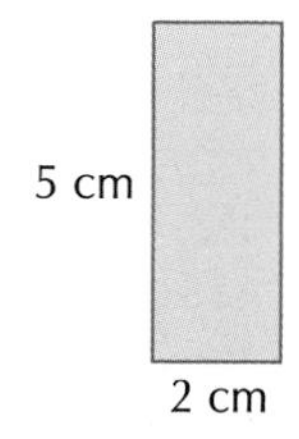

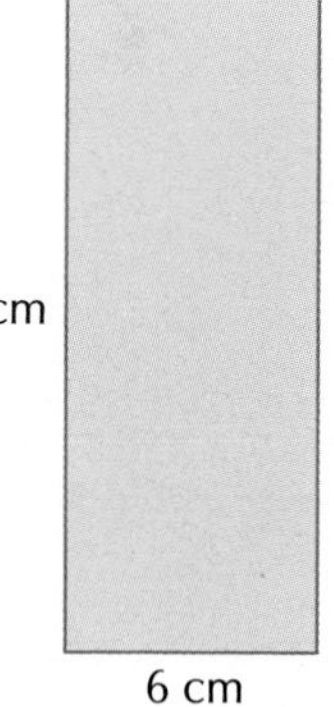

c

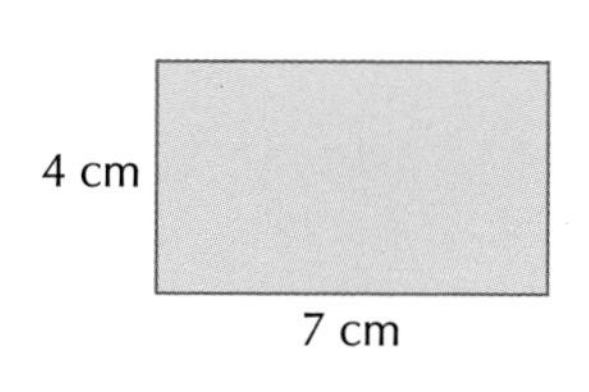

3 The isosceles triangle ABC is enlarged by a scale factor of 3 to give the isosceles triangle DEF.

Write down the lengths of the sides of triangle DEF.

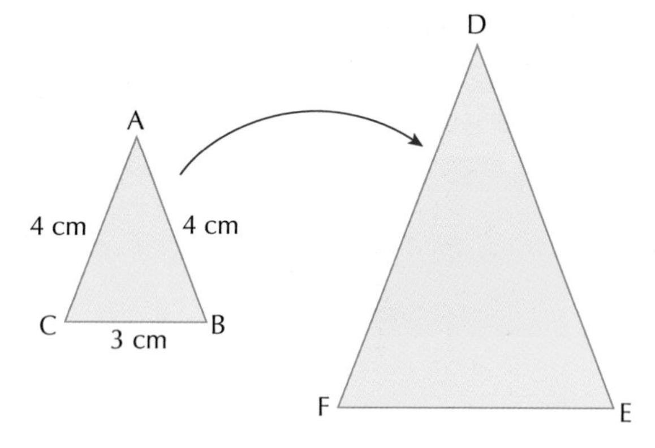

4 The trapezium ABCD is enlarged by a scale factor of 4 to give the trapezium EFGH.

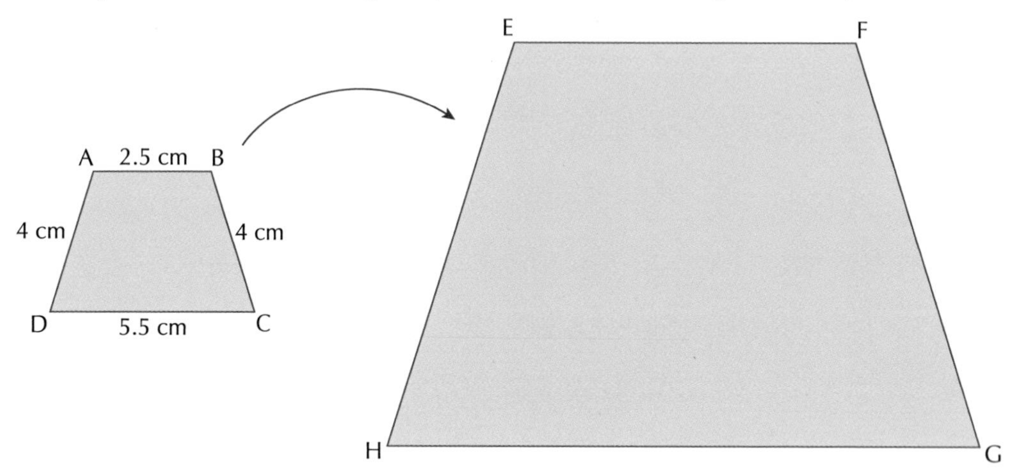

Write down the lengths of the sides of the trapezium EFGH.

5 Look at these four rectangles.

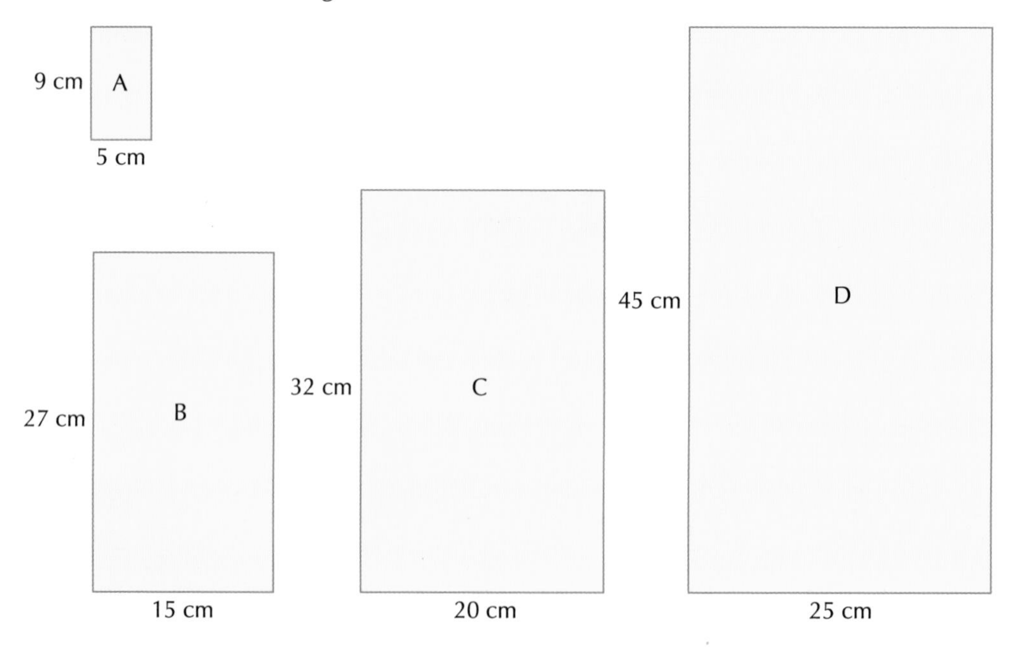

a Which rectangle is an enlargement of rectangle A, with a scale factor of 3?

b Which rectangle is not an enlargement of rectangle A? Explain your answer.

6 The diagram shows the length and height of a model car.

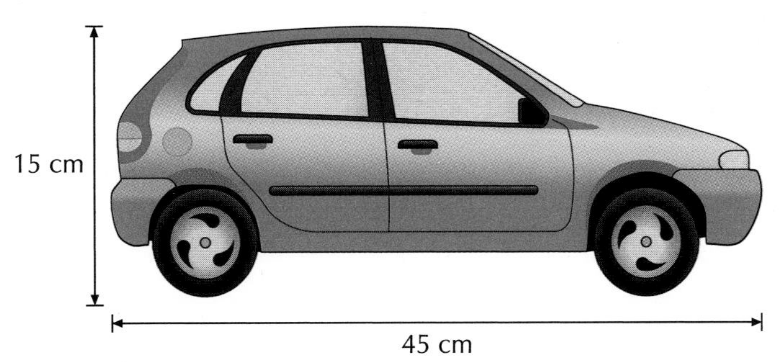

The real car has a length of 4.5 m and a height of 1.5 m.

What is the scale factor of the enlargement?

Challenge: Algebra with enlargements

A Rectangle B is an enlargement of rectangle A.

Work out the value of x.

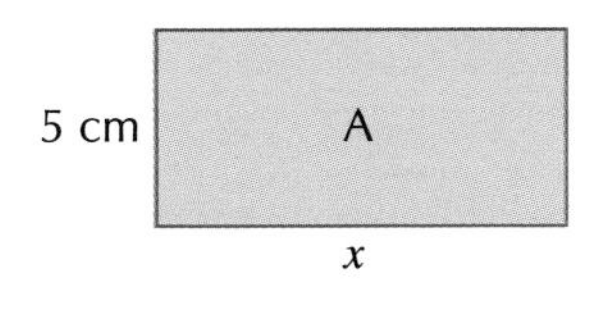

10 cm

B

$x + 8$

B Triangle Y is an enlargement of triangle X.

Work out the value of x.

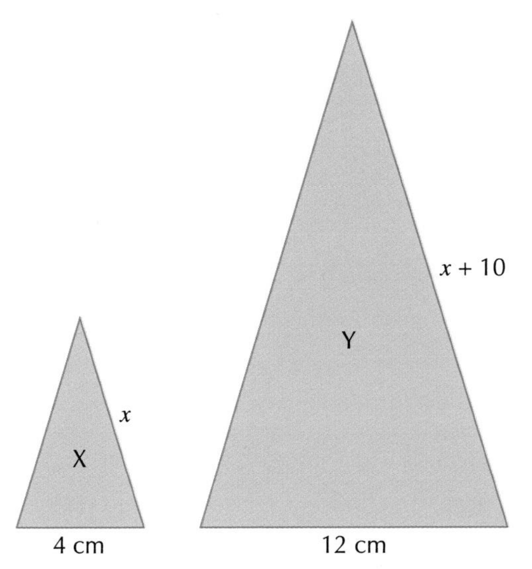

6.2 The centre of enlargement

Learning objective

- To enlarge a shape about a centre of enlargement

Key words
centre of enlargement
ray

The diagram shows triangle ABC enlarged by a scale factor of 2 to give triangle A′B′C′.

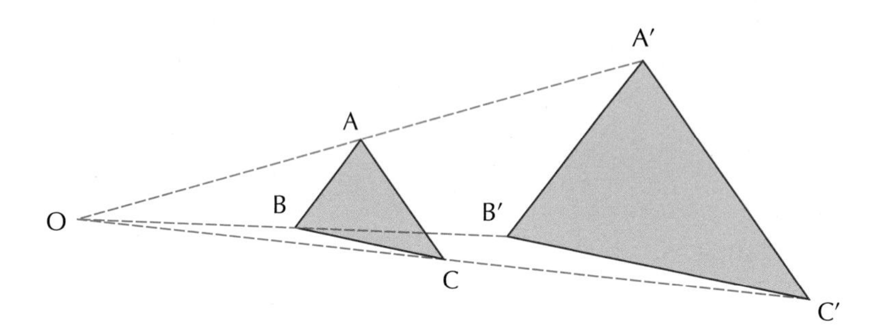

Each side of triangle A′B′C′ is twice as long as the corresponding side of triangle ABC.

The point O is called the **centre of enlargement.**

The dotted lines from O through the vertices of the triangles are called the guidelines or **rays** for the enlargement.

Notice that $OA' = 2 \times OA$, $OB' = 2 \times OB$ and $OC' = 2 \times OC$.

Triangle ABC is enlarged by a scale factor of 2 about the centre of enlargement O to give the triangle A′B′C′.

Example 3

Enlarge the triangle XYZ by a scale factor of 2 about the centre of enlargement O.

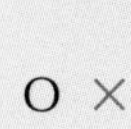

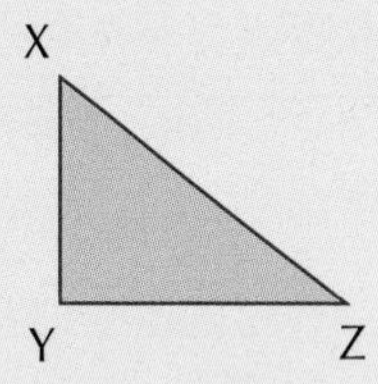

Draw rays OX, OY and OZ.

Measure the lengths of the three rays and multiply each of these lengths by two.

Then extend each of the rays to these new lengths, measured from O, and plot the points X′, Y′ and Z′.

Join X′, Y′ and Z′.

Triangle X′Y′Z′ is the enlargement of triangle XYZ by a scale factor of 2 about the centre of enlargement O.

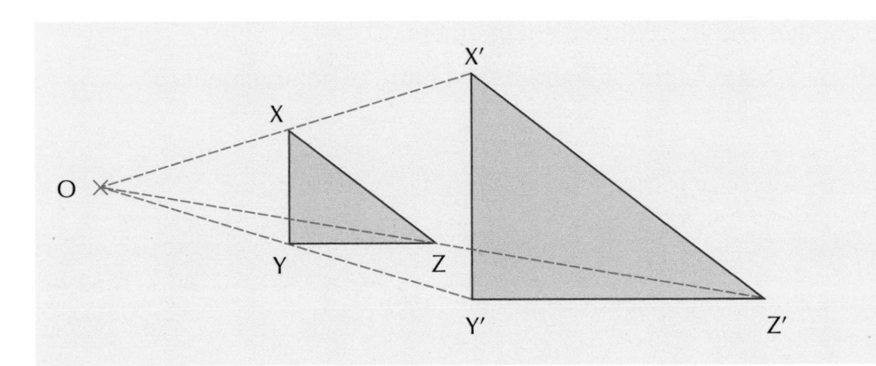

Example 4

Rectangle B is an enlargement of rectangle A by a scale factor of 2.

Find the position of the centre of enlargement, O.

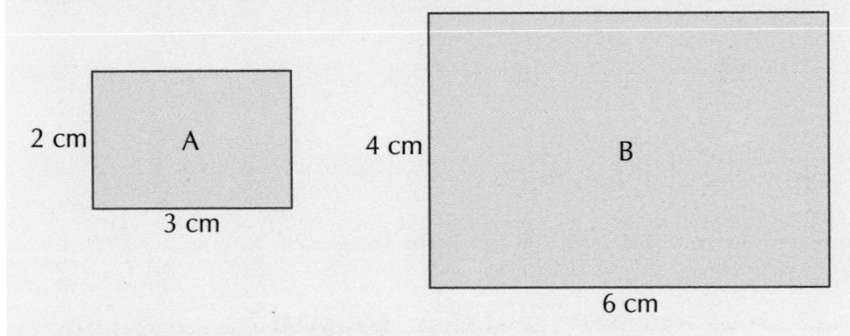

Draw rays through the four vertices as shown.

The point where the rays intersect is O, the centre of enlargement.

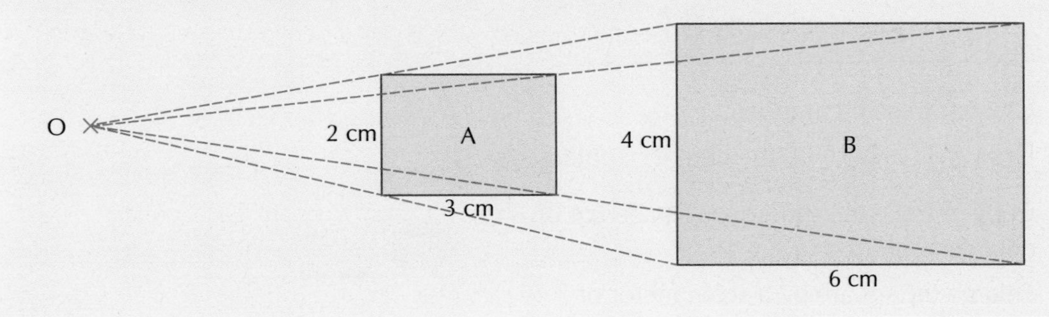

Exercise 6B

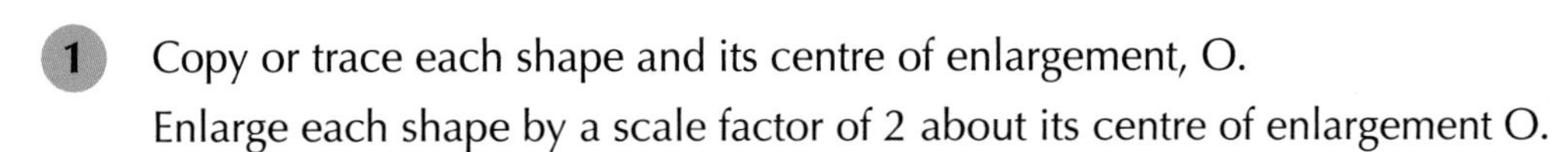

1 Copy or trace each shape and its centre of enlargement, O.

Enlarge each shape by a scale factor of 2 about its centre of enlargement O.

a

b

c

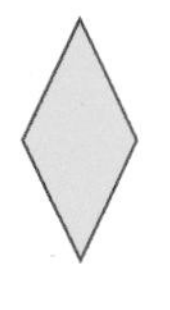

2 Copy or trace the triangle and its centre of enlargement, O.

Enlarge the triangle by a scale factor of 3 about the centre of enlargement O.

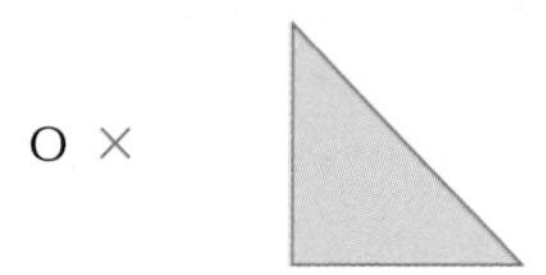

3 Triangle Y is an enlargement of triangle X by a scale factor of 2.

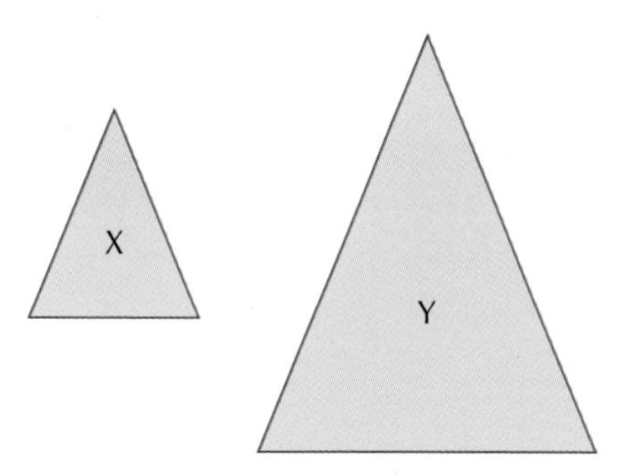

Copy or trace the triangles.

Draw rays to find, O, the centre of enlargement.

4 Trapezium Q is an enlargement of trapezium P by a scale factor of 2.

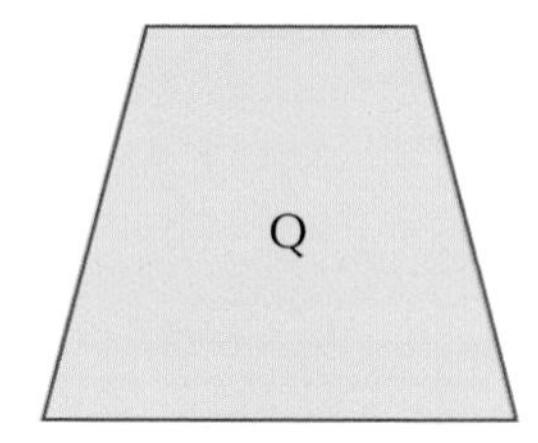

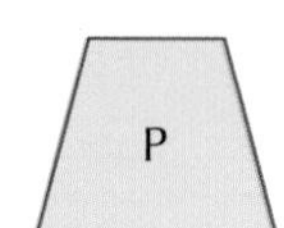

Copy or trace the trapezii.

Draw rays to find, O, the centre of enlargement.

5 Copy or trace the square and its centre of enlargement, O.

Enlarge the square by a scale factor of 3 about the centre of enlargement O.

× is the centre of the square.

Reasoning: Reductions

When the right-angled triangle A is enlarged by a scale factor of $\frac{1}{2}$ about the centre of enlargement O, it gives triangle B.

You will notice that, in fact, the triangle has been reduced! This is because the scale factor is less that 1. However, in mathematics this is still called an enlargement.

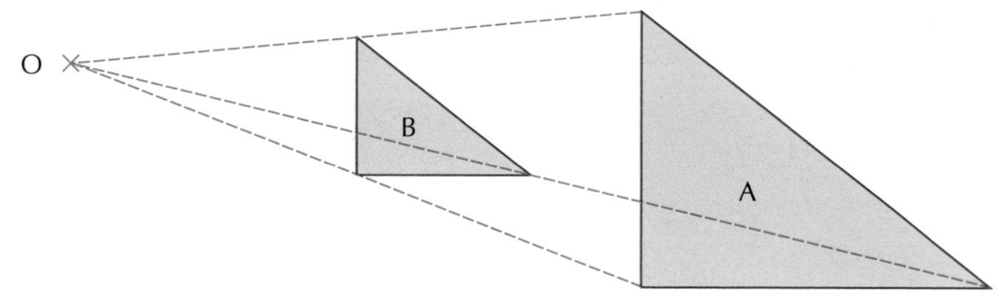

Notice that the reduced triangle B is between the original triangle A and the centre of enlargement.

The images will be smaller, for the reason explained above.

Now enlarge each shape by a scale factor $\frac{1}{2}$ about the centre of enlargement O.

A

O ×

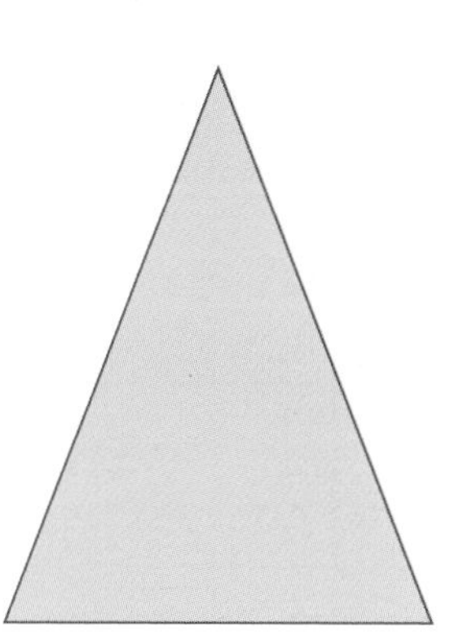

B O ×

6.3 Enlargements on grids

Learning objective

- To enlarge a shape on a coordinate grid

It can be helpful to draw enlargements on square grids.

Example 5

This rectangle is drawn on a centimetre-square grid.

Enlarge the rectangle by a scale factor of 3.

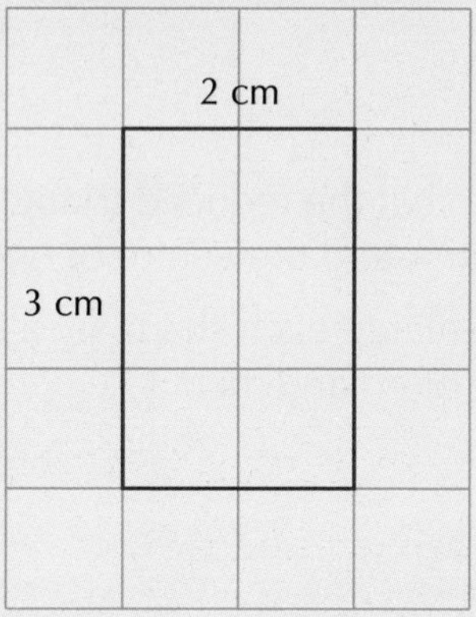

2 cm

3 cm

Original length = 3 cm, so the length on the enlargement will be 3 cm × 3 = 9 cm.

Original width = 2 cm, so the width on the enlargement will be 2 cm × 3 = 6 cm.

So the enlargement looks like this.

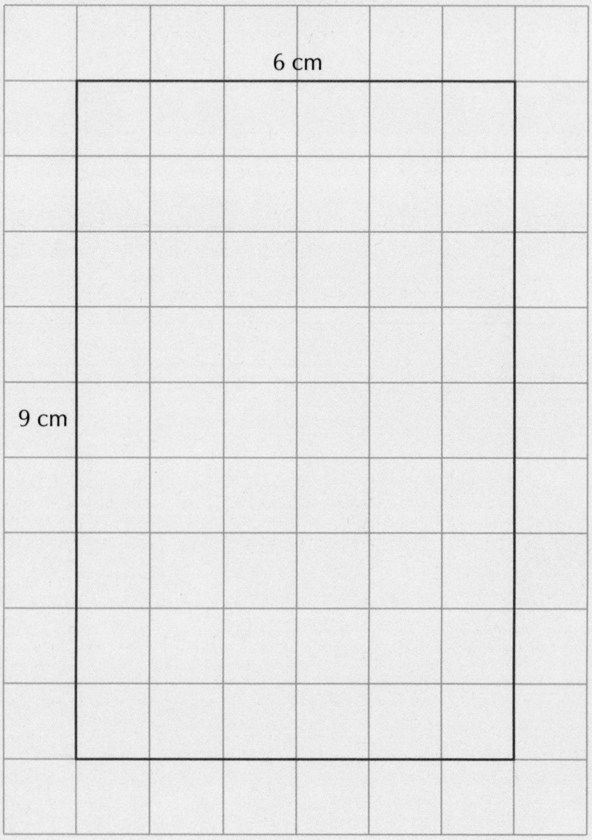

6 cm

9 cm

Example 6

The rectangle ABCD on this coordinate grid has been enlarged by a scale factor 3 about the origin O to give the image rectangle A′B′C′D′.

Compare the coordinates of the vertices of the object and the image.

What do you notice?

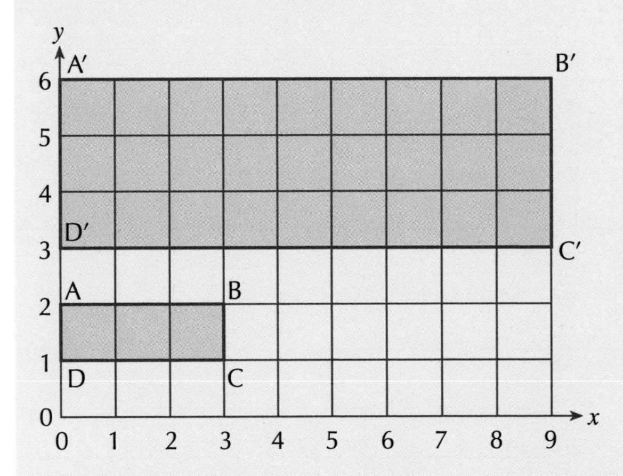

The coordinates of the vertices of the object are: A(0, 2), B(3, 2), C(3, 1) and D(0, 1).

The coordinates of the vertices of the image are: A′(0, 6), B′(9, 6), C′(9, 3) and D′(0, 3).

You should notice that if a shape is enlarged by a scale factor about the origin on a coordinate grid, the coordinates of a vertex on the enlarged shape are the coordinates of the corresponding vertex on the original shape, multiplied by the scale factor.

Exercise 6C

1 Copy each shape onto centimetre-squared paper.

Enlarge each one by a scale factor of 2.

a 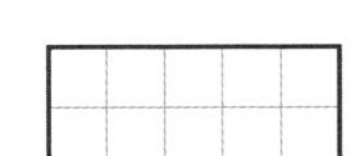 **b** 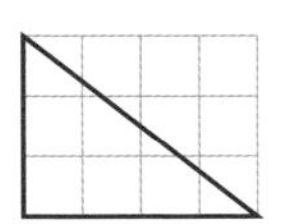 **c** 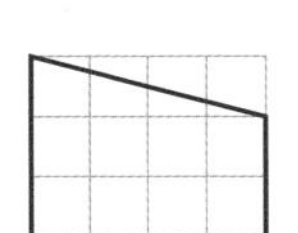 **d** 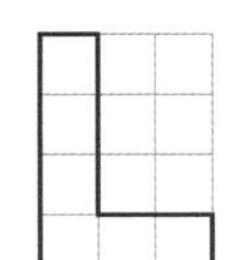

2 Copy each shape onto centimetre-squared paper.

Enlarge each one by a scale factor of 3.

a 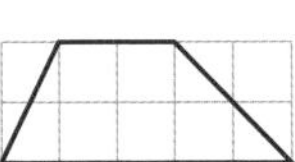 **b** 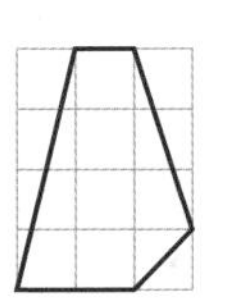

3 Copy each diagram onto centimetre-squared paper.

Enlarge each shape by the given scale factor, about the origin O.

a

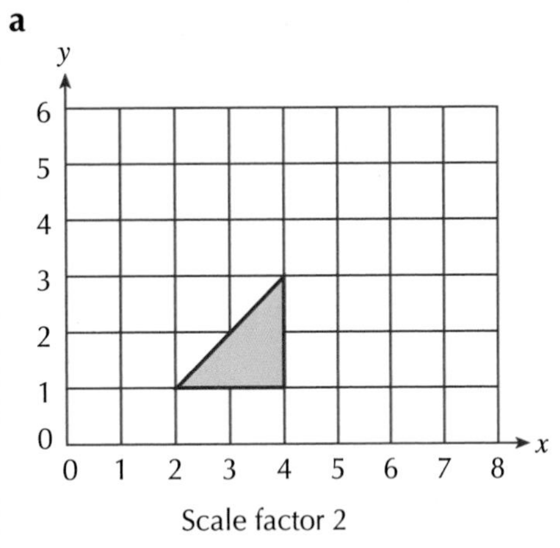

Scale factor 2

b

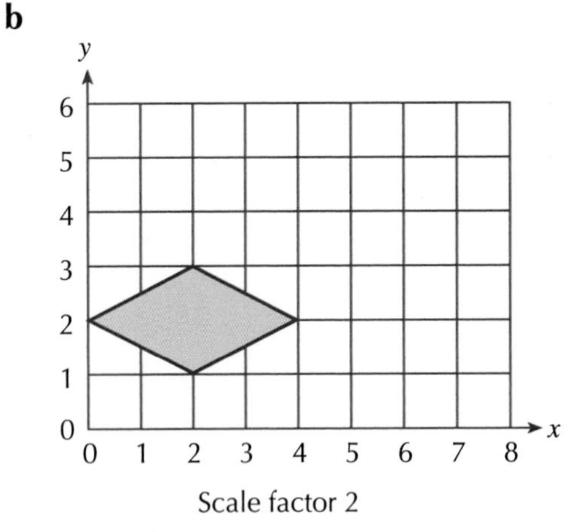

Scale factor 2

c

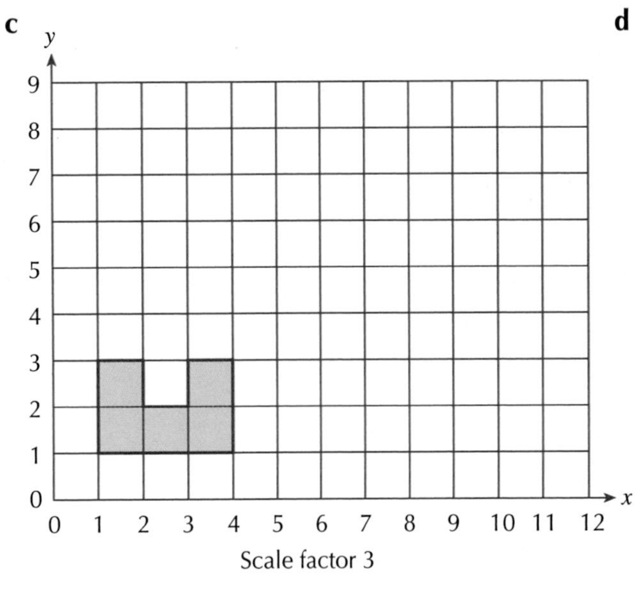

Scale factor 3

d

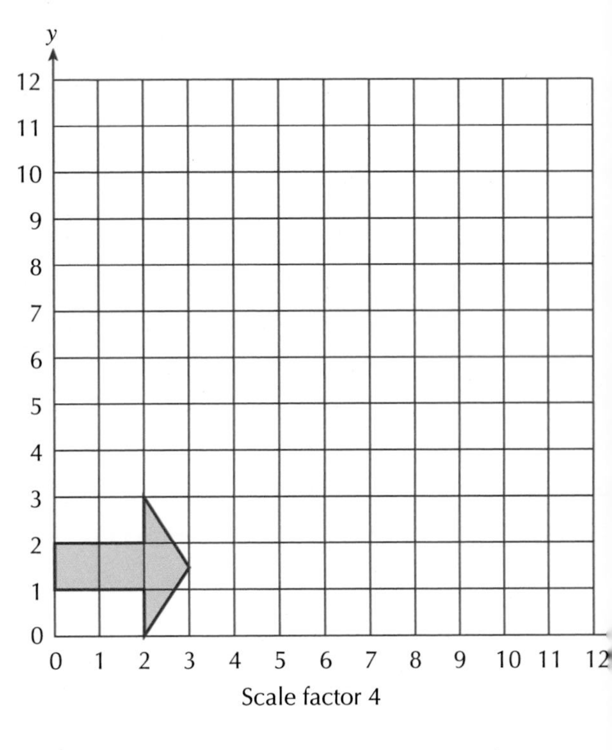

Scale factor 4

4 Draw a coordinate grid on centimetre-squared paper.

Label the x-axis and the y-axis from 0 to 12.

Plot the points A(2, 4), B(4, 4), C(3, 1) and D(1, 1).

Join them, in order, to form the parallelogram ABCD.

Enlarge the parallelogram by a scale factor of 3 about the origin O.

5 Copy these squares onto centimetre-squared paper.

Square A is enlarged to give square B.

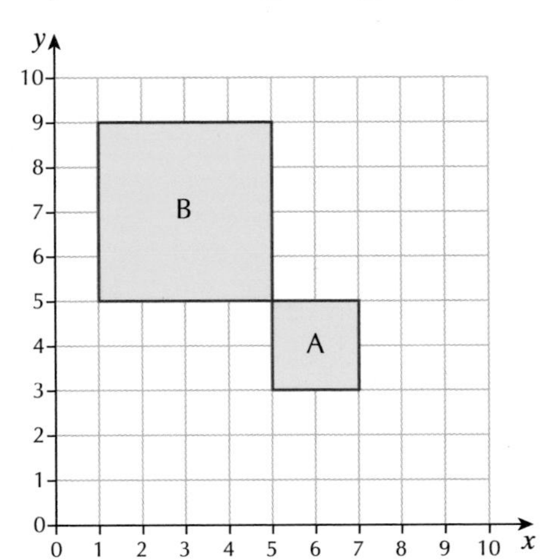

a What is the scale factor of the enlargement?

b By adding suitable rays to your diagram, find the coordinates of the centre of enlargement.

6 **a** Draw a coordinate grid on centimetre-squared paper.

Label the x-axis and the y-axis from 0 to 12.

Plot the points A(1, 3), B(3, 3), C(3, 1) and D(1, 1).

Join them, in order, to form the square ABCD.

b Write down the area of the square.

c Enlarge the square ABCD by a scale factor of 2 about the origin O.

What is the area of the enlarged square?

d Enlarge the square ABCD by a scale factor of 3 about the origin O.

What is the area of the enlarged square?

e Enlarge the square ABCD by a scale factor of 4 about the origin O.

What is the area of the enlarged square?

f Write down anything you notice about the increase in area of the enlarged squares.

Try to write down a rule to explain what is happening.

Activity: Enlarged stickmen

Work as part of a pair or a group.

Design a poster to show how the 'stickman' shown can be enlarged by different scale factors about any convenient centre of enlargement.

Ready to progress?

I can enlarge a shape by a given scale factor.

I can enlarge a shape about a centre of enlargement.
I can draw rays to find a centre of enlargement.

Review questions

1 Write down the scale factor for each enlargement.

a

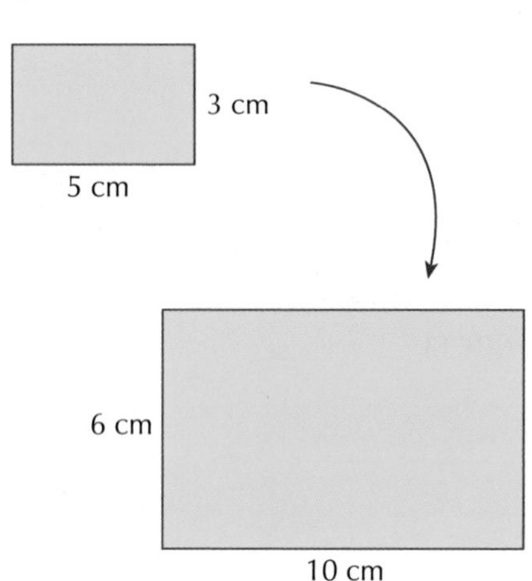

b

5 cm
4 cm
3 cm
20 cm
16 cm
12 cm

MR 2 Look at these four isosceles triangles.

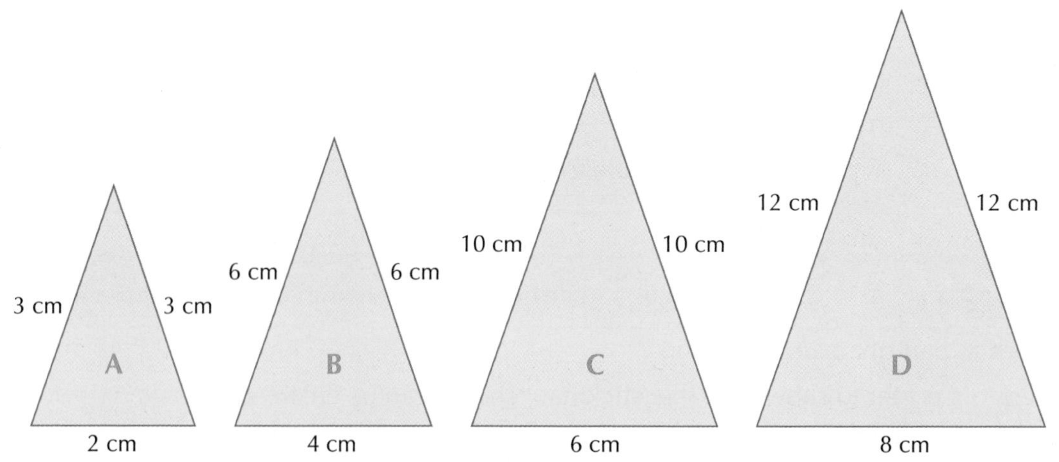

Which is the odd one out? Give a reason for your answer.

3 Copy each diagram onto centimetre-squared paper.

Enlarge each one by a scale factor of 2.

a

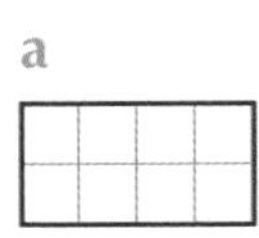

b

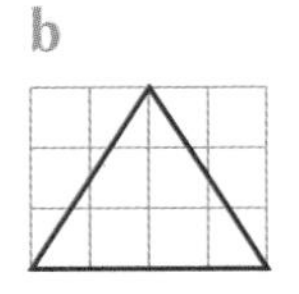

4 Square B is an enlargement of square A by a scale factor of 3.

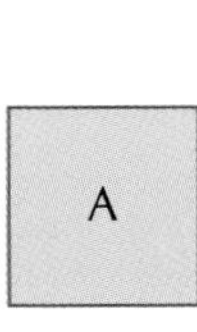

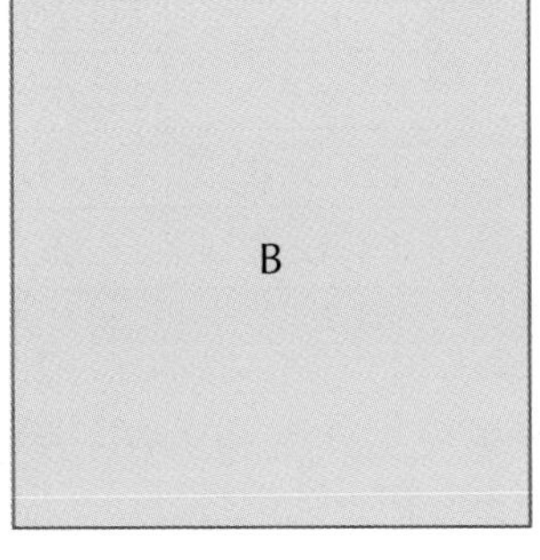

Copy or trace the squares. Draw rays to find, O, the centre of enlargement.

5 Copy this letter T onto centimetre-squared paper and enlarge it by a scale factor of 2 about the origin O.

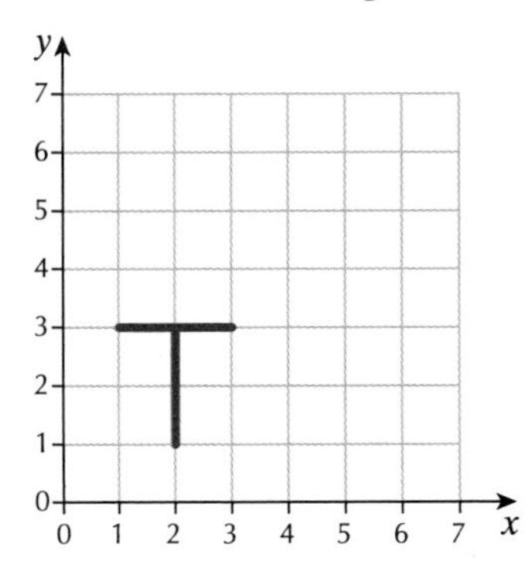

6 Copy the diagram onto centimetre-squared paper.

a ABCD is a square.

Write down the coordinates of A, B, C and D.

b Enlarge the square ABCD by a scale factor of 2 about the origin O.

Label the square A′B′C′D′.

c Write down the coordinates of A′, B′, C′ and D′.

d The square ABCD is enlarged by a scale factor of 3 about the origin O to give the square A″B″C″D″.

Without drawing the square, write down the coordinates of A″, B″, C″ and D″.

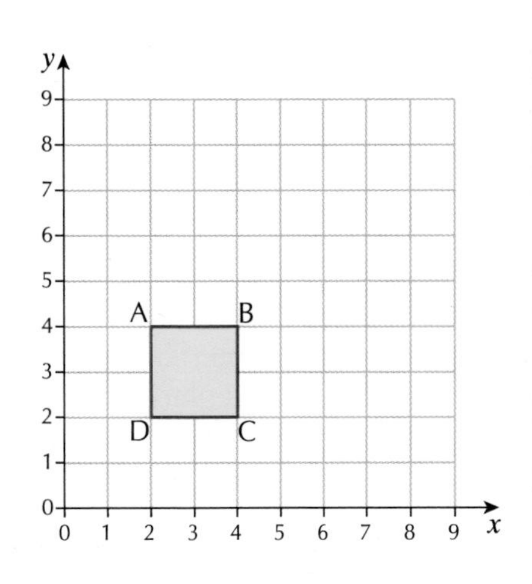

Problem Solving

Photographs

FastPrint advertises the cost of printing photographs in their shop.

The ″ symbol means inches.

Print size	Quantity	Price
6″ × 4″	1–49	£0.15 each
	50–99	£0.12 each
	100–249	£0.09 each
	250–499	£0.08 each
	500–750	£0.06 each
	751+	£0.05 each
7″ × 5″	£0.29	
8″ × 6″	£0.45	
12″ × 8″	£1.10	
13″ × 8″	£1.20	

1 A school decides to use FastPrint to buy prints of a year-group photograph. Students can choose the size of the prints they want. They can also choose to buy more than one size.

This is the school's order.

128	6″ × 4″ prints	75	10″ × 8″ prints
87	7″ × 5″ prints	60	13″ × 8″ prints

What is the total cost of buying these prints?

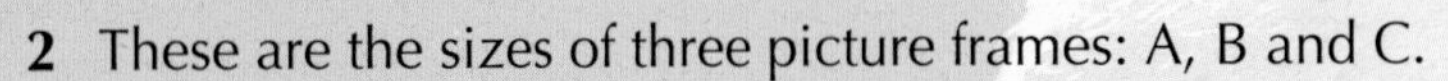

2 These are the sizes of three picture frames: A, B and C.

a The 7″ × 5″ print will fit inside frame A. What will be the area of the outside border?

b The 6″ × 4″ print will fit inside frame B. What will be the area of the outside border?

c The 13″ × 8″ print will fit inside frame C. What will be the area of the outside border?

A: 8″ × 6″
B: 7″ × 5″
C: 14″ × 9″

3 Some of the prints are actual mathematical enlargements of each other.

Write down the sizes of the prints that are exact enlargements of each other. State the scale factor of the enlargement.

13″ × 8″

12″ × 8″

8″ × 6″

7″ × 5″

6″ × 4″

4 EasyPrint also advertises the cost of photograph prints in their shop.

Print size	Price each
6″ × 4″	£0.12
7″ × 5″	£0.20
8″ × 6″	£0.42
12″ × 8″	£1.20
13″ × 8″	£1.32

a If you order one of each print size from EasyPrint, which prints are cheaper than FastPrint?

b If a school ordered 150 6″ × 4″ prints, which shop would they choose? How much would they save?

5 Any rectangle with length and width in the ratio 1.618 : 1 is known as a golden rectangle. The golden rectangle is said to be one of the most visually pleasing rectangular shapes. Many artists and architects have used the shape within their work.

a Copy and complete the table. You will need to work out the ratio of the length to the width, in the form n : 1, for each print size.

Print size	length : width	length ÷ width	length : width
6″ × 4″	6 : 4	6 ÷ 4 = 1.5	1.5 : 1
7″ × 5″	7 : 5	7 ÷ 5 = 1.4	
8″ × 6″	8 : 6		
12″ × 8″			
13″ × 8″			

b Which print is closest to being a golden rectangle?

7

Fractions

This chapter is going to show you:

- how to subtract any two fractions
- how to multiply any two fractions
- how to divide any two fractions.

You should already know:

- how to multiply and divide a fraction by an integer.

About this chapter

If you could count all the pips on this board, you would see that the fraction of the square that is green is $\frac{36}{121}$ or about 0.30.

The first value is exact. The second is a decimal approximation. A more accurate approximation is 0.298.

You can write numbers that are not integers in two different ways: as decimals or as fractions. But why do you need two different ways to write numbers?

In many cases, such as money or measuring, decimals are the better choice. However, some simple fractions, such as one third, cannot be written down exactly as decimals. In those cases it may be better to use fractions.

In other cases, such as common formulae involving simple fractions, writing them with decimals would make them more difficult to remember. It would not be easy to calculate accurately. An example is the volume of a sphere, which you will meet later in your mathematics course. It involves the fraction $\frac{4}{3}$ which, as a decimal, is 1.333 333…. In this case, it is easier to write the number as a fraction.

7.1 Adding and subtracting fractions

Learning objective

- To add or subtract any two fractions

In this section you will review the addition and subtraction of fractions. You need to be able to do this without using a calculator.

Before you can add or subtract two fractions you need to make sure they are written with the same number in the denominator. This example will illustrate the method.

Example 1

Work out these.

a $\frac{1}{2}+\frac{3}{8}$ **b** $\frac{3}{5}-\frac{1}{2}$

a Remember that before you can add or subtract fractions, you must make sure they have the same denominator (bottom number).

$\frac{1}{2}=\frac{4}{8}$ 8 is a multiple of both 2 and 8, so change the first fraction to eighths.

$\frac{1}{2}+\frac{3}{8}=\frac{4}{8}+\frac{3}{8}$

$\frac{4}{8}+\frac{3}{8}=\frac{7}{8}$ Add the numerators, keeping the same denominator.

b First, make sure that both fractions have the same denominator.

Since 5 and 2 have a common multiple of 10, change both fractions to tenths.

$\frac{3}{5}=\frac{3\times 2}{5\times 2}$

$=\frac{6}{10}$

$\frac{1}{2}=\frac{1\times 5}{2\times 5}$

$=\frac{5}{10}$

$\frac{6}{10}-\frac{5}{10}=\frac{1}{10}$

Exercise 7A

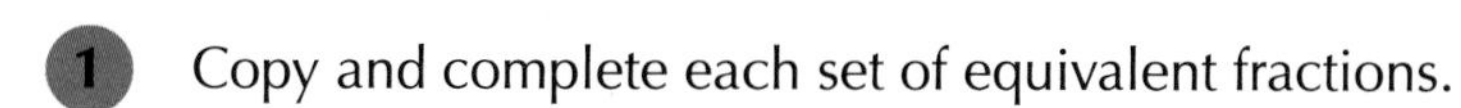

1 Copy and complete each set of equivalent fractions.

a $\frac{3}{4}=\frac{\square}{12}$ **b** $\frac{1}{5}=\frac{\square}{10}$ **c** $\frac{2}{3}=\frac{\square}{9}$ **d** $\frac{1}{2}=\frac{\square}{10}$

e $\frac{9}{10}=\frac{\square}{40}$ **f** $\frac{6}{7}=\frac{\square}{28}$ **g** $\frac{1}{8}=\frac{\square}{32}$ **h** $\frac{7}{9}=\frac{\square}{63}$

2 Copy and complete each addition.

a $\frac{1}{7}+\frac{1}{7}=\frac{\square}{7}$

b $\frac{1}{9}+\frac{1}{9}=\frac{\square}{9}$

c $\frac{1}{5}+\frac{2}{5}=\frac{\square}{5}$

d $\frac{1}{3}+\frac{1}{3}=\frac{\square}{3}$

e $\frac{3}{10}+\frac{3}{10}=\frac{\square}{10}=\frac{\square}{5}$

f $\frac{1}{8}+\frac{3}{8}=\frac{\square}{8}=\frac{\square}{2}$

g $\frac{2}{9}+\frac{4}{9}=\frac{\square}{9}=\frac{\square}{3}$

h $\frac{1}{12}+\frac{5}{12}=\frac{\square}{12}=\frac{\square}{2}$

i $\frac{1}{8}+\frac{5}{8}=\frac{\square}{8}=\frac{\square}{4}$

3 Copy and complete each subtraction.

a $\frac{5}{7}-\frac{1}{7}=\frac{\square}{7}$

b $\frac{8}{9}-\frac{4}{9}=\frac{\square}{9}$

c $\frac{4}{5}-\frac{2}{5}=\frac{\square}{5}$

d $\frac{2}{3}-\frac{1}{3}=\frac{\square}{3}$

e $\frac{7}{10}-\frac{3}{10}=\frac{\square}{10}=\frac{\square}{5}$

f $\frac{5}{8}-\frac{3}{8}=\frac{\square}{8}=\frac{\square}{4}$

g $\frac{7}{9}-\frac{4}{9}=\frac{\square}{9}=\frac{\square}{3}$

h $\frac{11}{12}-\frac{5}{12}=\frac{\square}{12}=\frac{\square}{2}$

i $\frac{11}{15}-\frac{1}{15}=\frac{\square}{15}=\frac{\square}{3}$

4 Copy and complete each addition.

a $\frac{1}{10}+\frac{1}{5}=\frac{1}{10}+\frac{2}{10}=\frac{\square}{10}$

b $\frac{1}{2}+\frac{1}{6}=\frac{\square}{6}+\frac{1}{6}=\frac{\square}{6}=\frac{\square}{3}$

c $\frac{1}{8}+\frac{1}{2}=\frac{1}{8}+\frac{\square}{8}=\frac{\square}{8}$

d $\frac{1}{2}+\frac{1}{4}=\frac{\square}{4}+\frac{1}{4}=\frac{\square}{4}$

e $\frac{3}{4}+\frac{1}{8}=\frac{\square}{8}+\frac{1}{8}=\frac{\square}{8}$

f $\frac{3}{5}+\frac{1}{10}=\frac{\square}{10}+\frac{1}{10}=\frac{\square}{10}$

g $\frac{5}{12}+\frac{1}{6}=\frac{5}{12}+\frac{\square}{12}=\frac{\square}{12}$

h $\frac{1}{7}+\frac{1}{14}=\frac{\square}{14}+\frac{1}{14}=\frac{\square}{14}$

i $\frac{3}{5}+\frac{4}{15}=\frac{\square}{15}+\frac{\square}{15}=\frac{\square}{15}$

5 Copy and complete each subtraction.

a $\frac{9}{10} - \frac{1}{5} = \frac{9}{10} - \frac{\square}{10}$
$= \frac{\square}{10}$

b $\frac{1}{3} - \frac{1}{6} = \frac{\square}{6} - \frac{1}{6}$
$= \frac{\square}{6}$

c $\frac{1}{2} - \frac{1}{8} = \frac{\square}{8} - \frac{1}{8}$
$= \frac{\square}{8}$

d $\frac{1}{2} - \frac{1}{4} = \frac{\square}{4} - \frac{1}{4}$
$= \frac{\square}{4}$

e $\frac{3}{4} - \frac{1}{8} = \frac{\square}{8} - \frac{1}{8}$
$= \frac{\square}{8}$

f $\frac{4}{5} - \frac{1}{10} = \frac{\square}{10} - \frac{1}{10}$
$= \frac{\square}{10}$

g $\frac{11}{12} - \frac{5}{6} = \frac{11}{12} - \frac{\square}{12}$
$= \frac{\square}{12}$

h $\frac{1}{7} - \frac{1}{14} = \frac{\square}{14} - \frac{1}{14}$
$= \frac{\square}{14}$

i $\frac{4}{5} - \frac{2}{15} = \frac{\square}{15} - \frac{2}{15}$
$= \frac{\square}{15}$
$= \frac{\square}{3}$

6 Convert the fractions to equivalent fractions with a common denominator, then work out the answers.

a $\frac{1}{5} + \frac{1}{4}$ **b** $\frac{1}{8} + \frac{1}{2}$ **c** $\frac{3}{4} + \frac{1}{5}$ **d** $\frac{1}{6} + \frac{2}{9}$

e $\frac{1}{4} - \frac{1}{5}$ **f** $\frac{5}{8} - \frac{1}{3}$ **g** $\frac{3}{4} - \frac{1}{5}$ **h** $\frac{5}{6} - \frac{2}{3}$

PS **7** A magazine fills $\frac{1}{4}$ of its pages with advertisements, $\frac{1}{12}$ with letters and the rest with articles.

a What fraction of the pages is taken up with articles?

b If the magazine has 150 pages, how many are used for articles?

PS **8** A survey showed that $\frac{1}{2}$ of the pupils walked to school, $\frac{1}{3}$ came by bus and the rest came by car.

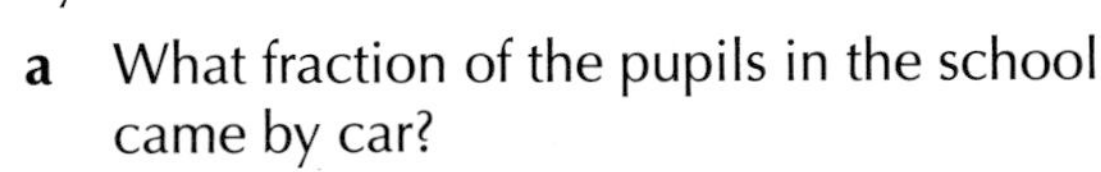

a What fraction of the pupils in the school came by car?

b If there were 1200 pupils in the school, how many came by car?

9 Here are two fractions.

$\frac{1}{4}$ $\frac{5}{8}$

Work out:

a the sum of the two fractions

b the difference between the two fractions.

10 Eve and Sophia were talking about this rectangle.

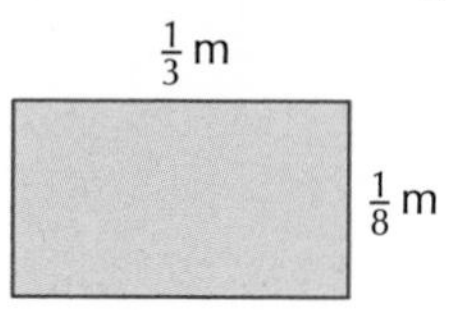

The perimeter of this rectangle is more than 1 m.
No it's not.
Sophia
Eve

Who is correct?

Explain your answer.

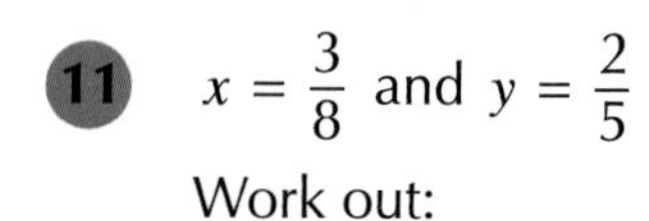

11 $x = \frac{3}{8}$ and $y = \frac{2}{5}$

Work out:

a $x + y$ **b** $y - x$ **c** $2x - y$ **d** $x - 2y$.

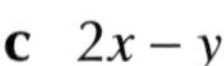

Challenge: Interesting fractions

You may use a calculator for this work.

A Work these out.

a $\frac{1}{2} + \frac{1}{4}$ **b** $\frac{1}{4} + \frac{1}{8}$ **c** $\frac{1}{8} + \frac{1}{16}$

B Use the pattern to help you write the answer to this one.

$\frac{1}{16} + \frac{1}{32}$

C Now work these out.

a $\frac{1}{3} + \frac{1}{9}$ **b** $\frac{1}{9} + \frac{1}{27}$ **c** $\frac{1}{27} + \frac{1}{81}$

D Write down the next pair of fractions and its answer.

7.2 Multiplying fractions

Learning objective

- To multiply two fractions

To multiply a fraction by an integer, you just multiply the numerator by the integer.

For example:

- $5 \times \frac{2}{3} = \frac{10}{3}$
- $\frac{3}{4}$ of $7 = \frac{3}{4} \times 7$

  $= \frac{21}{4}$

To multiply two fractions you multiply the numerators and multiply the denominators.

Example 2

Work out **a** $\frac{3}{4}$ of $\frac{1}{2}$ **b** $\frac{2}{3} \times \frac{3}{5}$

Solution

a $\frac{3}{4} \text{ of } \frac{1}{2} = \frac{3}{4} \times \frac{1}{2}$

$= \frac{3 \times 1}{4 \times 2}$ The numerator is 3×1 and the denominator is 4×2.

$= \frac{3}{8}$

b $\frac{2}{3} \times \frac{3}{5} = \frac{6}{15}$ $2 \times 3 = 6$ and $3 \times 5 = 15$

$= \frac{2}{5}$ Simplify the fraction as much as possible.

These rules are just an extension of the method for integers.

5 can be written as $\frac{5}{1}$ so:

$$5 \times \frac{2}{3} = \frac{5}{1} \times \frac{2}{3}$$
$$= \frac{10}{3}$$

as before.

Exercise 7B

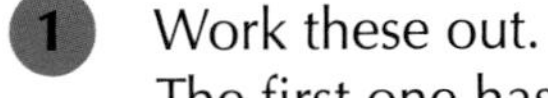

1 Work these out.
The first one has been done for you.

a $\frac{1}{2}$ of 18 **b** $\frac{1}{2}$ of 46 **c** $\frac{1}{2}$ of 60 **d** $\frac{1}{3}$ of 21

$18 \div 2 = 9$

e $\frac{1}{3}$ of 33 **f** $\frac{1}{4}$ of 28 **g** $\frac{1}{4}$ of 36 **h** $\frac{1}{5}$ of 35

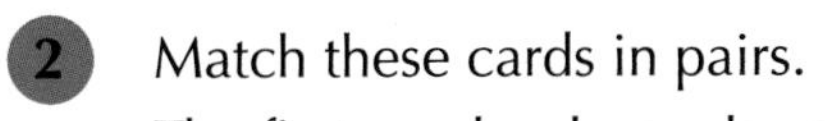

2 Match these cards in pairs.
The first one has been done for you.

$\frac{1}{3}$ of $30 = \frac{1 \times 30}{3}$
$= \frac{30}{3}$
$= 30 \div 3$
$= 10$

So $\frac{1}{3}$ of $30 \rightarrow 10$

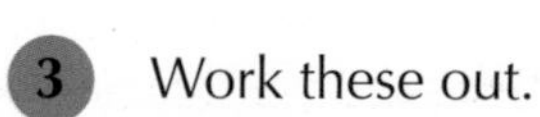

3 Work these out.

a $\frac{1}{3} \times 12$ **b** $\frac{2}{3} \times 12$ **c** $\frac{1}{4} \times 40$ **d** $\frac{3}{4} \times 40$

e $\frac{1}{8} \times 16$ **f** $\frac{5}{8} \times 16$ **g** $\frac{1}{12} \times 24$ **h** $\frac{7}{12} \times 24$

PS **4** A man earns £300. He pays out two-thirds on his rent. How much rent does he pay?

PS **5** Kathy has 24 pairs of shoes. Three-quarters of them are light colours. How many pairs of light-coloured shoes she does she have?

6 Work out each multiplication.
The first one has been done for you.

a $\frac{1}{4} \times \frac{1}{3} = \frac{1}{12}$ **b** $\frac{1}{5} \times \frac{1}{2}$ **c** $\frac{3}{4} \times \frac{1}{4}$ **d** $\frac{2}{5} \times \frac{2}{3}$

e $\frac{3}{8} \times \frac{1}{2}$ **f** $\frac{3}{4} \times \frac{2}{5}$ **g** $\frac{5}{8} \times \frac{2}{3}$ **h** $\frac{3}{4} \times \frac{7}{10}$

MR **7** Look at these two multiplications.

$\frac{3}{4} \times \frac{5}{8}$ $\frac{3}{8} \times \frac{4}{5}$

Which gives the bigger answer?

Explain how you worked it out.

MR **8** Andrew and Oliver are talking about fractions.

Who is correct? Explain your answer.

Investigation: Multiplication of fractions

Copy each multiplication and fill in the missing numbers.

A **a** $\frac{1}{5} \times 2 = \frac{\square}{5}$ **b** $\frac{2}{5} \times 2 = \frac{\square}{5}$ **c** $\frac{1}{5} \times 3 = \frac{\square}{5}$

B **a** $\frac{4}{5} \times 3 = \frac{\square}{\square}$ **b** $\frac{3}{7} \times 2 = \frac{\square}{\square}$ **c** $\frac{1}{5} \times 4 = \frac{\square}{\square}$

C **a** $\frac{4}{5} \times 4 = \frac{\square}{\square}$ **b** $\frac{4}{7} \times 5 = \frac{\square}{\square}$ **c** $\frac{1}{5} \times 8 = \frac{\square}{\square}$

D **a** $\frac{4}{5} \times 8 = \frac{\square}{\square}$ **b** $\frac{5}{7} \times 3 = \frac{\square}{\square}$ **c** $\frac{2}{7} \times 3 = \frac{\square}{\square}$

E **a** $\square \times 10 = 2$ **b** $\square \times 20 = 8$ **c** $\square \times 4 = \frac{24}{7}$

F **a** $\frac{1}{5} \times \square = 200$ **b** $\frac{4}{5} \times \square = 800$ **c** $\frac{\square}{7} \times 35 = 50$

7.3 Dividing fractions

Learning objective

- To divide one fraction by another

Key word

invert

You already know how to divide a fraction by integer.

You would answer both of these questions in the same way.

- What is a half of $\frac{2}{5}$?
- Work out $\frac{2}{5} \div 2$.

The answers are:

- $\frac{2}{5} \times \frac{1}{2} = \frac{2}{10} = \frac{1}{5}$
- $\frac{2}{5} \div 2 = \frac{1}{5}$ or

  $\frac{2}{5} \times \frac{1}{2} = \frac{2}{10} = \frac{1}{5}$

To divide by a fraction you **invert** it or 'turn it upside down'. This means that you swap the numerator and the denominator. You then multiply by the new fraction. Some examples will help to make the method clear.

Example 3

Work these out.

a $\frac{5}{8} \div 3$ **b** $\frac{2}{5} \div \frac{3}{4}$

a Write 3 as $\frac{3}{1}$ then invert it to get $\frac{1}{3}$. Then multiply.

$\frac{5}{8} \times \frac{1}{3} = \frac{5}{24}$

b Invert $\frac{3}{4}$ to get $\frac{4}{3}$ and then multiply.

$\frac{2}{5} \times \frac{4}{3} = \frac{8}{15}$

Exercise 7C

1 Copy and complete each calculation.

a $\frac{3}{4} \div 2 = \frac{3}{4} \times \frac{1}{2} = \frac{\square}{8}$

b $\frac{4}{5} \div 3 = \frac{4}{5} \times \frac{1}{3} = \frac{\square}{15}$

c $\frac{3}{8} \div 5 = \frac{3}{8} \times \frac{1}{5} = \frac{\square}{\square}$

d $\frac{3}{5} \div 4 = \frac{3}{5} \times \frac{1}{\square} = \frac{\square}{\square}$

e $\frac{5}{6} \div 2 = \frac{5}{6} \times \frac{1}{\square} = \frac{\square}{\square}$

f $\frac{3}{10} \div 4 = \frac{3}{10} \times \frac{1}{\square} = \frac{\square}{\square}$

g $\frac{4}{9} \div 3 = \frac{4}{9} \times \frac{\square}{\square} = \frac{\square}{\square}$

h $\frac{4}{5} \div 7 = \frac{4}{5} \times \frac{\square}{\square} = \frac{\square}{\square}$

i $\frac{7}{10} \div 5 = \frac{7}{10} \times \frac{\square}{\square} = \frac{\square}{\square}$

2 Copy and complete each calculation.

a $\frac{3}{4} \div \frac{2}{5} = \frac{3}{4} \times \frac{5}{2} = \frac{\square}{8}$

b $\frac{4}{5} \div \frac{2}{3} = \frac{4}{5} \times \frac{3}{2} = \frac{\square}{10}$

c $\frac{3}{8} \div \frac{5}{6} = \frac{3}{8} \times \frac{6}{5} = \frac{\square}{\square}$

d $\frac{3}{5} \div \frac{2}{3} = \frac{3}{5} \times \frac{3}{\square} = \frac{\square}{\square}$

e $\frac{5}{6} \div \frac{2}{5} = \frac{5}{6} \times \frac{5}{\square} = \frac{\square}{\square}$

f $\frac{3}{10} \div \frac{5}{8} = \frac{3}{10} \times \frac{\square}{\square} = \frac{\square}{\square}$

g $\frac{3}{8} \div \frac{2}{5} = \frac{3}{8} \times \frac{\square}{\square} = \frac{\square}{\square}$

h $\frac{5}{8} \div \frac{2}{3} = \frac{5}{8} \times \frac{\square}{\square} = \frac{\square}{\square}$

i $\frac{7}{10} \div \frac{4}{9} = \frac{7}{10} \times \frac{\square}{\square} = \frac{\square}{\square}$

3 Copy and complete each calculation.
The first one has been done for you.

a $5 \div \frac{1}{2} = 5 \times \frac{2}{1} = \frac{10}{1} = 10$

b $4 \div \frac{1}{3}$

c $3 \div \frac{1}{5}$

d $6 \div \frac{1}{4}$

e $8 \div \frac{1}{5}$

f $3 \div \frac{1}{4}$

g $7 \div \frac{1}{6}$

h $5 \div \frac{1}{3}$

4 Copy and complete each calculation.
Leave each answer as a top-heavy fraction.
The first one has been done for you.

a $5 \div \frac{2}{3} = 5 \times \frac{3}{2} = \frac{15}{2}$

b $7 \div \frac{2}{3}$

c $4 \div \frac{3}{5}$

d $8 \div \frac{2}{5}$

e $10 \div \frac{2}{3}$

f $4 \div \frac{3}{4}$

g $6 \div \frac{5}{8}$

h $9 \div \frac{5}{8}$

5 Copy and complete each calculation.

Leave each answer as a top-heavy fraction.

The first one has been done for you.

a $7 \div \frac{3}{5} = 7 \times \frac{5}{3} = \frac{35}{3}$ **b** $4 \div \frac{2}{5}$ **c** $8 \div \frac{2}{3}$ **d** $9 \div \frac{2}{5}$

e $5 \div \frac{2}{5}$ **f** $5 \div \frac{3}{8}$ **g** $4 \div \frac{5}{9}$ **h** $7 \div \frac{3}{8}$

6 Which calculation has:

a the smallest answer **b** the largest answer?

i $5 \div \frac{3}{4}$ **ii** $3 \div \frac{4}{5}$ **iii** $4 \div \frac{3}{5}$

PS **7** How many lengths of wire, each $\frac{4}{5}$ cm long, can Alex cut from a piece of wire 10 cm long?

PS **8** Mr Bishop has a piece of copper wire 25 cm long.

He needs 30 pieces, each $\frac{9}{10}$ cm long. Does he have enough copper wire?

Explain your answer.

MR **9** Joe and James are talking about fractions.

Five divided by one third is the same as three divided by five.
No it can't be, one of them must be bigger.
Joe
James

Is Joe or James correct?

Explain your answer.

Challenge: Algebra with fractions

You are told that:

$a \times b \times c = 100$

A **a** What would the answer be if a were doubled?

b What would the answer be if b were trebled?

c What would the answer be if c were halved?

B **a** What would the answer be if a were doubled, b were trebled and c were halved at the same time?

b What would the answer be if a were doubled, b were doubled and c were doubled at the same time?

c What would the answer be if a were halved, b were halved and c were halved at the same time?

Ready to progress?

I can add or subtract two simple fractions.

I can add or subtract two fractions with different denominators.
I can multiply two simple fractions.
I can divide two simple fractions.

Review questions

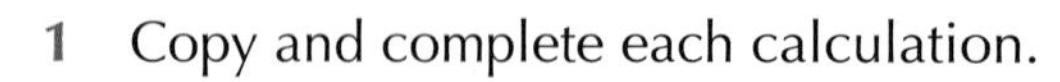

1 Copy and complete each calculation.

a $\frac{1}{5} + \frac{3}{5} = \frac{\square}{5}$

b $\frac{3}{10} + \frac{1}{10} = \frac{\square}{10} = \frac{\square}{5}$

c $\frac{1}{8} + \frac{3}{8} = \frac{\square}{8} = \frac{\square}{2}$

d $\frac{7}{8} - \frac{3}{8} = \frac{\square}{\square} = \frac{\square}{\square}$

e $\frac{7}{10} - \frac{3}{10} = \frac{\square}{10} = \frac{\square}{\square}$

f $\frac{5}{9} - \frac{2}{9} = \frac{\square}{9} = \frac{\square}{\square}$

2 Work these out.

a $\frac{1}{2}$ of 14 **b** $\frac{1}{2}$ of 48 **c** $\frac{1}{2}$ of 64 **d** $\frac{1}{3}$ of 30

e $\frac{1}{3}$ of 36 **f** $\frac{1}{4}$ of 20 **g** $\frac{1}{4}$ of 48 **h** $\frac{1}{5}$ of 45

3 Match these cards in pairs.

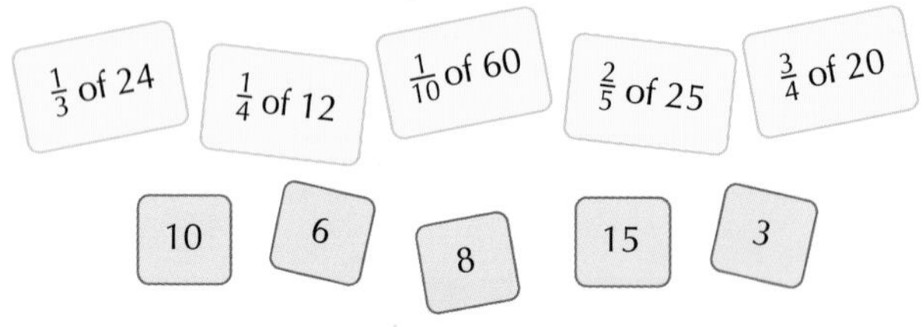

4 Copy and complete each calculation.

a $\frac{4}{5} \div 7 = \frac{4}{5} \times \frac{1}{7} = \frac{\square}{35}$

b $\frac{5}{6} \div 2 = \frac{5}{6} \times \frac{1}{2} = \frac{\square}{12}$

c $\frac{3}{4} \div 5 = \frac{3}{4} \times \frac{1}{5} = \frac{\square}{\square}$

d $\frac{4}{5} \div \frac{3}{4} = \frac{4}{5} \times \frac{4}{3} = \frac{\square}{\square}$

e $\frac{7}{10} \div \frac{2}{3} = \frac{7}{10} \times \frac{3}{2} = \frac{\square}{\square}$

f $\frac{5}{8} \div \frac{3}{5} = \frac{5}{8} \times \frac{5}{\square} = \frac{\square}{\square}$

g $\frac{4}{7} \div \frac{3}{5} = \frac{4}{7} \times \frac{\square}{\square} = \frac{\square}{\square}$

h $\frac{5}{6} \div \frac{8}{9} = \frac{5}{6} \times \frac{\square}{\square} = \frac{\square}{\square}$

i $\frac{7}{8} \div \frac{4}{5} = \frac{7}{8} \times \frac{\square}{\square} = \frac{\square}{\square}$

5 Copy and complete each calculation.

Leave your answers as top-heavy fractions.

The first one has been done for you.

a $8 \div \frac{3}{5} = 8 \times \frac{5}{3} = \frac{40}{3}$ **b** $2 \div \frac{5}{9}$ **c** $3 \div \frac{4}{7}$ **d** $6 \div \frac{5}{8}$

e $11 \div \frac{3}{4}$ **f** $5 \div \frac{3}{7}$ **g** $9 \div \frac{7}{8}$ **h** $10 \div \frac{11}{12}$

6 Work out each of these.

The first one has been done for you.

a $\frac{3}{5} \times \frac{1}{2} = \frac{3}{10}$ **b** $\frac{4}{5} \times \frac{1}{3}$ **c** $\frac{3}{10} \times \frac{2}{5}$ **d** $\frac{3}{8} \times \frac{5}{8}$

e $\frac{5}{6} \times \frac{1}{3}$ **f** $\frac{2}{5} \times \frac{3}{4}$ **g** $\frac{7}{8} \times \frac{2}{3}$ **h** $\frac{3}{5} \times \frac{9}{10}$

7 Look at these two multiplications.

$\frac{2}{3}$ of $\frac{5}{8}$ $\frac{2}{5}$ of $\frac{3}{8}$

Which gives the bigger answer?

Explain how you worked it out.

8 A survey of pupils in a school showed that $\frac{3}{5}$ of them enjoyed maths, $\frac{1}{4}$ didn't like maths and the rest weren't sure.

a What fraction weren't sure?

b If there were 1400 pupils in the school, how many enjoyed maths?

9 Jess needed 120 pieces of metal wire, each $\frac{3}{4}$ m long, to complete a wire sculpture. She has 150 m of metal wire on a roll.

Is this long enough for her to cut 120 lengths each $\frac{3}{4}$ m long?

Explain your answer.

10 Helen and David are discussing fractions.

A quarter of a third is bigger than a third of a quarter.
No it's not, they're the same size.

Is Helen or David correct?

Explain your answer.

Problem solving

The 2016 Olympic Games in Rio

The Rio 2016 Games will provide the best possible environment for peak performances. Athletes will enjoy world-class facilities, including a superb village, all located in one of the world's most beautiful cities, in a compact layout for maximum convenience.

The competition venues will be clustered in four zones and connected by a high-performance transport ring. Of the 34 competition venues, of which 18 are already operational, eight will undergo some permanent works, seven will be totally temporary and nine are being constructed as permanent legacy venues.

The Rio Games will also celebrate and showcase sport, thanks to the city's stunning setting and a desire to lift event presentation to new heights. At the same time, Rio 2016 will be an opportunity to deliver the broader aspirations for the long-term future of the city, region and country – an opportunity to hasten the transformation of Rio de Janeiro into an even greater global city.

The Rio 2016 Games will provide an opportunity for some wonderful performances by many athletes. These Games will celebrate with an opening ceremony on 5 August 2016 and a closing ceremony on 21 August.

The Olympic Village

The athletes will enjoy world-class facilities. The Olympic Village will be located in Bara de Tijuree, set to become another of the world's most romantic cities. It will be laid out for the maximum convenience of the 10 500 athletes expected to be representing around 200 nations.

Nearly half of the athletes will be able to reach their venues in less than 10 minutes, and almost 75% will be able to do so in less than 25 minutes.

Competition venues

The competition venues will be in:

Barra de Tijuree

Copacabana

Deodoro

Maracanã.

Tickets

There will be 6 million tickets available for spectators who wish to see events.

The average price of the tickets is US$36.

55% of the tickets will cost less than US$30

Over 2 million tickets will cost less than US$20

1 How long will the actual Olympic games last In Rio?

2 How many of the 10 500 athletes will be expected to:

  a reach their venues in under 10 minutes

  b reach their venues in under 25 minutes

  c reach their destination in 25 minutes or over?

3 What fraction of the athletes will take longer than 25 minutes to reach their venue?

4 What fraction of the venues are to be permanent legacy venues?

5 Of the 36 venues hosting the competition:

- half are in Barra de Tijuree
- one-twelfth are in Deodoro
- one-third are in Copacabana
- the rest are in Maracanã.

  a How many of the venues are in:

    i Barra de Tijuree ii Copacabana iii Deodora?

  b What fraction of the venues are in Maracanã?

5 How much is the revenue from the six million tickets?

6 What fraction of tickets cost less than US$20?

7 How many of the tickets cost less than US$30?

8

Algebra

This chapter is going to show you:

- more about expanding brackets and factorising algebraic expressions
- how to simplify more complicated expressions.

You should already know:

- how to collect like terms in an expression
- how to multiply out a simple bracket
- how to simplify simple expressions.

About this chapter

When the wind catches a sail, or a kite, it takes on a curved shape like those shown on this page. Mathematicians have been interested in the shapes that form naturally, and have used algebra to investigate them. They have used their skills in rearranging expressions and solving equations that involve powers of a variable, such as x^2 or x^3.

These and higher powers occur in many areas of mathematics and science. You have met formulae that involve powers, such as the formulae for the area of a circle and the volume of a cube. The equations of curved lines often involve powers.

In this chapter you will extend your algebraic skills to deal with expressions involving powers.

8.1 Expanding brackets

Learning objective

- To multiply out brackets with a variable outside them

Do you remember how to multiply out a term that includes brackets? If there is a number outside the brackets, you must multiply each term inside the brackets by that number.

This is also called expanding brackets.

Example 1

Multiply out the brackets in these expressions.

a $2(x + 3)$ **b** $4(t - 3)$ **c** $3(2p + 6)$

a $2(x + 3) = 2x + 6$ $2 \times x = 2x$ and $2 \times 3 = 6$

b $4(t - 3) = 4t - 12$ $4 \times t = 4t$ and $4 \times -3 = -12$

c $3(2p + 6) = 6p + 18$ $3 \times 2p = 6p$ and $3 \times 6 = 18$

If there is a variable (letter) outside the brackets then you multiply what is inside the brackets by that variable.

Example 2

Expand $m(4p + 2)$.

Multiply each term by m.

$m(4p + 2) = m \times 4p + m \times 2$

$= 4mp + 2m$

Example 3

Expand $t(5t - 3)$.

Multiply each term by t.

$t(5t - 3) = t \times 5t - t \times 3$

$= 5t^2 - 3t$

Example 4

Write down the expanded expression for the area of this rectangle.

$4x + 3$

x

To find the area, multiply the two different sides.

Area $= x \times (4x + 3)$

$= x(4x + 3)$

$= 4x^2 + 3x$

Now try this exercise. It starts with a few questions to get you thinking algebraically.

Exercise 8A

1 Simplify each of these expressions by collecting like terms.

a $2x + 5x$ **b** $6a + 4a$ **c** $8t + t$ **d** $3y + y + 5y$

e $9m - 3m$ **f** $6k - 2k$ **g** $7n - n$ **h** $4p - 8p$

2 Simplify each of these expressions.

a $5m + m + 4m$ **b** $3y + 5y + y$ **c** $7t + 4t + t$ **d** $6p + 3p + 5p$

e $7n + 3n + 4n$ **f** $6p + 2p + p$ **g** $8t - t + 4t$ **h** $5e - 3e + 4e$

i $8k + 3k - 5k$ **j** $5h + h - 3h$ **k** $8m - 5m - m$ **l** $4t + 5t - 3t$

3 Write each expression as simply as possible.

a $3 \times 3x$ **b** $3 \times 4a$ **c** $4 \times 5t$ **d** $4 \times 3y$ **e** $5 \times 3k$

f $2t \times 6$ **g** $3x \times 7$ **h** $5m \times 2$ **i** $8t \times 3$ **j** $4y \times 6$

4 Multiply out the brackets.

a $4(t + 3)$ **b** $2(x + 6)$ **c** $4(4m - 2)$ **d** $5(2k - 3)$

e $3(3 + 2x)$ **f** $4(5 - 3k)$ **g** $2(7 - 3y)$ **h** $3(5 - x)$

5 Multiply out each expression.

a $x(y + 2)$ **b** $m(3a + 2)$ **c** $k(2p + 4)$ **d** $n(6m + 3)$

e $t(5 + 4q)$ **f** $g(3 + 4h)$ **g** $h(7 + 5g)$ **h** $k(3 + 2d)$

i $a(4b - 3)$ **j** $c(5 - 4d)$ **k** $f(2 - 3m)$ **l** $b(5 - 4a)$

m $d(5a + 3)$ **n** $e(7f + 3)$ **o** $y(3x + 2)$ **p** $p(2q + 5)$

q $q(3 - 4p)$ **r** $t(6 - 3s)$ **s** $w(8 - 5k)$ **t** $n(3 - 2m)$

6 Write down an expression, in expanded form, for the area, A, of each rectangle.

a $x + 5$, y

b $2x + 3$, m

c $6 + 3a$, d

d $2a + 3$, k

e $3 + 5y$, n

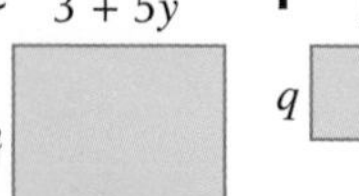

f $5p + 6$, q

7 Write each expression as simply as possible.

a $x \times 4x$ **b** $a \times 5a$ **c** $t \times 6t$ **d** $y \times 4y$ **e** $k \times 2k$

f $t \times 5t$ **g** $x \times 8x$ **h** $m \times 3m$ **i** $t \times 4t$ **j** $y \times 5y$

8 Multiply out each expression.

a $x(x + 2)$ **b** $m(3m + 2)$ **c** $k(4k + 1)$ **d** $n(4n + 3)$

e $t(6 + 2t)$ **f** $g(1 + 4g)$ **g** $h(3 + 5h)$ **h** $d(2 + 3d)$

i $a(5a - 2)$ **j** $c(3 - 4c)$ **k** $t(5 - 3t)$ **l** $b(7 - 4b)$

m $d(8d + 7a)$ **n** $e(5e + 3)$ **o** $y(2x + 3y)$ **p** $p(5 + 4p)$

q $q(7q - 5)$ **r** $t(2t - 5)$ **s** $w(3w - 4)$ **t** $n(8n - 5)$

9 Write down an expression, in expanded form, for the area, A, of each rectangle.

a $4m + 3$, m **b** $6 + 3t$, t **c** $3k + 1$, k **d** $4 + 3x$, x **e** $2g + 7$, g **f** $3 + 2n$, n

MR **10** James has asked Jess for some help multiplying out brackets.

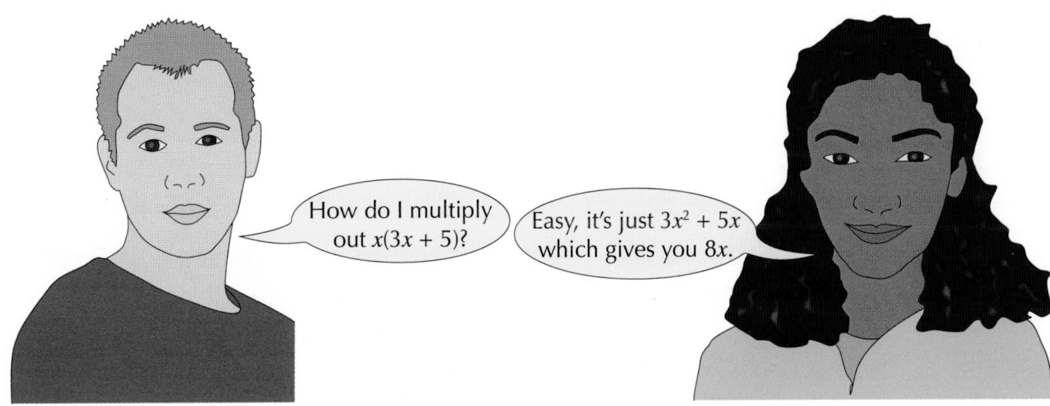

How do I multiply out x(3x + 5)?
Easy, it's just 3x² + 5x which gives you 8x.

Explain why what Jess has told James is wrong.

Challenge: Mixed letters

Joy had written some letters.

Just as she sealed the last one, she realised that she had put some of the letters in the wrong envelopes.

She had to open them all and rematch them.

A If there were ten letters, what is the minimum number of envelopes Joy would need to open, to be sure of having all the right letters in the right envelopes.

B If there were x letters, what is the minimum number of envelopes Joy would need to open, to be sure of having all the right letters in the right envelopes.

8.2 Factorising algebraic expressions

Learning objective

- To factorise expressions

To factorise an algebraic expression, you do the opposite of multiplying out the brackets. You look for a factor that is common to all the terms, then take it outside the brackets.

- If you multiply out $3(x - 2)$ you get $3x - 6$.
- If you factorise $3x - 6$ you get $3(x - 2)$.

In this case, 3 is a common factor of $3x$ and 6 so you take it outside the brackets.

You have just been expanding expressions such as $t(5t - 3) = 5t^2 - 3t$.

Therefore, you should be able to see that you can factorise $5t^2 - 3t$ to get back to $t(5t - 3)$.

In this case t is a common factor of $5t^2$ and $3t$.

Example 5

Factorise each expression.

a $2xy + 3y$ **b** $5k^2 - 4k$

a The HCF of $2xy + 3y$ is y. You can take y outside the brackets.

$2xy + 3y = y(2x + 3)$ The terms inside are $2xy \div y = 2x$ and $3y \div y = 3$.

b The common factor of $5k^2$ and $4k$ is k. You can take k outside the brackets.

$5k^2 - 4k = k(5k - 4)$ The terms inside are $5k^2 \div k = 5k$ and $-4k \div k = -4$.

Exercise 8B

1 Write down all the factors of each number.

a 24 **b** 35 **c** 40 **d** 28

e 36 **f** 18 **g** 28 **h** 50

2 Write down all the factors of each expression.

a $2x$ **b** $3m$ **c** $4t$ **d** $5y$

e $3x^2$ **f** $2m^2$ **g** $5t^2$ **h** $3k$

3 Write down the highest common factor of the terms in each pair.

a $4mt$ and $3t$ **b** $2pq$ and $5q$ **c** $3x$ and $5xy$ **d** $3a$ and $4am$

e $5t^2$ and $3t$ **f** $4q^2$ and $7q$ **g** $2x$ and $3x^2$ **h** $3a$ and $7a^2$

4 Copy and complete each factorisation.

a $mx + 2m = \ldots(x + 2)$ **b** $mt + 3m = \ldots(t + 3)$ **c** $np + 2p = \ldots(n + 2)$

d $x^2 - 3x = \ldots(x - 3)$ **e** $p^2 - p = \ldots(p - 1)$ **f** $y^2 - 2y = \ldots(y - 2)$

g $4k + xk = \ldots(4 + x)$ **h** $3k + k^2 = \ldots(3 + k)$ **i** $2x - xt = \ldots(2 - t)$

5 Copy and complete each factorisation.

a $3t + mt = t(\ldots + \ldots)$ **b** $2x + xy = x(\ldots + \ldots)$ **c** $5p + pq = p(\ldots + \ldots)$

d $6k - k^2 = k(\ldots - \ldots)$ **e** $n^2 - 5n = n(\ldots - \ldots)$ **f** $x^2 - 8x = x(\ldots - \ldots)$

g $5x + x^2 = x(\ldots)$ **h** $h + h^2 = h(\ldots)$ **i** $2t - 3t^2 = t(\ldots)$

6 Write down the missing length in each rectangle below.

a

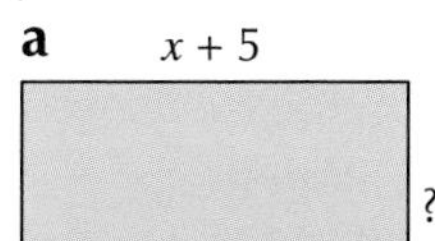

Area = $x^2 + 5x$

b

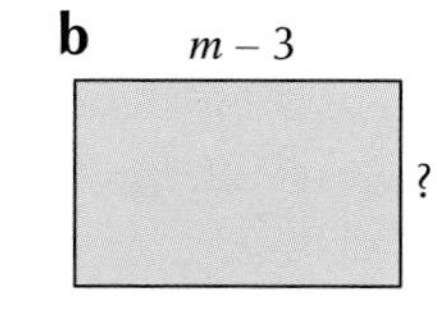

Area = $mp + 3p$

c

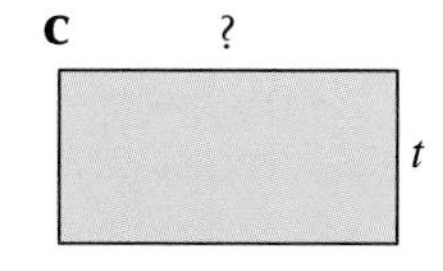

Area = $t^2 + 5t$

d

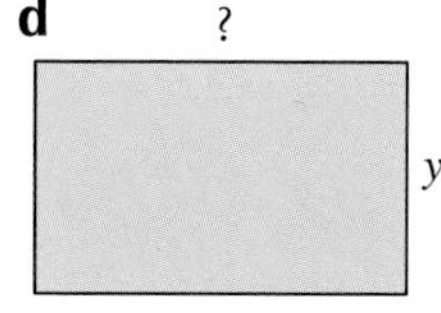

Area = $4y - y^2$

7 When Ian's mum checked his homework, she said: 'I think these are wrong.'

Look at Ian's work.

Write it out again and correct all the errors.

a $x^2 + 5x = x(x - 5)$

b $6m + mt = m(6 + m)$

c $3y^2 + 2y = y(2y + 3)$

8 Copy and complete each factorisation.

a $4t^2 + 5t = t(\ldots + \ldots)$ **b** $6x^2 + x = \ldots(6x + 1)$ **c** $6t + 4 = 2(\ldots + \ldots)$

d $3x^2 - mx = \ldots(3x - m)$ **e** $5t^2 + kt = t(\ldots + \ldots)$ **f** $9x + 6 = \ldots(3x + 2)$

g $5t + 8t^2 = t(\ldots)$ **h** $3x^2 - 2x = \ldots(3x - 2)$ **i** $10t + 15 = 5(\ldots)$

9 Al has asked Chris for some help.

How should Chris explain to Al how to factorise $12p^2 + 5p$?

How can I factorise 12p² + 5p?
That's easy, you just …

Investigation: An age-old problem

Andrew reversed the digits in his dad's age and the result was the same as his grandpa's age.

In one year's time, the ages of his dad and his grandpa will add up to 101.

How old was Andrew's grandpa when Andrew's dad was born?

8.3 Expand and simplify

Learning objective

- To expand expressions with two brackets and simplify them

You already know how to expand simple expressions with one set of brackets.

Sometimes, you need to expand two brackets and add the results together. Then you need to simplify the result, to find the answer.

You have met both of these processes – expanding brackets and simplifying – before. Now you are going to put them together.

Example 6

Expand and simplify $4(5 + 2y) + 2(5y - 6)$.

$4(5 + 2y) + 2(5y - 6) = 20 + 8y + 10y - 12$ — Multiply out both brackets.

$= 8y + 10y + 20 - 12$ — Bring like terms together.

$= 18y + 8$ — Simplify to obtain the final answer.

Example 7

Expand and simplify $4(2u + 3i) - 2(u - 2i)$.

$4(2u + 3i) - 2(u - 2i) = 8u + 12i - 2u + 4i$ — Multiply out both brackets.

$= 8u - 2u + 12i + 4i$ — Bring like terms together.

$= 6u + 16i$ — Simplify to obtain the final answer.

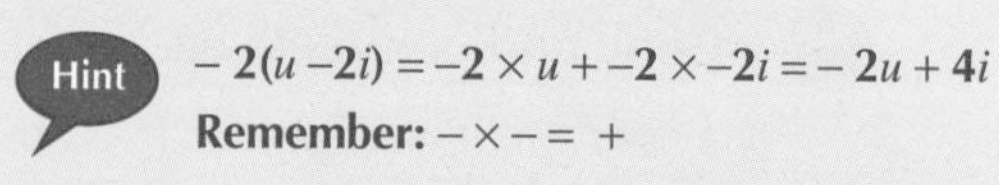

Hint
$-2(u - 2i) = -2 \times u + -2 \times -2i = -2u + 4i$
Remember: $- \times - = +$

Example 8

Expand and simplify $x(3x + 4) - x(x - 5)$.

$x(3x + 4) - x(x - 5) = 3x^2 + 4x - x^2 + 5x$ — Multiply out both brackets.

$= 3x^2 - x^2 + 4x + 5x$ — Bring like terms together.

$= 2x^2 + 9x$ — Simplify to obtain the final answer.

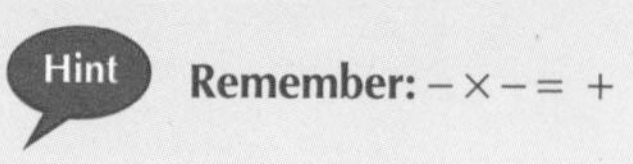

Exercise 8C

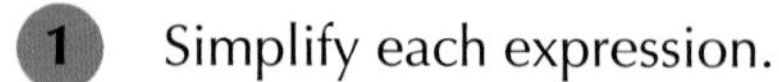

1 Simplify each expression.

a	$5m + m + 4m$	**b**	$6y + 5y + y$	**c**	$2t + 5t + t$	**d**	$8p + 3p + 5p$
e	$n + 5n + 6n$	**f**	$7p + 5p + p$	**g**	$7t - t + 2t$	**h**	$e - 3e + 6e$
i	$k + 4k - 5k$	**j**	$2h + h - 4h$	**k**	$7m - 5m - 4m$	**l**	$4t + 5t - 3t$

2 Simplify each expression.

a	$4t + 5g + 6t + 3g$	**b**	$6x + y + 3x + 4y$	**c**	$2m + k + 4m + 3k$
d	$9x + 6y - 3x + 2y$	**e**	$6m + p - 3m + 2p$	**f**	$5n + 3t - n + 6t$
g	$12k + 4g - 3k - 2g$	**h**	$6d + 8b - 3d - 2b$	**i**	$6q + 4p - 3q - 2p$
j	$6g - 2k + 2g + 4k$	**k**	$6x - 9y + 2x + 3y$	**l**	$4d - 2e - 8d + 4e$

3 Multiply out each expression.

a $4(y + 3)$ **b** $2(3a + 4)$ **c** $5(2p + 3)$ **d** $3(2m + 3)$

e $t(4 + 3q)$ **f** $g(2 + 5h)$ **g** $h(3 + 7g)$ **h** $k(4 + 3d)$

i $a(3a - 3)$ **j** $c(4 - c)$ **k** $f(2 - 3f)$ **l** $b(5 - 4b)$

m $5(3a + 2)$ **n** $e(5f + 1)$ **o** $y(y + 4)$ **p** $p(2p + 3)$

4 Expand and simplify each expression.

a $2(3x + 4) + 3(x + 2)$ **b** $4(2k + 3) + 3(4k + 7)$

c $5(2t + 3) + 2(3t + 4)$ **d** $4(3q + 2) + 3(2q + 1)$

e $6(3h + 2) + 4(2h - 1)$ **f** $5(6 + 3f) + 2(2 - 3f)$

g $4(3 - 2y) + 3(2 + 3y)$ **h** $6(2t - 5) + 3(5t - 2)$

5 Expand and simplify each expression.

a $3(2x + 5) - 2(x + 3)$ **b** $5(2k + 4) - 2(4k + 1)$

c $6(3t + 4) - 3(2t + 5)$ **d** $7(2q + 3) - 4(3q + 4)$

e $8(2h + 5) - 3(4h - 2)$ **f** $7(w + 4) - 3(2w - 3)$

g $5(4x - 3) - 3(3x - 2)$ **h** $9(2t - 3) - 2(6t - 3)$

6 Expand and simplify each expression.

a $x(2x + 5) + x(4x + 3)$ **b** $p(3p + 4) + p(2p + 1)$

c $k(5k + 3) + k(2k + 4)$ **d** $d(3d + 5) + d(2d + 3)$

e $n(5n + 6) + n(3n - 5)$ **f** $f(5f + 3) + f(3f - 2)$

g $p(p - 5) + p(2p - 4)$ **h** $y(5y - 2) + y(4y - 3)$

7 Expand and simplify each expression.

a $x(8x + 5) - x(4x + 1)$ **b** $p(5p + 4) - p(2p + 1)$

c $k(4k + 4) - k(2k + 3)$ **d** $d(3d + 7) - d(2d + 4)$

e $n(7n + 5) - n(3n - 2)$ **f** $f(6f + 5) - f(3f - 4)$

g $p(3p - 1) - p(p - 5)$ **h** $y(4y - 3) - y(2y - 7)$

8 Helen has asked Dean for some help.

Is what Dean says correct? Explain how you know.

I have to add together 3(x + 5) and x(x + 5) then simplify it.
That's easy, its going to be x² + 8x + 15.

Challenge: All legs and heads

A farmer has cows and chickens on his farm.

Looking at all of his animals, he can see a total of 50 heads and 172 legs.

How many more cows than chickens does the farmer have?

Ready to progress?

I can expand simple brackets.
I can expand simple brackets and simplify.

I can factorise simple algebraic expressions.
I can expand brackets that give rise to powers.

Review questions

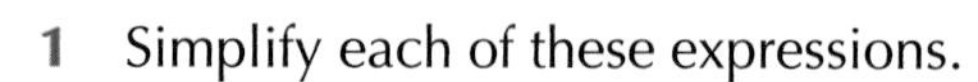

1 Simplify each of these expressions.

a $3p + p + 4p$ **b** $x + 7x + 5x$ **c** $5q + 4q + 3q$ **d** $3t + 5t + t$

e $4n + n + 5n$ **f** $4p + 7p - p$ **g** $5m - m + 2m$ **h** $3a - 5a + 6a$

i $7h + 2h - 8h$ **j** $4g + g - 7g$ **k** $5n - 3n - n$ **l** $2t + 3t - 8t$

2 Write down an expression for the perimeter, P, of each shape.

Write your answers as simply as possible.

a

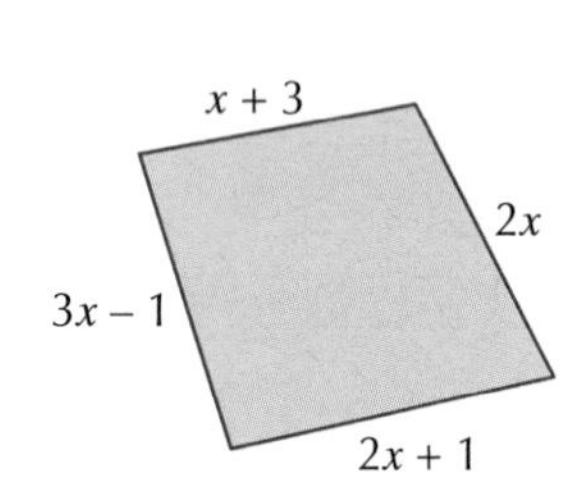

b

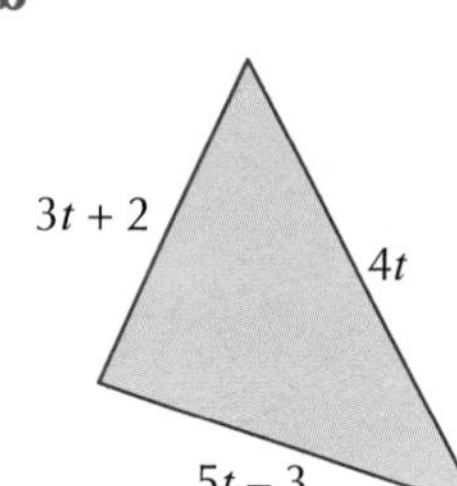

c

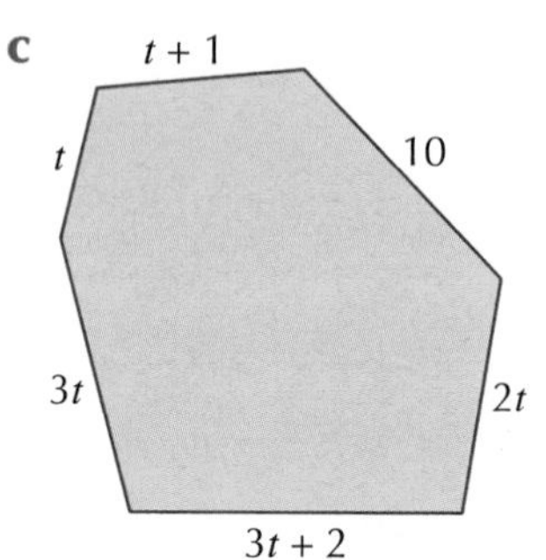

3 Multiply out the brackets.

a $2(m + 4)$ **b** $4(t + 5)$ **c** $3(x - 7)$ **d** $5(t - 2)$

e $m(3 + y)$ **f** $t(3 - h)$ **g** $x(5 - 2t)$ **h** $k(2 - 3t)$

4 Write down an expression, in expanded form, for the area, A, of each rectangle.

a

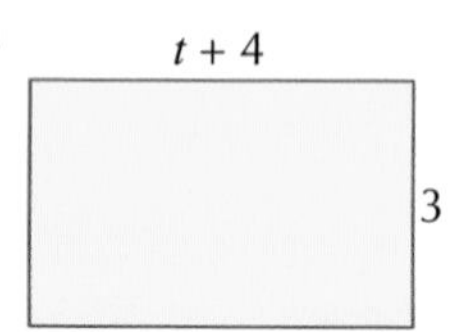

b

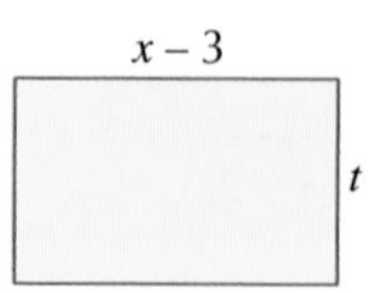

c

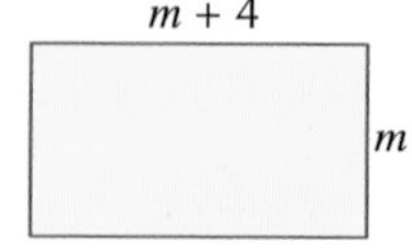

d

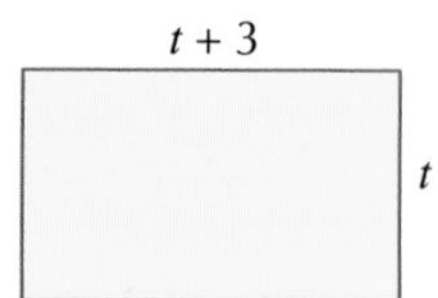

5 Copy and complete each factorisation.

a $6t + 10 = 2(\ldots + \ldots)$ **b** $3x + 6 = \ldots(x + 2)$ **c** $5t - 2mt = t(\ldots - \ldots)$

d $6x - yx = \ldots(6 - y)$ **e** $4t - 5pt = t(\ldots)$ **f** $3x + mx = \ldots(3 + m)$

g $5t^2 + 3t = t(\ldots)$ **h** $2x^2 + 7x = \ldots(2x + 7)$ **i** $4t^2 - mt = t(\ldots)$

6 Write down the missing length in each rectangle.

a

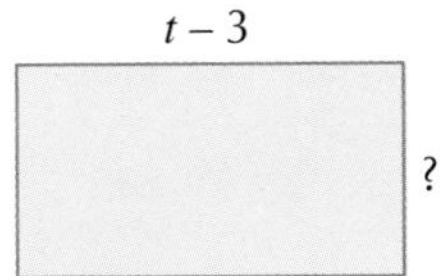

Area = $t^2 - 3t$

b

Area = $x^2 + 4x$

c

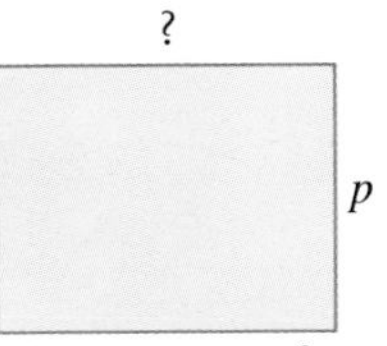

Area = $mp - p^2$

7 This is David's homework.

His older sister, Maisie, decided to check his homework.

She said that he had made some mistakes.

Explain what David has done wrong and correct any errors.

a $8t - t^2 = t(t - 8)$

b $3mp + 2p^2 = m(3p + 2p^2)$

c $4x^2 + 3xy = x(4x + 3y)$

8 Expand and simplify each expression.

a $3(3x + 5) + 2(x + 4)$ **b** $3(2k + 5) + 2(4k + 3)$

c $3(2t + 4) + 4(3t + 1)$ **d** $3(3q + 5) + 2(2q + 3)$

e $5(2h + 3) - 2(4h + 1)$ **f** $5(w + 3) - 3(2w - 1)$

g $4(4x - 2) - 2(5x - 3)$ **h** $7(2t - 1) - 3(4t - 5)$

9 Expand and simplify each expression.

a $x(3x + 5) + x(5x + 3)$ **b** $p(4p + 3) + p(5p + 2)$

c $k(3k + 2) + k(4k + 5)$ **d** $d(2d + 3) + d(3d + 5)$

e $n(5n + 1) - n(2n - 1)$ **f** $f(5f + 4) - f(2f - 3)$

g $p(2p - 3) - p(p - 4)$ **h** $y(5y - 2) - y(3y - 5)$

10 Jenna has asked her dad for help with her homework.

How can I factorise $5t^2 - 2t$?
That's easy, you just …

How should her dad explain how to factorise $5t^2 - 2t$?

Challenge
Calfornia Gold

While walking their dog on their own land, a couple suddenly noticed a coin sticking out of the ground. Naturally, they picked it up. It was then that they realised this was no ordinary coin, but an old coin dated 1886 – it would be worth a lot of money.

So they went back with a spade and dug up 1427 coins, all dated between 1847 and 1894.

An expert has told them that each coin could be worth as much as £600 000.

We'll be billionaires now my dear.
Well almost.

1 The coins were found in 2014. What was the age of:

a the first coin they found

b the oldest of the coins when they were found?

2 Each coin had a mass of 15 grams.

What is the mass of:

a 2 coins b x coins c all 1427 coins?

3 When the coins were found, the approximate price of gold was £28 per gram.

How much money:

a was 2 grams of gold worth

b was y grams of gold worth

c was the 1427 coins worth?

4 The expert said that each coin could be worth £600 000.

What could be the worth of:

a 2 coins b w coins c all 1427 coins?

5 Who is correct, Mr T or Mrs T ?

Explain your answer.

6 Are the coins worth more as they are or sold just as gold?

Explain if the couple should sell them as coins or as gold.

9 Decimal numbers

This chapter is going to show you:

- how to extend your ability to work with powers of 10
- how to know when to make suitable rounding and to use rounded numbers to estimate the results of calculations.

You should already know:

- how to multiply and divide by 10 and 100.

About this chapter

One of the earliest aids to arithmetic was the abacus. To use it, you move beads to represent numbers. There are many different types. The Chinese abacus has a number of rods, each holding seven beads. A bar across the rods separates the beads into two sections, with two on one side of the bar and five on the other. The five beads each represent one unit, each of the separate two beads represents five units. So the number seven is represented by one 'five-unit' bead and two 'single-unit' beads.

The Japanese abacus is different, with only four 'single-unit' beads and one 'five-unit' bead on each rod.

Skilful abacus users can calculate very quickly and accurately. There are resources on the internet to show you how to do it.

More recently, though, computers and calculators have been developed to help you work out quite complicated calculations. Modern calculators can do so much more than simple arithmetic. In this chapter you will learn more about the decimal system of counting and practise your skills in using a calculator.

9.1 Multiplication of decimals

Learning objective

- To practise multiplying decimal numbers

Work through this section to review what you already know about the multiplication of decimals. It is very important that you are able to carry out this sort of multiplication.

Example 1

Write down the answer to each multiplication.

Use the fact that $27 \times 4 = 108$.

a 27×0.4 **b** 2.7×4 **c** 2.7×0.4

a There is one decimal place in the multiplication 27×0.4, so there will be one decimal place in the answer.

So $27 \times 0.4 = 10.8$.

b There is one decimal place in the multiplication 2.7×4, so there must be one decimal place in the answer.

So $2.7 \times 4 = 10.8$.

c There are two decimal places in the multiplication 2.7×0.4, so there are two decimal places in the answer.

So $2.7 \times 0.4 = 1.08$.

Example 2

Find the answer to each multiplication.

a 0.3×0.04 **b** 900×0.3 **c** 50×0.05

a There are three decimal places in the multiplication, so there must be three decimal places in the answer.

$3 \times 4 = 12$ Remove the decimal places from the original question.

$0.3 \times 0.04 = 0.012$ Put the decimal places back in the answer.

b There is one decimal place in the multiplication, so there must be one decimal place in the answer.

It is important to include the zero until you decide where to put the decimal point. Once you have done that you can leave the zero out.

$900 \times 3 = 2700$ Remove the decimal places from the original question.

$900 \times 0.3 = 270.0$ Put the decimal places back in the answer.

$900 \times 0.3 = 270$ You can leave out the zero after the decimal point.

c There are two decimal places in the multiplication, so there must be two decimal places in the answer.

$50 \times 5 = 250$ Remove the decimal places from the original question.

$50 \times 0.05 = 2.50$ Put the decimal places back in the answer.

$50 \times 0.05 = 2.5$ You can leave out the zero after the decimal point.

Example 3

Without using a calculator, work out the answer to 134×0.6.

First, note the one decimal place in the calculation.

Ignore the decimal point and calculate 134×6 by the column method.

$$\begin{array}{r} 134 \\ \times \quad 6 \\ \hline 804 \\ \hline {\scriptstyle 2\,2} \end{array}$$

Now put one decimal place back into the answer.

$134 \times 0.6 = 80.4$

Exercise 9A

Do not use a calculator to answer any of these questions.

1 $83 \times 24 = 1992$

Use this fact to write down the answer to each multiplication.

a 8.3×24 **b** 83×2.4 **c** 8.3×2.4 **d** 0.83×0.24

2 $25 \times 32 = 800$

Use this fact to write down the answer to each multiplication.

a 2.5×32 **b** 25×3.2 **c** 2.5×3.2 **d** 2.5×0.32

3 Work out each multiplication.

a 2.6×5 **b** 3.4×6 **c** 4.91×4 **d** 6.12×5

e 31.5×7 **f** 22.4×8 **g** 14.6×6 **h** 19.1×4

4 Work out each multiplication.

a 10×0.5 **b** 0.7×10 **c** 0.3×100 **d** 0.6×10

e 10×0.7 **f** 0.8×100 **g** 100×0.1 **h** 0.4×100

5 Work out each multiplication.

a 40×0.5 **b** 0.7×20 **c** 0.3×200 **d** 0.6×50

e 40×0.7 **f** 0.8×300 **g** 400×0.1 **h** 0.4×500

6 Work out the answer to each multiplication.

a 0.3×0.6 **b** 0.5×0.5 **c** 0.9×0.7 **d** 0.6×0.6

e 0.9×0.8 **f** 0.7×0.6 **g** 0.5×0.8 **h** 0.4×0.4

i 0.7×0.7 **j** 0.9×0.3 **k** 0.4×0.8 **l** 0.3×0.2

7 **a** Work out the answer to 123×4.

b Use your answer to part **a** to write down the answers to these multiplications.

i 1.23×0.4 **ii** 12.3×0.04

8 Calculate the area of each rectangle.

a

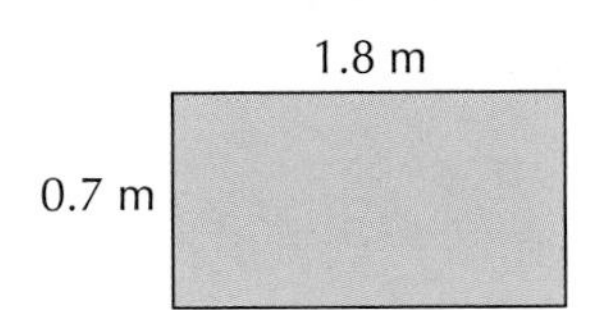

b

2.3 m

1.1 m

9 Kieron calculated that he needed 4 m^2 of carpet for his hall.
He saw some carpet priced at £5.15 per m^2.
How much will it cost him if he chooses this carpet?

10 Mrs Bishop took all her class of 30 pupils to a theme park.
It cost £6.25 for each pupil to get in.
Mrs Bishop got in free.
How much did it cost Mrs Bishop to get all 30 pupils into the theme park?

Investigation: Mystical multiplication

A Complete each calculation.

a $0.7 + 0.3$ **b** 0.7×0.7 **c** 0.3×0.3

d $(0.7 \times 0.7) - (0.3 \times 0.3)$ **e** $0.7 - 0.3$

B What do you notice about the answers to parts **d** and **e** of question **A**?

C Repeat question A with more pairs of decimals that add up to 1.
e.g. $0.8 + 0.2$, $0.6 + 0.4$

D Explain what you have found out.

9.2 Powers of 10

Learning objective

- To understand and work with both positive and negative powers of ten

Key words

negative power

What do you remember about positive powers of 10? This section will remind you how to use them to solve problems. It will also introduce you to **negative powers** of ten.

This table shows some powers of 10. Look at it carefully and see if you can spot any patterns.

Power	10^4	10^3	10^2	10^1	10^0	10^{-1}	10^{-2}	10^{-3}
Value	10 000	1000	100	10	1	0.1	0.01	0.001

Can you see how the negative powers of ten move you into the decimal numbers 0.1, 0.01 and so on?

Note that:

- multiplying by 10^{-1} is the same as multiplying by 0.1
- multiplying by 10^{-2} is the same as multiplying by 0.01
- multiplying by 10^{-3} is the same as multiplying by 0.001, and so on.

Example 4

Calculate the value of each expression.

a 34.5×10^{-1} **b** 0.89×10^{-2} **c** 7632×10^{-3}

a $34.5 \times 10^{-1} = 34.5 \times 0.1$, which has two decimal places.
So put two decimal places in the answer to $345 \times 1 = 345$.
$34.5 \times 10^{-1} = 3.45$

b $0.89 \times 10^{-2} = 0.89 \times 0.01$, which has four decimal places.
So put four decimal places in the answer to $89 \times 1 = 89$.
$0.89 \times 10^{-2} = 0.0089$

c $7632 \times 10^{-3} = 7632 \times 0.001$, which has three decimal places.
So put three decimal places in the answer to $7632 \times 1 = 7632$.
$7632 \times 10^{-3} = 7.632$

You may have noticed a pattern in the example.

This can help you to work out problems like these quite quickly.

In part **a** you can see that $34.5 \times 10^{-1} = 3.45$, but notice also that $34.5 \div 10 = 3.45$.

In part **b** you can see that $0.89 \times 10^{-2} = 0.0089$, but notice also that $0.89 \div 100 = 0.0089$.

In part **c** you can see that $7632 \times 10^{-3} = 7.632$, but notice also that $7632 \div 1000 = 7.632$.

Spotting the pattern will help you to calculate multiplications such as these, as you can:

- multiply by 10^{-1} by dividing by 10 (move digits 1 place to the right)
- multiply by 10^{-2} by dividing by 100 (move digits 2 places to the right)
- multiply by 10^{-3} by dividing by 1000 (move digits 3 places to the right).

Example 5

Calculate the value of each expression.

a 45.6×10^{-1} **b** 3.73×10^{-2} **c** 152×10^{-3}

a $45.6 \times 10^{-1} = 45.6 \div 10$
$= 4.56$ The digits have moved 1 place to the right.

b $3.73 \times 10^{-2} = 3.73 \div 100$
$= 0.0373$ The digits have moved 2 places to the right.

c $152 \times 10^{-3} = 152 \div 1000$
$= 0.152$ The digits have moved 3 places to the right.

Example 6

Calculate the value of each expression.

a 0.752×10^1 **b** 1.508×10^2 **c** 0.0371×10^3

a $0.752 \times 10^1 = 0.752 \times 10$

$= 7.52$ The digits have moved 1 place to the left.

b $1.508 \times 10^2 = 1.508 \times 100$

$= 150.8$ The digits have moved 2 places to the left.

c $0.0371 \times 10^3 = 0.0371 \times 1000$

$= 37.1$ The digits have moved 3 places to the left.

Summary

Remember that the number in the power tells you how many places the digits move when you multiply.

- Multiply by a positive power of ten → number gets bigger, digits move to the left.
- Multiply by a negative power of ten → number gets smaller, digits move to the right.

Exercise 9B

1 Multiply these numbers.

a 57×10 **b** 69×100 **c** 78×1000

d 714×1000 **e** 802×100 **f** 315×10

2 Divide these numbers.

a $748 \div 10$ **b** $329 \div 100$ **c** $473 \div 1000$

d $58 \div 1000$ **e** $85 \div 100$ **f** $17 \div 10$

3 Multiply these numbers.

a 14.3×10 **b** 36.2×100 **c** 57.3×1000

d 32.14×1000 **e** 12.85×100 **f** 39.17×10

4 Divide these numbers.

a $6.34 \div 10$ **b** $47.3 \div 100$ **c** $66.3 \div 1000$

d $2.7 \div 1000$ **e** $3.76 \div 100$ **f** $71.93 \div 10$

5 Complete these calculations.

a 11.5×10 **b** $63.7 \div 100$ **c** 42.3×1000

d $3.65 \div 1000$ **e** 1.07×100 **f** $9.14 \div 10$

g 4.1×10 **h** $3.8 \div 100$ **i** 7.4×1000

6 Change each number to decimal form.

The first one has been done for you.

a $10^{-1} = 0.1$ **b** 10^{-2} **c** 10^{-3} **d** 10^{-5} **e** 10^{-6}

7 Change each number to index form.

The first one has been done for you.

a $0.01 = 10^{-2}$ **b** 0.001 **c** 0.1 **d** 0.0001 **e** 0.000 000 1

8 Multiply each number by 10^2.

The first one has been done for you.

a $9.6 \times 10^2 = 9.6 \times 100$
$= 960$
b 0.18 **c** 204.6 **d** 12.97

9 Multiply each number by 10^3.

The first one has been done for you.

a $8.16 \times 10^3 = 8.16 \times 1000$
$= 8160$
b 0.71 **c** 824.6 **d** 29.66

10 Multiply each number by 10^{-1}.

The first one has been done for you.

a $7.7 \times 10^{-1} = 7.7 \div 10$
$= 0.77$
b 0.63 **c** 514.6 **d** 235.8

11 Multiply each number by 10^{-2}.

The first one has been done for you.

a $18.6 \times 10^{-2} = 18.6 \div 100$
$= 0.186$
b 1.84 **c** 218.5 **d** 34.6

12 The table shows some country populations, with the land mass (km^2) per person given in index form.

Copy and complete the table to write the decimal numbers in full.

The first one has been done for you.

	Population	Land mass per person (km^2)
Singapore	Five million	$14.2 \times 10^{-5} = 0.000\ 142$
Hong Kong	Seven million	15.7×10^{-5}
Belgium	Eleven million	27.3×10^{-4}
Japan	127 million	29.9×10^{-4}

Activity: Prefixes

You have already met three prefixes that you can use to make decimal multiples of units. They are:

- kilo-, as in kilogram (1000 grams)
- centi-, as in centilitre (one hundredth of a litre)
- milli-, as in millimetre (one thousandth of a metre).

This table gives the main prefixes and their equivalent multiples, written as powers of 10.

Prefix	giga	mega	kilo	centi	milli	micro	nano
Power	10^9	10^6	10^3	10^{-2}	10^{-3}	10^{-6}	10^{-9}

For example, 8 000 000 000 watts could be written as 8 gigawatts.

A Use suitable prefixes to write each quantity in a simpler form.

a 3 000 000 watts **b** 5000 metres **c** 3 000 000 000 bytes

d 0.07 grams **e** 0.004 metres **f** 0.000 0055 litres

B Use the internet or a reference book to find out how far light travels in 1 nanosecond.

9.3 Rounding suitably

Learning objective

- To round numbers, where necessary, to a suitable degree of accuracy

Key words

suitable degree of accuracy

There are two main uses of rounding, both of which you have met before.

- One reason for rounding is to give an answer to a **suitable degree of accuracy**.
- The other reason for rounding is to enable you to make an estimate of the answer to a problem.

Example 7

Round each number to the stated number of decimal places.

a 7.35 to one decimal place

b 3.764 to two decimal places

c 3.98 to one decimal place

a 7.35 = 7.4 to one decimal place The 5 after the 3 means you round up.

b 3.764 = 3.76 to two decimal places The 4 after the 6 means you round down.

c 3.98 = 4.0 to one decimal place The 8 after the 9 means you round the 3.9 up to 4.0.

Example 8

Round each number to one significant figure.

a 18.67 **b** 0.037 61 **c** 7.95

a 18.67 = 20 to one significant figure — The 8 means you round 18 up to 20.

b 0.037 61 = 0.04 to one significant figure — The 7 means you round up to 0.04.

c 7.15 = 7 to one significant figure — The 1 means you round down.

Example 9

Estimate the answer to each calculation.

a 23.7 + 69.3 **b** 3.1×5.2 **c** $13.9 \div 4.2$

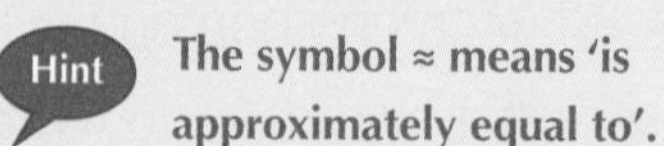

The symbol ≈ means 'is approximately equal to'.

Round the numbers to one significant figure each time, unless it seems more sensible to do something else.

a $23.7 + 69.3 \approx 20 + 70 = 90$

b $3.1 \times 5.2 \approx 3 \times 5 = 15$

c $13.9 \div 4.2 \approx 12 \div 4 = 3$ Approximate 13.9 to 12 instead of 10, so you can easily divide by 4.

Example 10

Which measurement has been rounded to the most suitable degree of accuracy?

a The distance from my house to the local post office

i 721.4 m **ii** 721 m **iii** 700 m

b The time in which an athlete runs a 100 m race

i 10.14 seconds **ii** 10.1 seconds **iii** 10 seconds

c The amount that a lorry can carry

i 3.48 tonnes **ii** 3 tonnes **iii** 3 481 783 kg

a Distances are usually rounded to one or two significant figures.
Hence, 700 m is the most suitable answer.

b 100-metre times need to be as accurate as possible.
Hence, 10.14 s is the most suitable answer.

c The mass a lorry can carry is a rough estimation.
Hence, 3 tonnes is the most suitable answer.

Exercise 9C

1 Round each number to one decimal place.

a 1.25	**b** 24.19	**c** 9.04	**d** 31.13
e 1.88	**f** 5.14	**g** 1.97	**h** 4.25

2 Round each number to two decimal places.

a 3.258	**b** 35.194	**c** 7.043	**d** 42.135
e 2.887	**f** 6.146	**g** 3.971	**h** 7.255

3 Round each number to:

i one decimal place **ii** two decimal places.

a 1.278 **b** 46.1724 **c** 6.836 **d** 17.138

e 3.997 **f** 7.0689 **g** 2.754 **h** 2.1528

4 Round each number to the nearest whole number.

Use your rounded numbers to estimate the value of each calculation.

a $3.9 + 8.2$ **b** $8.3 + 1.7$ **c** $7.1 - 1.8$ **d** $6.9 - 3.1$

e 6.1×7.6 **f** 9.2×8.9 **g** $7.8 \div 1.9$ **h** $9.2 \div 3.1$

5 Round each number to the nearest ten.

Use your rounded numbers to estimate the value of each calculation.

a $28 + 71$ **b** $72 + 29$ **c** $82 - 27$ **d** $58 - 21$

e 72×65 **f** 81×78 **g** $64 \div 31$ **h** $83 \div 19$

6 Round each number to one significant figure.

Use your rounded numbers estimate the value of each calculation.

a $281 + 92$ **b** $816 + 37$ **c** $758 - 117$ **d** $548 - 79$

e 182×85 **f** 273×37 **g** $193 \div 39$ **h** $413 \div 18$

7 State, with reasons, which is the most suitable degree of accuracy for:

a the average speed of a car journey

i 63.7 mph **ii** 60 mph **iii** 63.757 mph

b the size of an angle in a right-angled triangle

i 23.478° **ii** 23° **iii** 20°

c the mass of a sack of potatoes

i 50 kg **ii** 47 kg **iii** 46.89 kg

d the time taken to boil an egg

i 4 minutes 3 seconds **ii** 4 minutes **iii** 4 minutes 3.7 seconds

e the time to run 100 m

i 12.8 seconds **ii** 13 seconds **iii** 10 seconds

f the distance from home to school

i 1.7581 km **ii** 2 km **iii** 1.8 km.

Activity: Rounding

Below are four calculations and four exact answers.

Use estimations to match up each calculation to its answer.

Explain how you have matched each one.

8.3×3.9 $11.4 \div 1.5$ 9.3×6.1 $84 \div 3.2$

56.73 32.37 26.25 7.6

9.4 Dividing decimals

Learning objective

- To confirm ability to divide with decimals

This section will give you more practice in dividing decimal numbers by integers (whole numbers).
Remember the steps.

1. Count the decimal places in the decimal number.
2. Ignore the decimal point and just divide as if they are both whole numbers.
3. Put the decimal places back into your answer.

Example 11

Work out each of these.

a $0.8 \div 2$ **b** $0.12 \div 3$ **c** $27.5 \div 5$ **d** $1.4 \div 4$

a There is one decimal place in 0.8.
Work out $8 \div 2$ and ignore the decimal point. $8 \div 2 = 4$
Put one decimal place back into the answer of 4 to give 0.4.
$0.8 \div 2 = 0.4$

b There are two decimal places in 0.12.
Work out $12 \div 3$ and ignore the decimal point. $12 \div 3 = 4$
Put two decimal places back into the answer of 4 to give 0.04.
$0.12 \div 3 = 0.04$

c There is one decimal place in 27.5.
Work out $275 \div 5$ and ignore the decimal point. $275 \div 5 = 55$
Put one decimal place back into your answer of 55 to give 5.5.
$27.5 \div 5 = 5.5$

d You can see that 14 is not divisible exactly by 4 but 140 is.
So think of 1.4 as 1.40.
There are two decimal places in 1.40.
Work out $140 \div 4$ and ignore the decimal point. $140 \div 4 = 35$
Put two decimal places back into your answer of 35 to give 0.35.
$1.4 \div 4 = 0.35$

Exercise 9D

Do not use a calculator to answer any of these questions.

1 Work out each division.

a $0.6 \div 3$ **b** $0.9 \div 3$ **c** $2.4 \div 4$ **d** $3.5 \div 5$
e $2.1 \div 7$ **f** $4.8 \div 8$ **g** $5.4 \div 9$ **h** $6.4 \div 8$

2 Work out each division.

a 0.36 ÷ 2 **b** 0.45 ÷ 5 **c** 0.16 ÷ 4 **d** 0.45 ÷ 9

e 0.24 ÷ 6 **f** 0.81 ÷ 9 **g** 0.63 ÷ 7 **h** 0.56 ÷ 8

3 Work out each division.

a 2.42 ÷ 2 **b** 3.25 ÷ 5 **c** 6.44 ÷ 4 **d** 8.55 ÷ 9

e 5.22 ÷ 6 **f** 8.01 ÷ 9 **g** 4.27 ÷ 7 **h** 2.56 ÷ 8

4 Work out each division.

a 18.02 ÷ 2 **b** 37.95 ÷ 5 **c** 23.04 ÷ 4 **d** 49.14 ÷ 9

e 18.36 ÷ 6 **f** 99.18 ÷ 9 **g** 16.52 ÷ 7 **h** 10.24 ÷ 8

5 Work out each division by changing the first number.
The first one has been started for you.

a 16.1 ÷ 2 = 16.10 ÷ 2 … **b** 7.3 ÷ 5 **c** 13.2 ÷ 4 **d** 9.4 ÷ 5

e 17.3 ÷ 2 **f** 19.3 ÷ 4 **g** 26.2 ÷ 5 **h** 10.6 ÷ 4

PS **6** The perimeter of a square is 4.32 cm.
Work out the length of one side.

PS **7** Jenna paid £3.76 for eight cakes.
How much does one cake cost?

PS **8** Leon walks 8.32 km in four hours.
What is Leon's average speed?

PS **9** I have a length of cloth 2.3 m long. I want to cut it into five equal pieces.
How long will each piece be?

PS **10** I see a sign saying: '£1.08 for four grapefruit'.
How much should five grapefruit cost me?

Investigation: Spot the link

A Given that 548 ÷ 8 = 68.5, work out each division.

a 548 ÷ 80 **b** 5480 ÷ 8 **c** 54.8 ÷ 8 **d** 54.8 ÷ 80

B Given that 1380 ÷ 3 = 460, work out each division.

a 138 ÷ 3 **b** 13.8 ÷ 3 **c** 1.38 ÷ 3 **d** 1380 ÷ 30

9.5 Solving problems

Learning objective

- To solve real-life problems involving multiplication or division

Almost certainly, you use decimal calculations on many occasions, especially when you go to a shop. People at work often need to do some calculations while solving business problems.

Example 12

Which jar of jam offers the better value?

The larger jar is four times as big as the small jar.

Four small jars would cost $4 \times £0.89 = £3.56$.

One large jar costs £3.59.

So, the smaller jar is better value.

Example 13

A box contains 12 identical model cars.

Each model has 4 lights.

Each model has a mass of 200 g.

The box has a mass of 150 g.

a How many lights are there altogether?

b Three models are removed.

What is the total mass of the box and its remaining contents?

a 12 cars with 4 lights each have a total of $12 \times 4 = 48$ lights.

b If 3 cars are removed, there are 9 left.

The remaining cars have total mass of 9×200 g $= 1800$ g.

The box has a mass of 150 g.

So the total mass $= 1800 + 150 = 1950$ g.

Exercise 9E

PS **1** Find two odd numbers that add up to 48.

PS **2** The product of 2 and 3 is 6, because $2 \times 3 = 6$.

Work out the product of 6 and 7.

PS **3** The space inside a cupboard is 70 cm high. Tins are 15 cm high.

How many layers of tins will fit in the cupboard?

PS **4** Read this rule for the number walls.

Use the rule to fill in the missing numbers.

Some may have more than one correct answer.

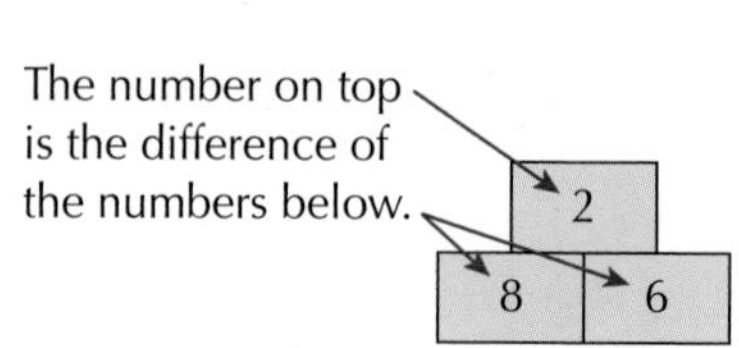

a

...	
22	17

b

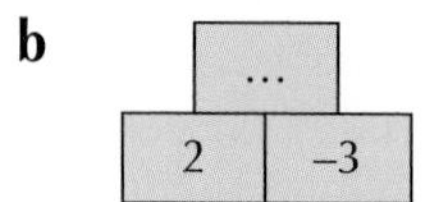

c

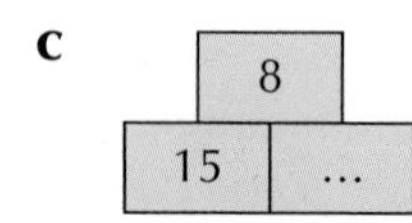

d

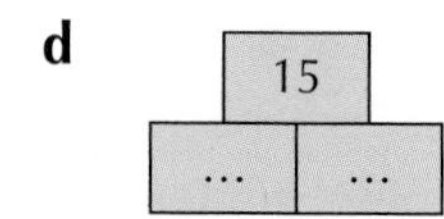

5 Yoghurts are sold individually for 45p.
They are also sold in a multipack of 6 for £2.45.
Which is the cheaper way to buy them?
Show your working.

6 The total of two brothers' ages is 110 years.
The difference in their ages is 4 years.
How old is the younger brother?

7 There are 5 blue, 3 green and 2 white counters in a bag.
The counters are numbered from 1 to 10.
Each counter has a mass of 6 g.
Match each statement to a correct calculation.

	10×6
The mass of the odd-numbered counters	$(5 + 2) \times 6$
The total mass of the counters	6×6
The mass of the counters that are not green	5×6

8 A photocopying company charges 5p per sheet.
How many sheets will they photocopy for £3?

9 A school buys some trophies for prizes.
The trophies cost £8.25 each.
The school had £100 to spend on prizes.
They buy as many trophies as possible.
How much money is left?

10 Oliver is twice as old as Rebecca.
The sum of their ages is 36 years.
How old are they?

11 Stickers cost £30.00 per box.
In the box there are 100 sheets.
Each sheet has 20 stickers on it.
What is the cost of one sticker?

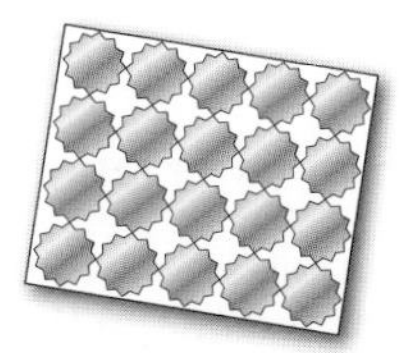

Investigation: Magic square

Here is a magic square.
All the rows and columns add up to the same value, called the magic value.

A Copy and complete the magic square.

B There are lots more ways of making that magic value, using groups of four numbers.
For example, the four corners all add up to the magic value too.
How many more ways can you find to make the magic value?

	2	3	
5			
9	7	6	12
4	14		1

Ready to progress?

I can round numbers to the nearest 10, 100 or 1000.
I can multiply and divide by 10, 100 and 1000.
I can solve simple number problems.

I can multiply and divide by simple powers of 10.
I can round numbers correct to one decimal place.

I can round numbers in order to make sensible estimations.
I can multiply and divide by any positive power of 10.

Review questions

All questions in this section, except question 6, should be attempted without a calculator.

1 Complete each of these calculations.

a 68×10 **b** 74×100 **c** 92×1000 **d** 713×100

e $859 \div 10$ **f** $118 \div 100$ **g** $584 \div 1000$ **h** $39 \div 10$

2 Find two odd numbers that add up to 56.

3 Round each number to one decimal place.

a 2.36 **b** 23.24 **c** 8.15 **d** 38.37 **e** 3.98

4 Round each number to two decimal places.

a 3.369 **b** 35.083 **c** 7.154 **d** 42.025 **e** 0.997

5 **a** Kath's height is 1.4 m. Brian is 0.2 m taller than Kath. How tall is Brian?

b Joe is 1.55 m tall. Hannah is 0.2 m shorter than Joe. What is Hannah's height?

c Helen's height is 1.6 m. What is Helen's height in centimetres?

6 A supermarket sells tins of beans.

They sell them in packs of four or six.

Which pack is the better value?

£1.30

7 I have compared the prices of two cleaning companies.

		Kleengo	Brushup
Costs	Weekly first hour	£8	£10
	Each extra hour	£7	£5

I am looking for a cleaner for 3 hours per week. Which is the cheaper company for me?

8 Work these out.

a 3.7×5 **b** 4.3×6 **c** 5.82×4 **d** 7.24×5

e 10×0.9 **f** 0.3×100 **g** 100×0.7 **h** 0.8×100

9 Change each number to index form.

The first one has been done for you.

a $0.001 = 10^{-3}$ **b** 0.1 **c** 0.01 **d** 0.000 01 **e** 0.000 001

10 Work out the answer to each of these.

a 0.4×0.7 **b** 0.8×0.8 **c** 0.9×0.5 **d** 0.3×0.7 **e** 0.9×1.1

11 The table lists the populations of some countries. The population density (number of people per km^2) for each country is given in index form.

Copy and complete the table, using decimal numbers.

The first one has been done for you.

	Population	Land mass per person (km^2)
Vatican City	826	$53.3 \times 10^{-5} = 0.000\ 533$
Gibraltar	31 000	21.9×10^{-5}
Monaco	33 000	59×10^{-6}
Bermuda	65 000	8.2×10^{-4}

12 State, with reasons, which is the most suitable degree of accuracy for:

a the average speed of a train
i 93.8 km/h **ii** 90 km/h **iii** 93.765 km/h

b the angle that a ladder is leaning against the wall
i 74.583° **ii** 75° **iii** 70°

c the mass of a packet of flour
i 1 kg **ii** 1.2 kg **iii** 1.187 kg

d the time taken to cook a pie
i 28 minutes 13 seconds **ii** 30 minutes **iii** 28 minutes.

Mathematical reasoning

Paper

Paper sizes

Standard paper sizes are called A0, A1, A2 and so on.
A sheet of A0 is approximately 1188 mm by 840 mm.
The next size, A1, is exactly half of A0.
This is a diagram of a piece of paper of size A0.

Paper for sale 1 ream = 500 sheets

Reams of paper – Special offer

Grade of paper	Prices		
	1 ream	5–9 reams	10+ reams
Standard	£2.70	SAVE $\frac{1}{3}$ per ream	EXTRA 10% discount
Special	£2.40		
Quality	£3.60		
Photo	£4.20		

Example for standard paper

1 ream costs £2.70

5 reams at full price = 5 × £2.70 = £13.50

Saving = $\frac{1}{3}$ of £13.50 = £4.50

Cost of 5 reams = £9.00

10 reams with $\frac{1}{3}$ off = 2 × £9.00 = £18.00

Extra 10% discount = £1.80

Cost of 10 reams = £16.20

Buy 5 reams

You save £4.50

Buy 10 reams

You save £10.80

Use the information on these pages to answer these questions.

1 How many sheets of paper are there in 5 reams of paper?

2 The thickness of a piece of paper is 0.1 mm.
How high would a stack of 2 reams (1000 sheets) of this paper be?
Give your answer in centimetres.

3 This is a table of paper sizes.

Paper size	A0	A1	A2	A3	A4	A5
Length (mm)	1188	840	594			
Width (mm)	840	594				

Copy and complete the table.
Check your result for A4 paper by measuring.

4 How many pieces of A5 paper would you need, to make a piece of A1 paper?

5 Look at the special offers and work out the cost of:

a 3 reams of quality grade paper

b 5 reams of photo grade paper

c 10 reams of special grade paper.

6 Work out the saving if you use the offers to buy 20 reams of photo paper.

10 Surface area and volume of 3D shapes

This chapter is going to show you:

- how to work out the surface areas of cubes and cuboids
- how to work out the volumes of cubes and cuboids
- how to work out the volumes of triangular prisms.

You should already know:

- how to work out the areas of squares and rectangles
- that the units for area are square centimetres (cm^2) and square metres (m^2).

About this chapter

Cubes and cuboids are very common shapes. You will see them used everywhere, from packaging for products such as stock cubes and shoes, to huge buildings.

Builders use small bricks in housing and massive stone blocks in cathedrals.

Just walk around any supermarket and you will see the enormous variety of different packets and containers, in all shapes and sizes.

Another interesting shape is the prism. Triangular prisms are not just used to wrap chocolate, but also in the study of light, in science. A clear prism can be positioned to show how a ray of light can be split into the different colours of a rainbow.

10.1 Surface areas of cubes and cuboids

Learning objective

- To work out the surface areas of cubes and cuboids

Key words	
cube	cuboid
surface area	

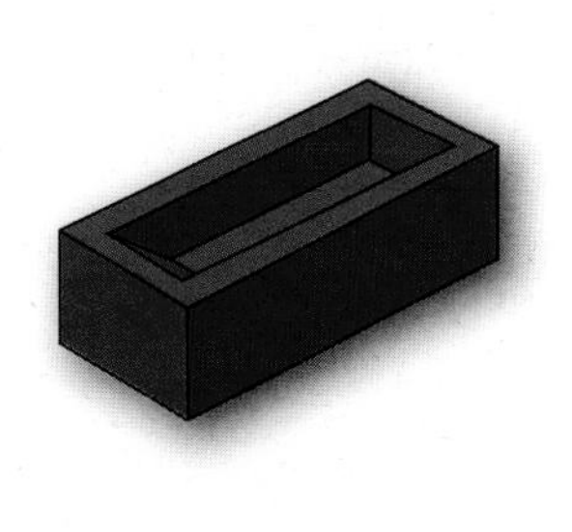

Shapes that are made from squares in 3D are called **cubes**. Their length, width and height (edge lengths) are all the same.

Shapes that are made from rectangles in 3D are called **cuboids**. Their length, width and height can all be different.

You can find the **surface area** of a cuboid by working out the total area of its six faces. They are all rectangles.

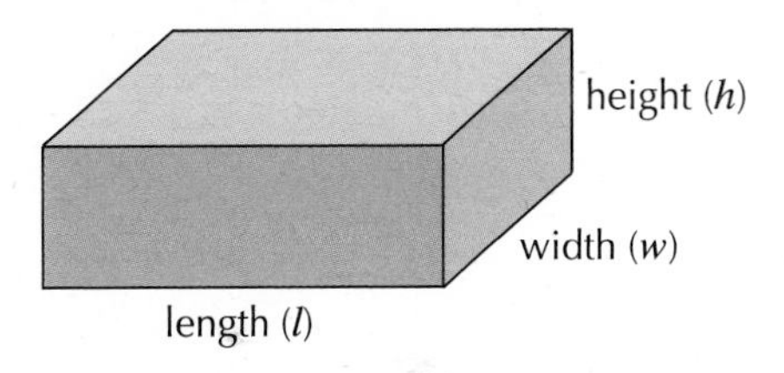

Area of top and bottom faces = 2 × length × width = $2lw$

Area of front and back faces = 2 × length × height = $2lh$

Area of the two sides = 2 × width × height = $2wh$

So the surface area of a cuboid = $S = 2lw + 2lh + 2wh$.

Note that the units for area are square centimetres (cm^2) or square metres (m^2).

Example 1

Work out the surface area of this cuboid.

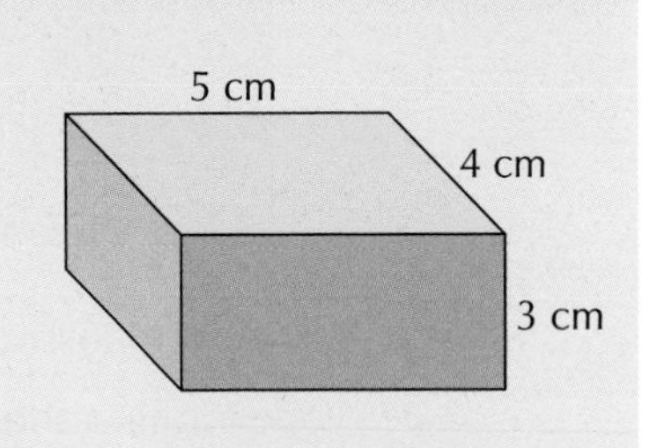

The formula for the surface area of a cuboid is:

$S = 2lw + 2lh + 2wh$

$= (2 \times 5 \times 4) + (2 \times 5 \times 3) + (2 \times 4 \times 3)$

$= 40 + 30 + 24$

$= 94 \text{ cm}^2$

Example 2

Work out the surface area of this cube.

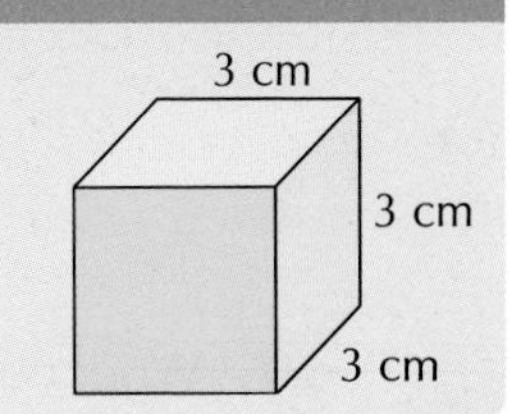

There are six square faces and each one of them has an area of $3 \times 3 = 9 \text{ cm}^2$.

So the surface area of the cube is $6 \times 9 = 54 \text{ cm}^2$.

Exercise 10A

1 This is a cuboid.

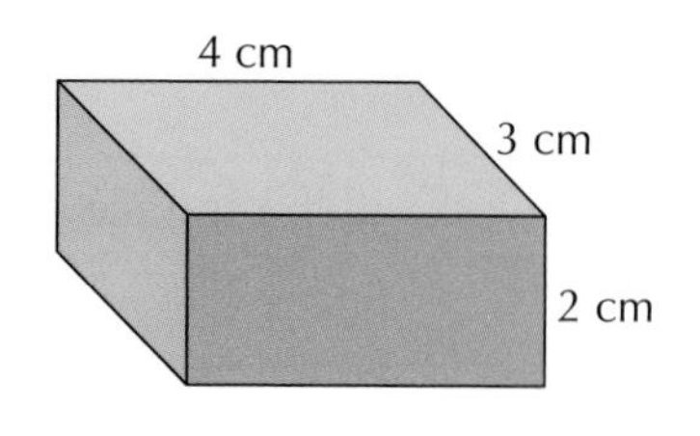

Copy and complete this calculation.

The formula for the surface area of the cuboid is:

$S = 2lw + 2lh + 2wh$

$= (2 \times \ldots \times \ldots) + (2 \times \ldots \times \ldots) + (2 \times \ldots \times \ldots)$

$= \ldots + \ldots + \ldots$

$= \ldots \text{ cm}^2$.

2 This is a cube.

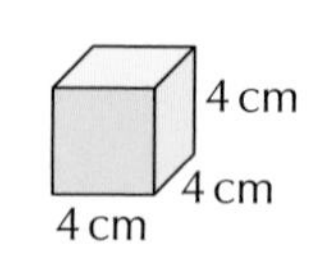

Copy and complete this calculation.

There are … square faces and each one has an area of $\ldots \times \ldots = \ldots \text{ cm}^2$.

So the surface area of the cube is $\ldots \times \ldots = \ldots \text{ cm}^2$.

3 Work out the surface area of each cuboid.

a

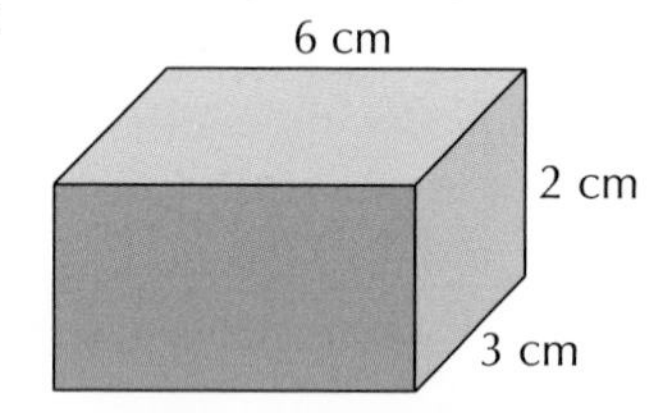

b

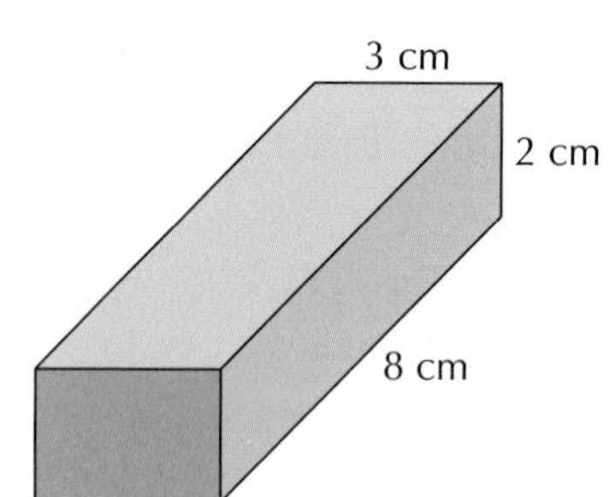

c

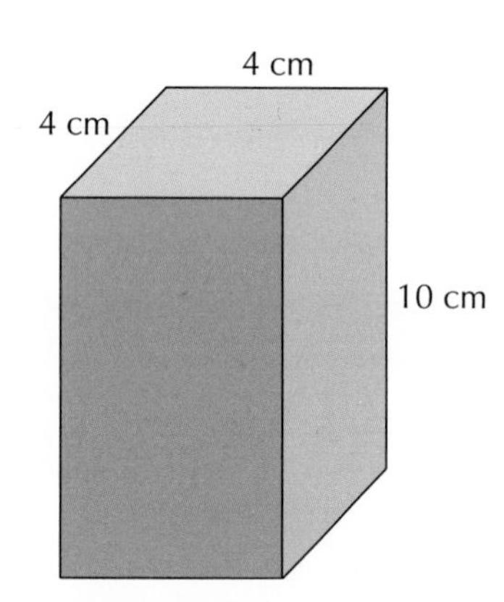

d

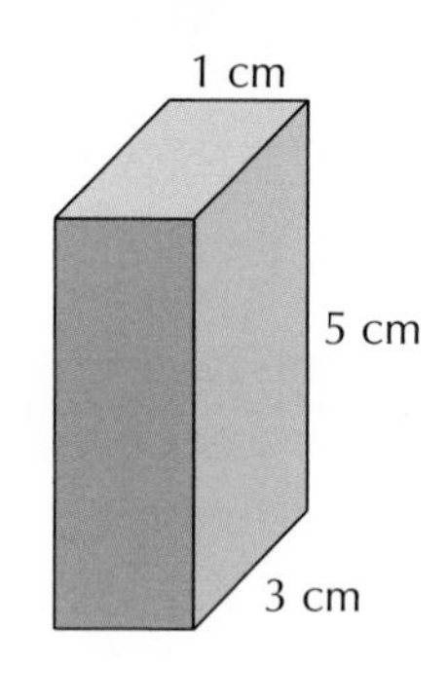

4 A cuboid measures 3 cm by 4 cm by 5 cm.

Work out its surface area.

5 Work out the surface area of each cube.

a

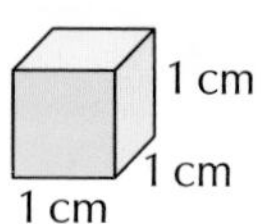

b

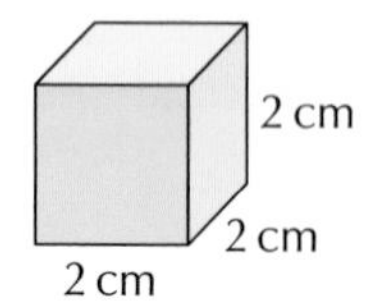

c

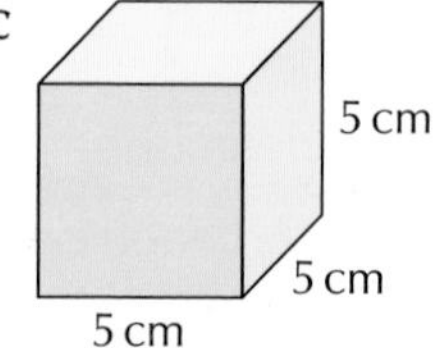

d

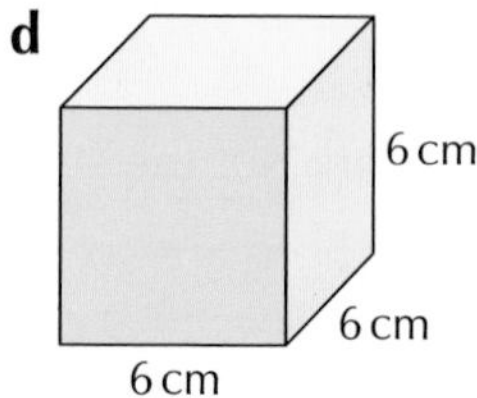

6 Work out the surface area of a cube with an edge length of 1.5 m.

7 Work out the surface area of this breakfast cereal packet.

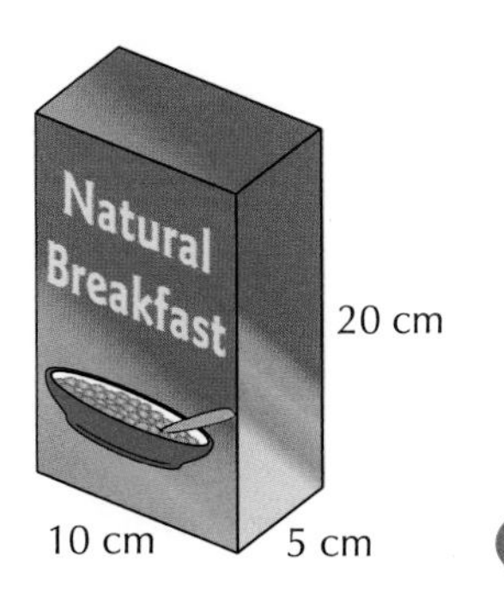

8 Work out the surface area of the outside of this open water tank.

Hint: Treat this as a cuboid without a top.

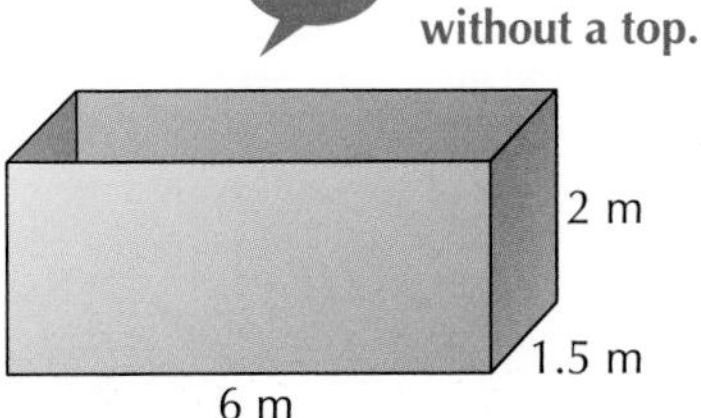

Investigation: An open box problem

An open box is made from a piece of card, measuring 16 cm by 12 cm, by cutting out a square from each corner.

Investigate the surface area of the open box formed when you cut out squares of different sizes from the corners of the card.

For example: cut off four squares with each one measuring 1 cm by 1 cm.

This will leave the net to make the open box.

The area of one square is $1 \times 1 = 1$ cm^2.

So the area of the four squares is $4 \times 1 = 4$ cm^2.

The area of the card is $16 \times 12 = 192$ cm^2.

The surface area of the box is $192 - 4 = 188$ cm^2.

A Now cut off four squares, each one measuring 2 cm by 2 cm.

Copy and complete this calculation.

The area of the square is $2 \times 2 = \ldots$ cm^2.

So the area of the four squares is $4 \times \ldots = \ldots$ cm^2.

The area of the card is $16 \times 12 = 192$ cm^2.

The surface area of the box is $192 - \ldots = \ldots$ cm^2.

B Copy and complete the table for squares of different sizes.

Size of squares cut out	Area of the four squares (cm^2)	Surface area of box (cm^2)
1 cm by 1 cm	4	188
2 cm by 2 cm		
3 cm by 3 cm		
4 cm by 4 cm		
5 cm by 5 cm		

10.2 Volume formulae for cubes and cuboids

Learning objectives

- To use simple formulae to work out the volume of a cube or cuboid
- To work out the capacity of a cube or cuboid

Key words	
capacity	height
litre	volume

Volume is the amount of space occupied by three-dimensional (3D) shapes like these.

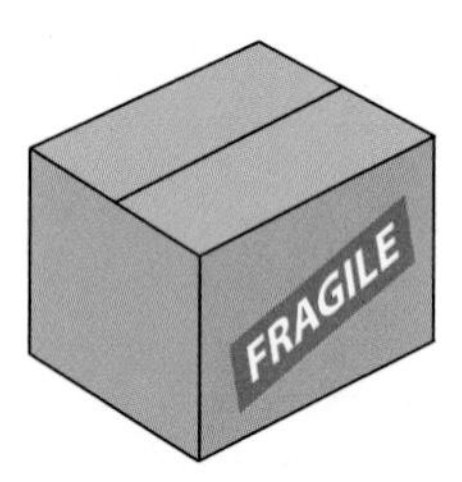

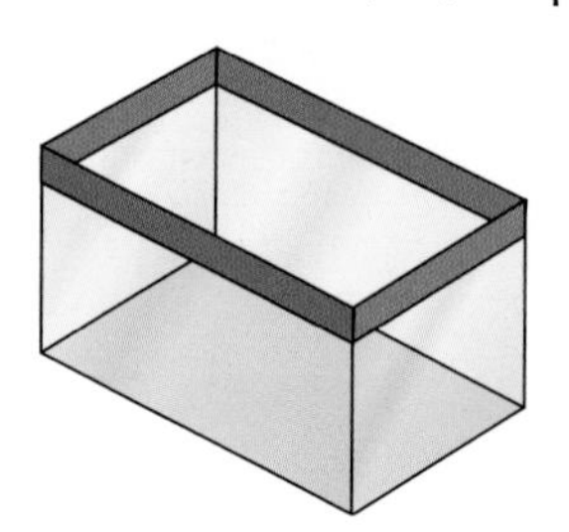

They have **height** as well as length and width.

This diagram shows a cuboid that measures 4 cm by 3 cm by 2 cm.

It is made up of cubes of edge length 1 cm.

The top layer has 12 cubes.

There are two layers so the cuboid has 24 cubes altogether.

This number is the same as you would get by multiplying all its edge lengths.

$$4 \times 3 \times 2 = 24$$

So you can find the volume of a cube or cuboid by multiplying its length by its width by its height.

Volume of a cube or cuboid = length × width × height

You can also use letters to write this as a formula:

$$V = l \times w \times h = lwh$$

where V = volume, l = length, w = width and h = height.

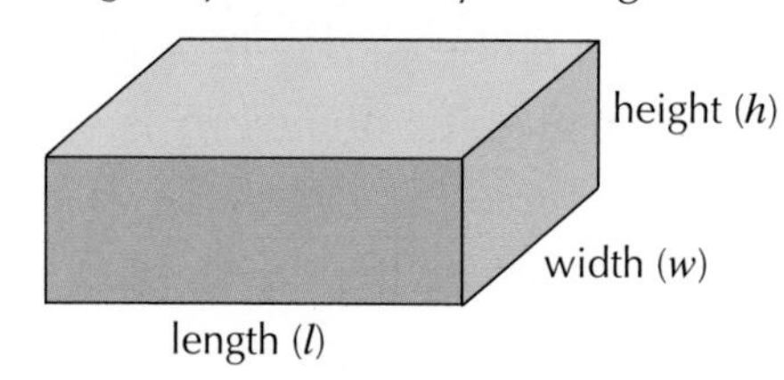

The metric units of volume in common use are:

- the cubic centimetre (cm^3)
- the cubic metre (m^3)

So, the volume of the cuboid above is 24 cm^3.

The cubes that make it up each have a volume of 1 cm^3.

Example 3

Work out the volume of this cuboid.

The formula for the volume of a cuboid is:

$V = lwh$

$= 5 \times 4 \times 3$

$= 60 \text{ cm}^3$

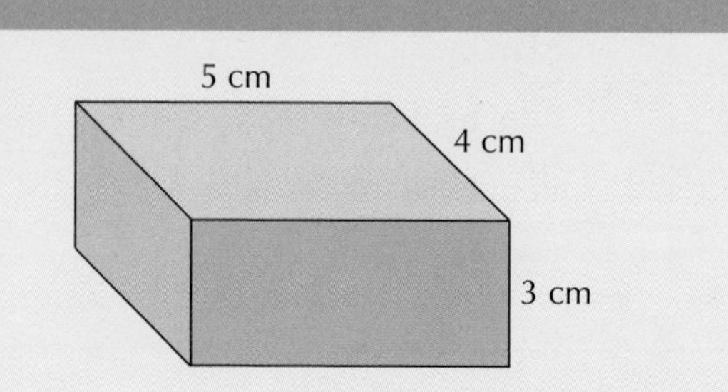

Example 4

Work out the volume of this cube.

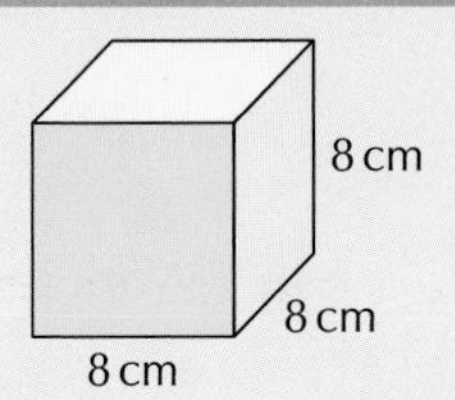

The formula for the volume of a cuboid is:

$V = lwh$ Here the length, width and height are all the same.

$= 8 \times 8 \times 8$

$= 512 \text{ cm}^3$

Example 5

The volume of this cuboid is 24 cm^3.

Work out the value of h.

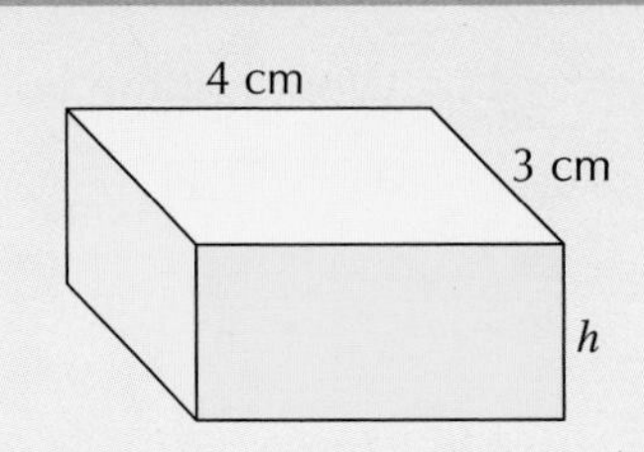

The formula for the volume of a cuboid is:

$V = lwh$

$24 = 4 \times 3 \times h$ Substitute known values into the formula.

$= 12h$

So $h = 2$ cm.

Capacity

The **capacity** of a 3D shape is the volume of liquid or gas it can hold.

The metric unit of capacity is the **litre** (l). One litre equals 1000 cm^3.

Example 6

Work out the volume, V, of the tank shown.

Then work out the capacity of the tank, in litres.

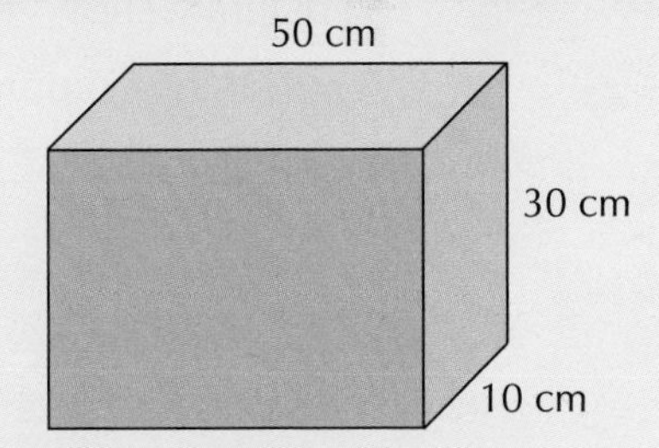

$V = lwh$

$= 50 \times 30 \times 10$

$= 15\,000 \text{ cm}^3$

Since $1000 \text{ cm}^3 = 1$ litre, the capacity of the tank is $15\,000 \div 1000 = 15$ litres.

Exercise 10B 

1 This is a cuboid.

Copy and complete this calculation.

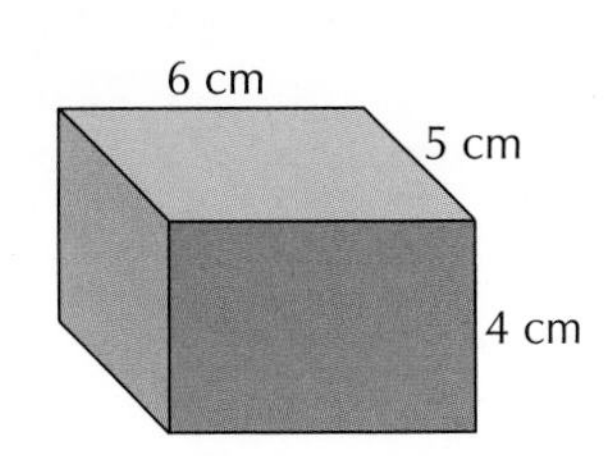

The formula for the volume of the cuboid is:

$V = lwh$

$= \ldots \times \ldots \times \ldots$

$= \ldots \text{ cm}^3$.

2 This is a cube.

Copy and complete this calculation.

The volume of the cube is … × … × … = … cm^3.

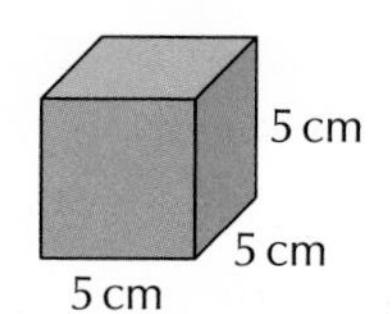

3 Work out the volume of each cuboid.

a

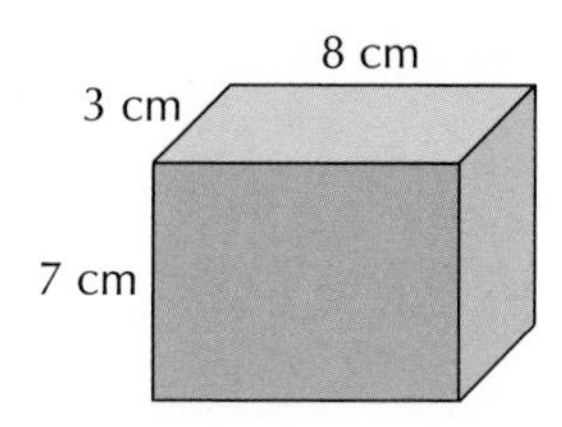

b

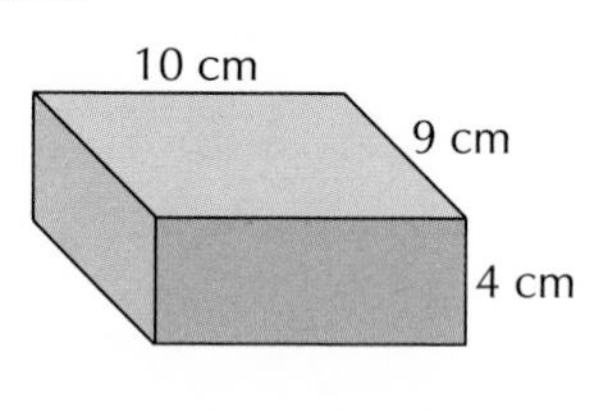

c

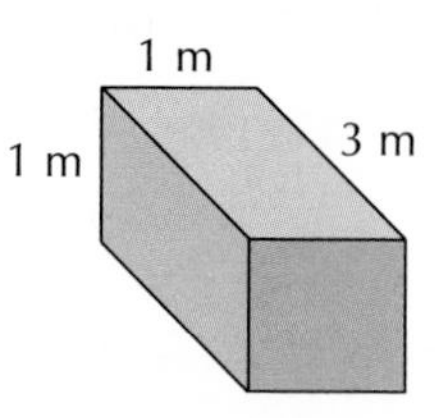

4 Work out the volume of each cube.

a

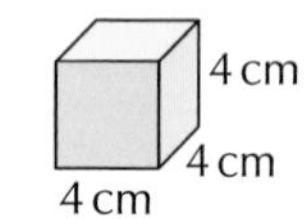

b

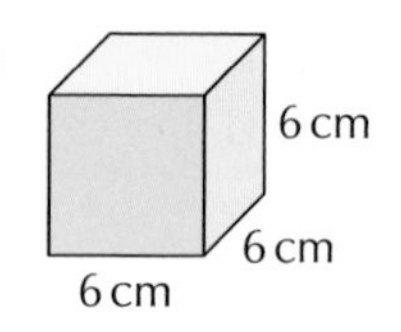

c

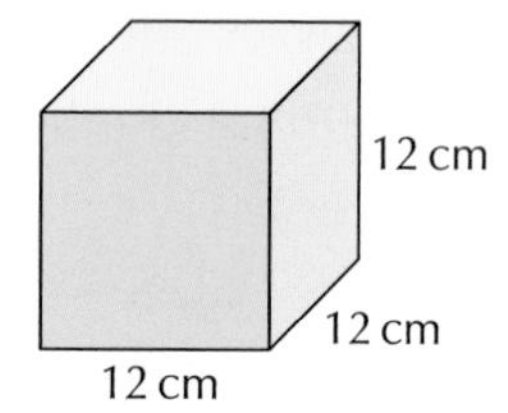

5 Work out the capacity, in litres, of each container.

a

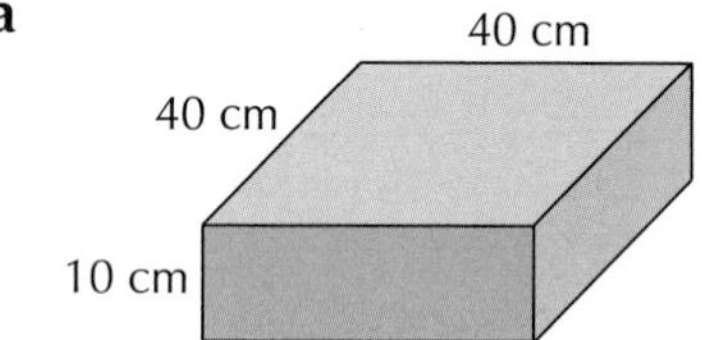

b

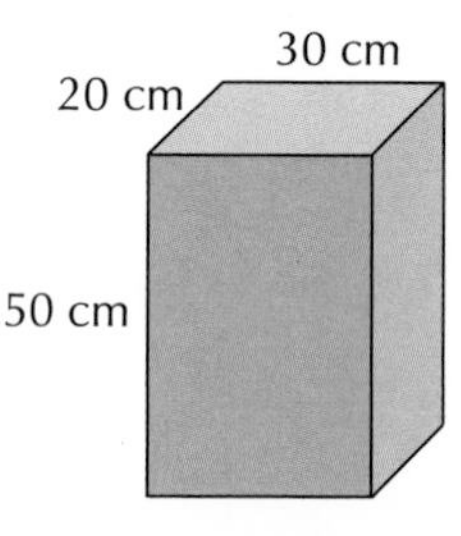

c

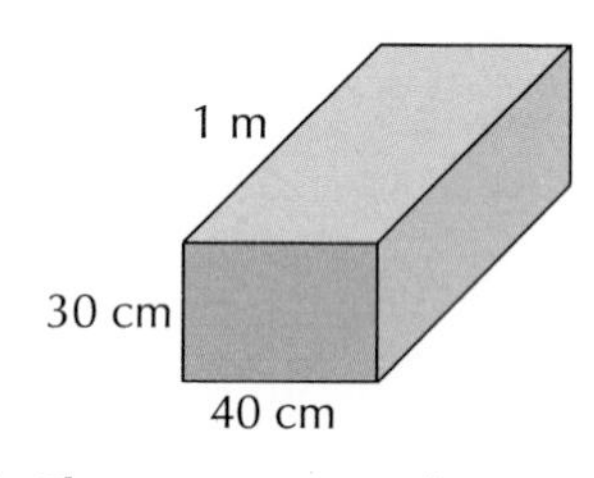

Hint Change 1 m to centimetres.

6 The diagram shows the dimensions of a swimming pool.

Work out the volume of the pool, giving the answer in cubic metres.

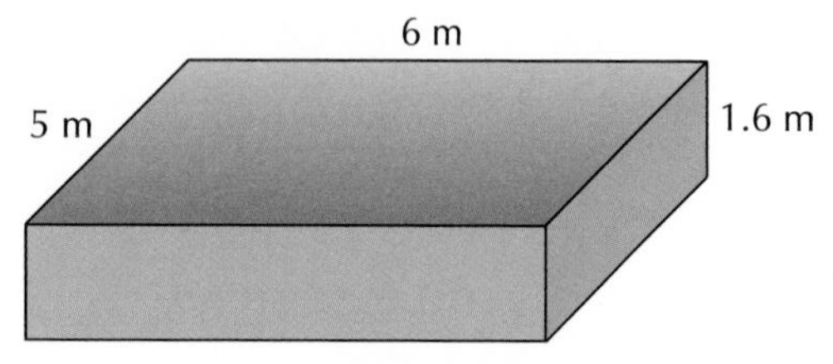

7 Work out the missing lengths for each cuboid.

a

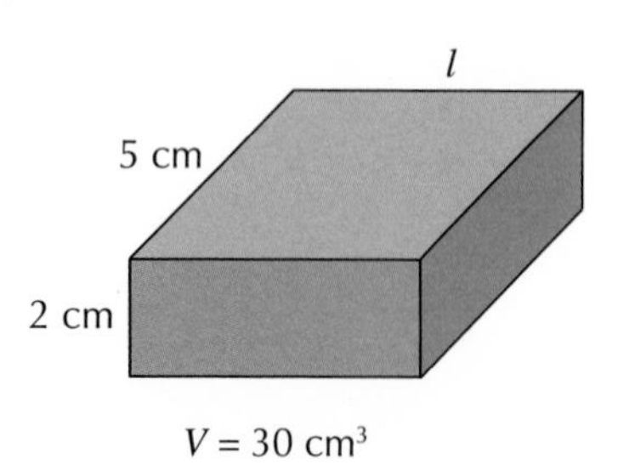

$V = 30$ cm^3

b

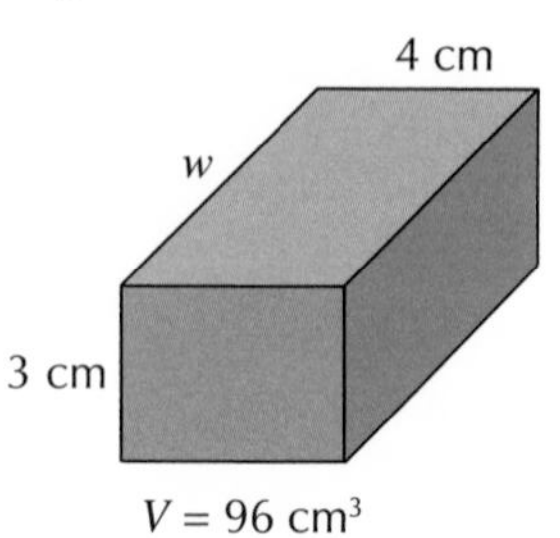

$V = 96$ cm^3

c

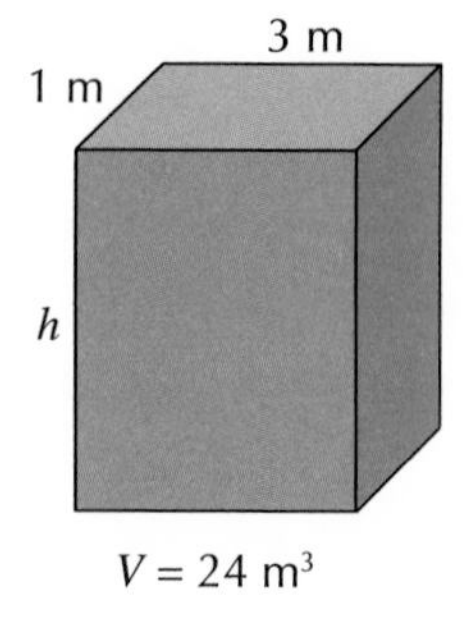

$V = 24$ m^3

MR **8** A cube has a capacity of 1 litre. What is the length of each side?

Investigation: Painted cubes

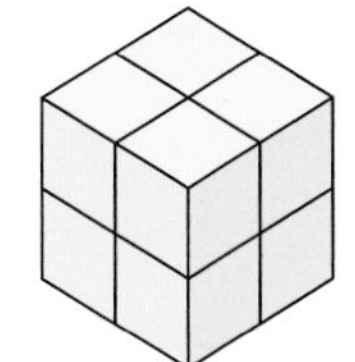

A 2 by 2 by 2 cube is made from eight smaller yellow cubes, as shown.

The outside of the large cube is painted red and then the large cube is taken apart.

A **a** How many of the smaller cubes have no faces painted red?

b How many of the smaller cubes have one face painted red?

c How many of the smaller cubes have two faces painted red?

d How many of the smaller cubes have three faces painted red?

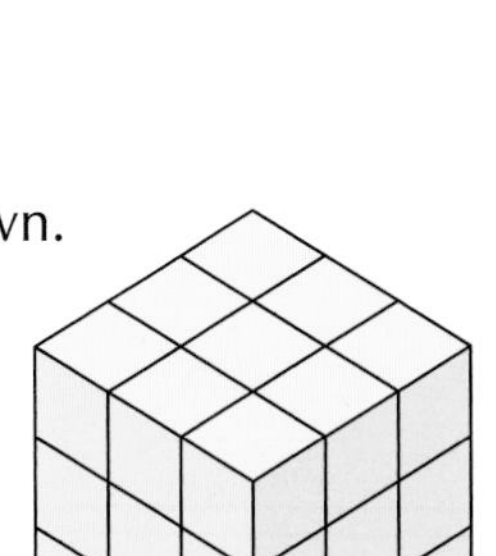

B Now a 3 by 3 by 3 cube is made from 27 smaller yellow cubes, as shown.

Again, the outside of the large cube is painted red and then the large cube is taken apart.

a How many of the smaller cubes have no faces painted red?

b How many of the smaller cubes have one face painted red?

c How many of the smaller cubes have two faces painted red?

d How many of the smaller cubes have three faces painted red?

C Now repeat for a 4 by 4 by 4 cube.

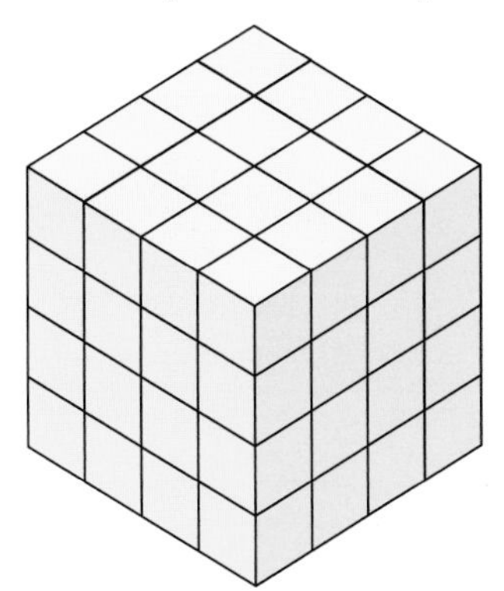

D Copy and complete this table.

Size of yellow cube	Number of cubes	No faces painted red	One face painted red	Two faces painted red	Three faces painted red
2 by 2 by 2	8				
3 by 3 by 3	27				
4 by 4 by 4					

10.3 Volumes of triangular prisms

Learning objective

- To work out the volume of a triangular prism

Key words

triangular prism

This is an example of a **triangular prism**.

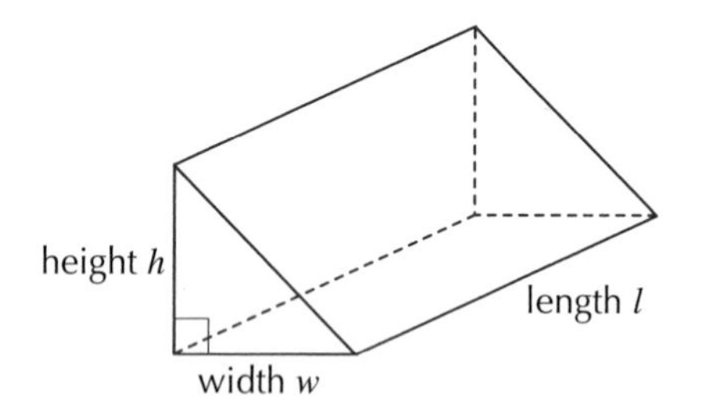

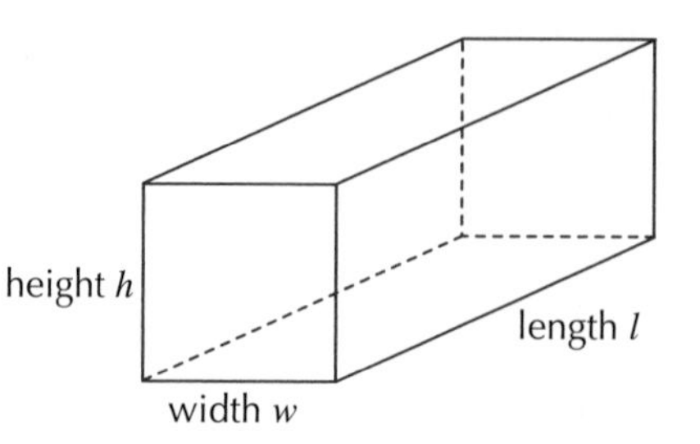

Notice that the triangular prism is half of a cuboid with the same measurements.

The volume of a cuboid is $V = lwh$, so the volume of a triangular prism is $V = \frac{lwh}{2}$.

Example 7

Work out the volume of this triangular prism.

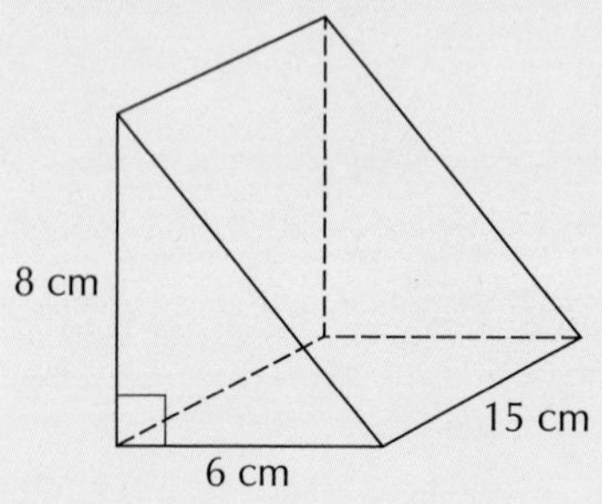

The volume is given by:

$$V = \frac{lwh}{2}$$

$$= \frac{15 \times 6 \times 8}{2}$$

$$= \frac{720}{2}$$

$$= 360 \text{ cm}^3.$$

Exercise 10C

1 This is a triangular prism.

Copy and complete this calculation.

The volume of the triangular prism is given by:

4 cm

7 cm

6 cm

$$V = \frac{lwh}{2}$$

$$= \frac{\ldots \times \ldots \times \ldots}{2}$$

$$= \frac{\ldots}{2}$$

$$= \ldots \text{ cm}^3.$$

2 This is a triangular prism.

Copy and complete this calculation.

The volume of the triangular prism is given by:

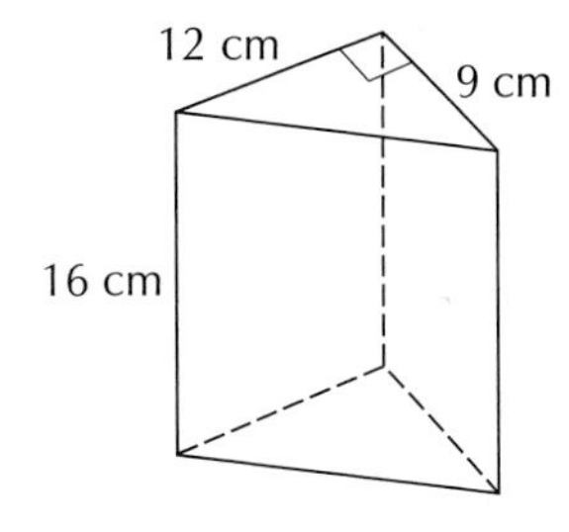

$$V = \frac{lwh}{2}$$

$$= \frac{\ldots \times \ldots \times \ldots}{2}$$

$$= \frac{\ldots}{2}$$

$$= \ldots \text{ cm}^3.$$

3 Work out the volume of each triangular prism.

a

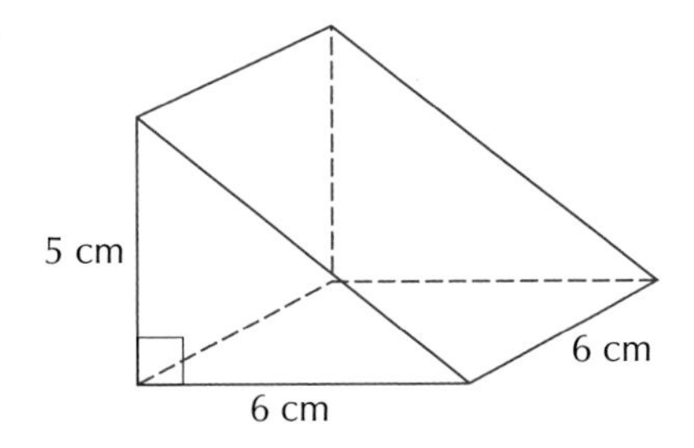

b

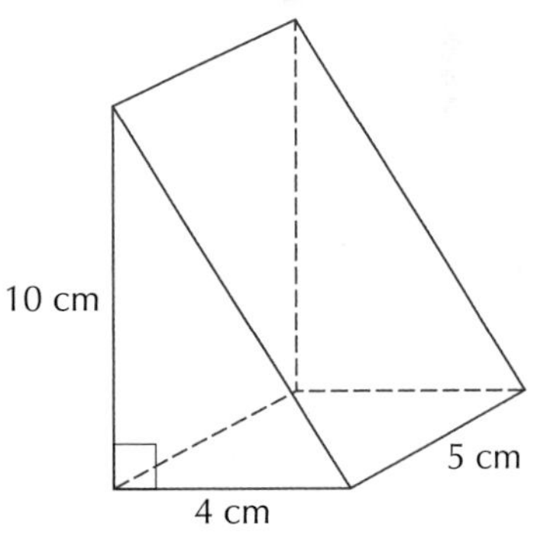

c

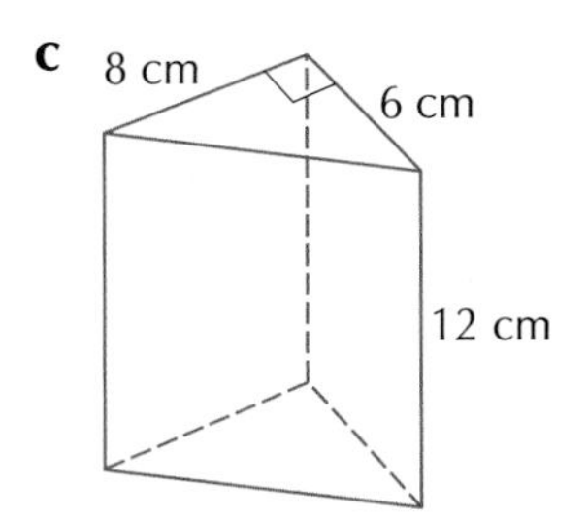

4 This gift box is in the form of a triangular prism.

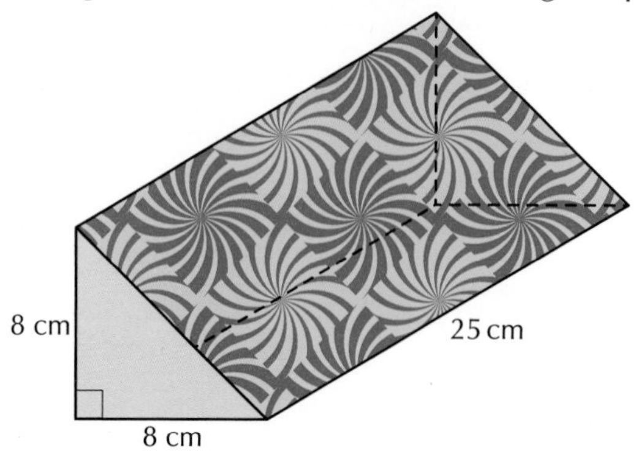

Work out the volume of the box.

5 Joe is making a solid concrete ramp for wheelchair-access to his house.

The dimensions of the ramp are shown on the diagram.

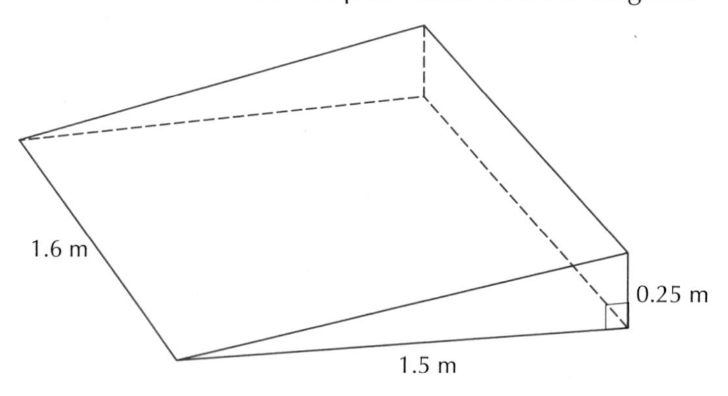

a Calculate the volume of the ramp, giving your answer in cubic metres.

b One cubic metre of cement weighs 2.4 tonnes. What is the mass of the concrete that Joe uses?

6 This triangular prism has a volume of 60 cm^3.

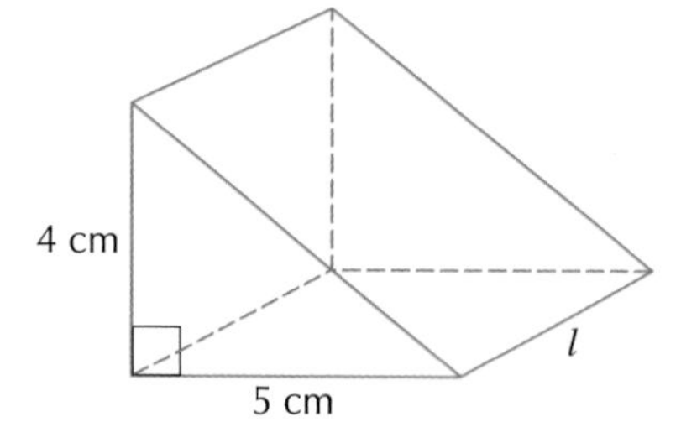

Work out the length of the triangular prism, labelled l on the diagram.

Problem solving: Surface area of triangular prisms

To work out the surface area of this triangular prism, you need to work out the area of two triangles and three rectangles.

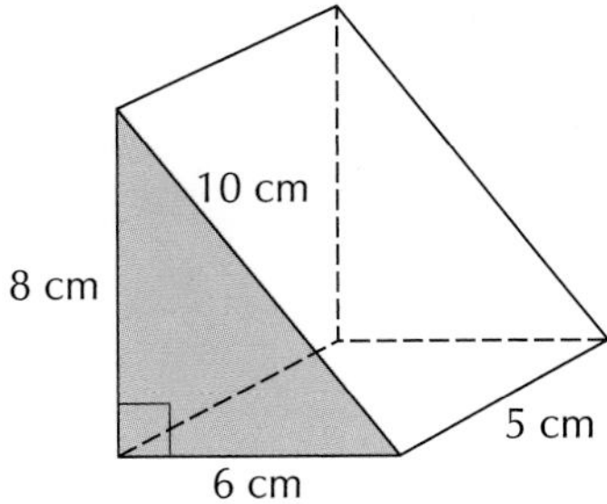

The areas of the shaded triangle $= \dfrac{8 \times 6}{2}$

$= \dfrac{48}{2}$

$= 24 \text{ cm}^2$.

So the area of the two triangles $= 2 \times 24 = 48 \text{ cm}^2$.

The areas of the three rectangles are:

$6 \times 5 = 30 \text{ cm}^2$

$8 \times 5 = 40 \text{ cm}^2$

$10 \times 5 = 50 \text{ cm}^2$.

So the surface area is $48 + 30 + 40 + 50 = 168 \text{ cm}^2$.

Work out the surface area of each triangular prism.

A

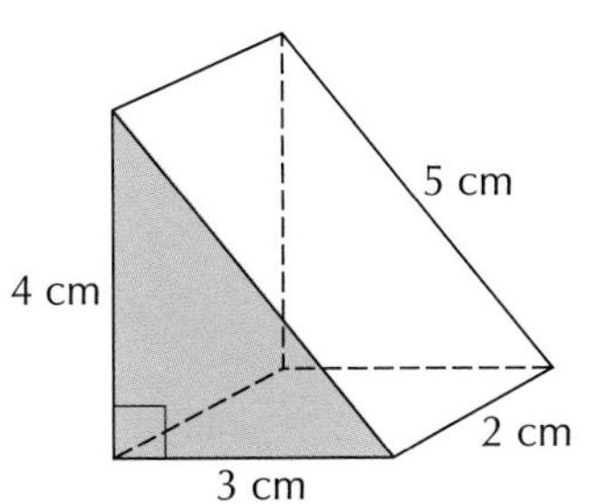

B

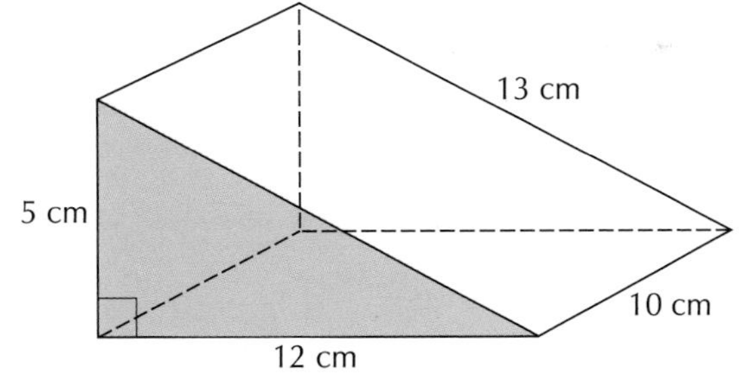

Ready to progress?

I can work out the surface areas of cubes and cuboids.
I can work out the volumes of cubes and cuboids.
I can work out the capacities of cubes and cuboids, measured in litres.
I can work out the volume of a triangular prism.

Review questions

1 This is a cuboid.

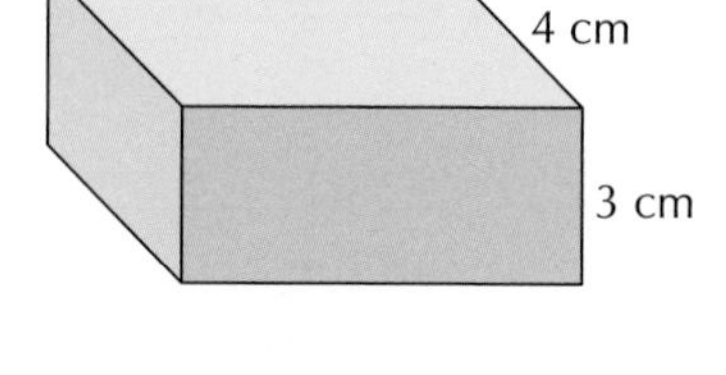

a Copy and complete this calculation.

The formula for the surface area of the cuboid is:

$S = 2lw + 2lh + 2wh$

$= (2 \times \ldots \times \ldots) + (2 \times \ldots \times \ldots) + (2 \times \ldots \times \ldots)$

$= \ldots + \ldots + \ldots$

$= \ldots \text{ cm}^2$.

b Copy and complete this calculation.

The formula for the volume of the cuboid is:

$V = lwh$

$= \ldots \times \ldots \times \ldots$

$= \ldots \text{ cm}^3$.

2 For each cuboid, work out:

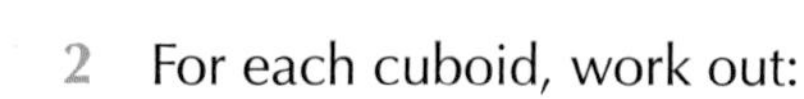

i the surface area ii the volume.

a

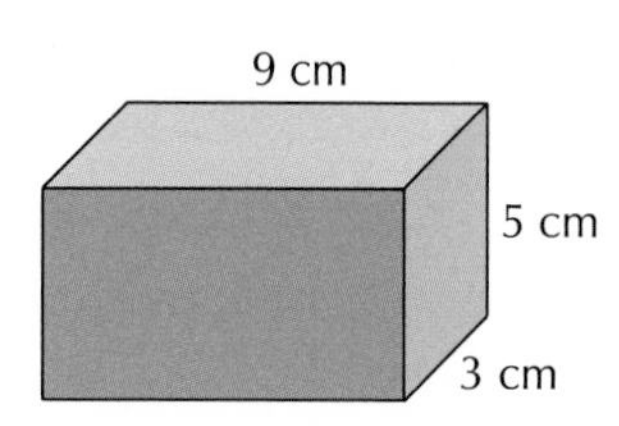

b

10 m
3 m
7 m

c

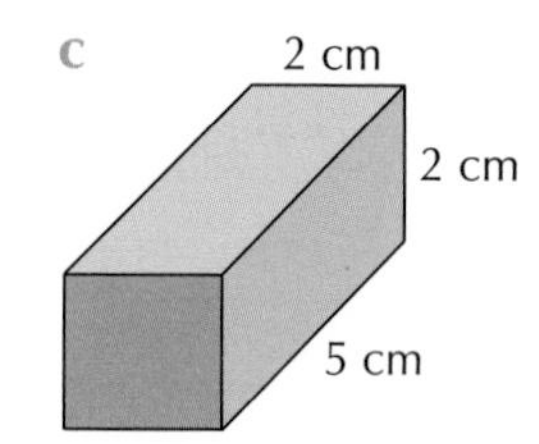

3 The diagram shows the dimensions of a carton of apple juice.

Work out the capacity of the carton.

Give your answer in litres.

4 This is a triangular prism.

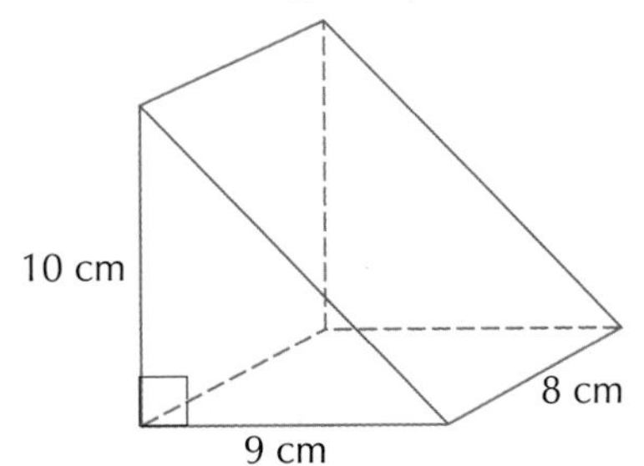

Copy and complete this calculation.

The volume of the triangular prism is given by:

$$V = \frac{lwh}{2}$$

$$= \frac{\ldots \times \ldots \times \ldots}{2}$$

$$= \frac{\ldots}{2}$$

$$= \ldots \text{ cm}^3.$$

5 This picture shows the dimensions of a skateboard ramp.

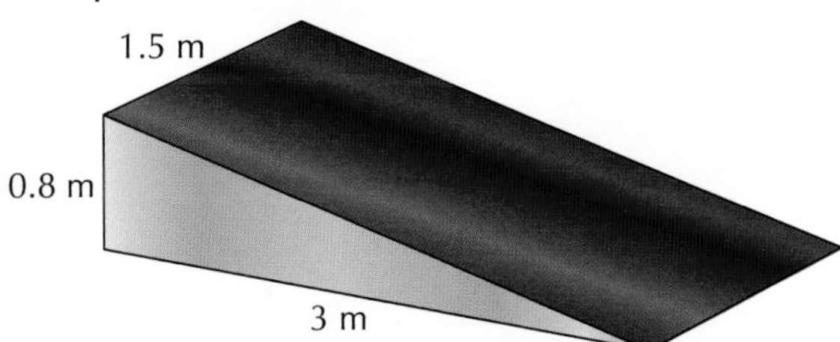

Work out the volume of the ramp.

Give your answer in cubic metres.

6 These two cuboids, A and B, have the same volume.

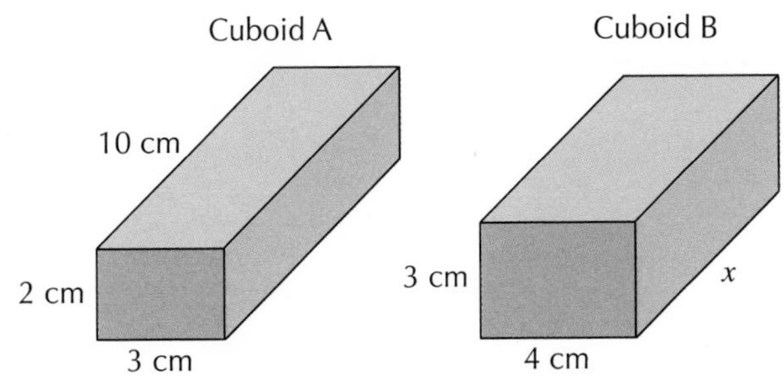

a Work out the value of the length marked x on cuboid B.

b Which cuboid has the greater surface area? Show how you decide.

Investigation

A cube investigation

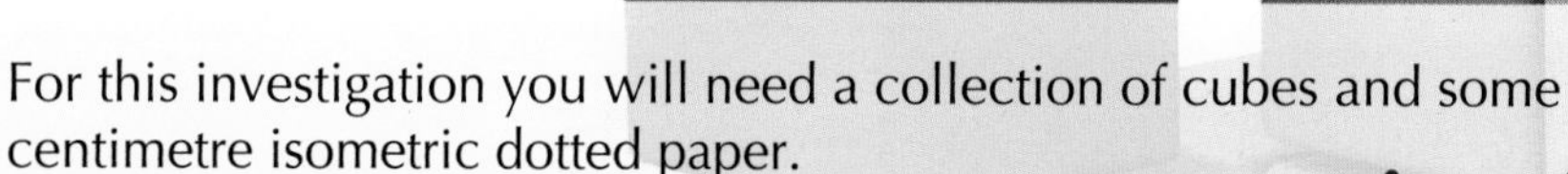

For this investigation you will need a collection of cubes and some centimetre isometric dotted paper.

Two cubes can only be arranged in one way to make a 3D shape, as shown.

Copy the diagram onto centimetre isometric dotted paper. The surface area of the shape is 10 cm^2.

Three cubes can be arranged in two different ways, as shown.

Copy the diagrams onto centimetre isometric dotted paper. The surface area of both 3D shapes is 14 cm^2.

Here is an arrangement of four cubes. The surface area of the 3D shape is 18 cm^2.

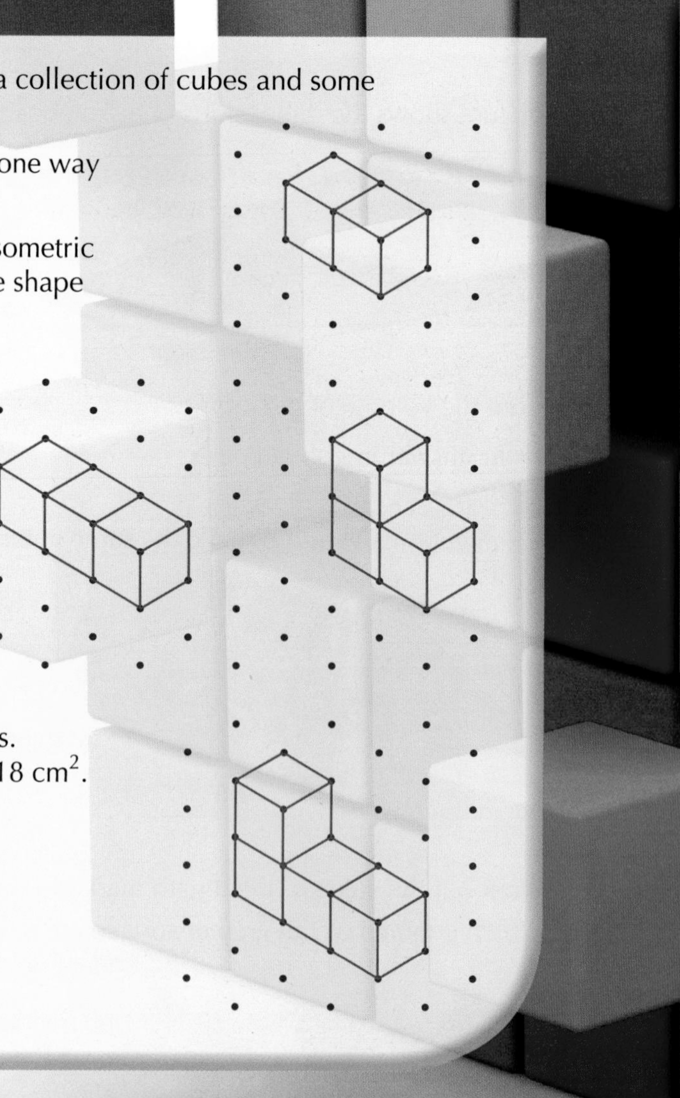

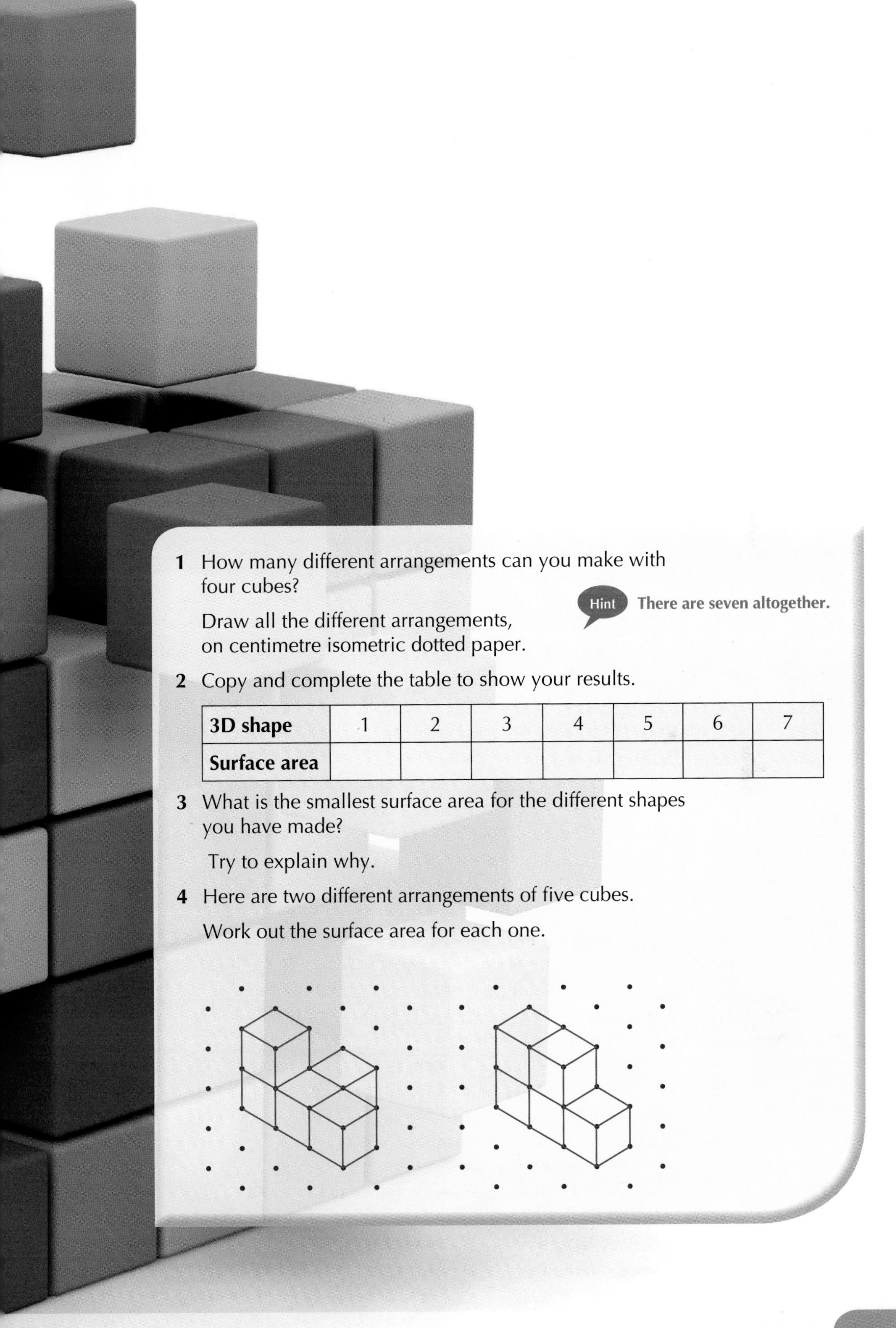

1 How many different arrangements can you make with four cubes?

Draw all the different arrangements, on centimetre isometric dotted paper.

Hint There are seven altogether.

2 Copy and complete the table to show your results.

3D shape	1	2	3	4	5	6	7
Surface area							

3 What is the smallest surface area for the different shapes you have made?

Try to explain why.

4 Here are two different arrangements of five cubes.

Work out the surface area for each one.

11

Solving equations graphically

This chapter is going to show you:

- how to solve linear equations graphically
- how to use straight-line graphs to solve problems
- how to solve simple quadratic equations
- how to use quadratic graphs to solve problems.

You should already know:

- how to draw linear graphs of the form $y = mx + c$
- how to draw a simple quadratic graph.

About this chapter

There are all sorts of different equations that arise from real situations. One example is the satellite dish.

The dish picks up waves and focuses them to a point. It can do this because of its shape. If you were to cut through a satellite dish vertically, across its diameter, you would see that its cross-section is a parabola.

The parabola is the curve formed by the graph of a quadratic equation. This is an equation of the form $y = ax^2 + bx + c$, and the graph always has the same basic shape. You will see parabolic curves used in other sorts of application, such as parabolic headlights.

Some equations are difficult to solve by algebraic methods and are more easily solved by drawing a graph. Such a graph can help you to see roughly what the solution should be and provides a way of checking if your algebraic solution is sensible.

11.1 Graphs from equations in the form $y = mx + c$

Learning objectives

- To draw a linear graph from any linear equation
- To solve a linear equation from a graph

Key words

linear equation

Think back to your earlier work on the **linear equation** $y = mx + c$.

This equation can be represented by a straight-line graph.

The equation $y = mx + c$ is a general equation, which means that m and c can be any numbers.

Example 1

Draw the graph of $y = 3x + 1$.

First, construct a table of values.

Use values for x that make the working easy, as shown here.

x	0	1	2	3
$y = 3x + 1$	1	4	7	10

Next, draw the axes on graph paper, as a coordinate grid.

Choose values for both variables that will fit in with the values in your table.

Here, the values for x, on the horizontal axis, are from 0 to 3 and those for y, on the vertical axis, are from 0 to 10.

Then plot the points given in the table, and join them with a straight line.

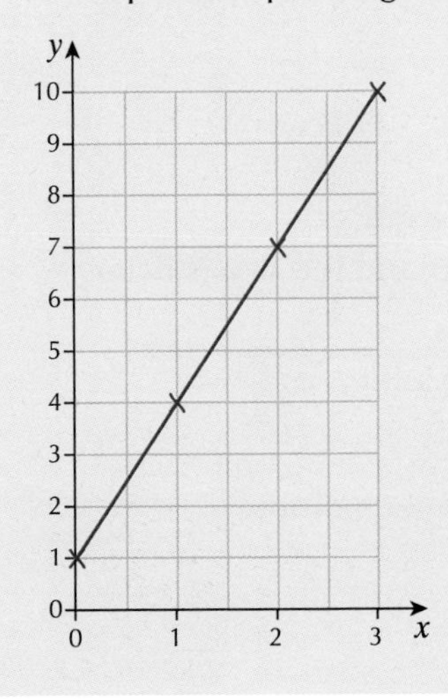

Remember that you plot x on the horizontal axis and y on the vertical axis.

Exercise 11A

Each question in this exercise is a short investigation into the positions of the graphs of equations written in the form $y = mx + c$.

After completing the investigations, you should find out something very important and useful about the values of m and c.

What you discover will help you see where the straight-line graph lies for each equation.

1 **a** Copy and complete this table for the four equations shown.

The table has been started for you.

x	0	1	2	3
$y = x + 1$	1			
$y = x + 2$		3		
$y = x + 3$			5	
$y = x + 4$				7

b Draw a coordinate grid. Number the x-axis from 0 to 3 and the y-axis from 0 to 7.

c On your grid, draw the graph for each equation in the table.

d What do you notice about each graph?

e Use what you have noticed to draw the graphs of these equations.

i $y = x + 5$ **ii** $y = x + 0.5$

2 **a** Copy and complete this table for the four equations shown.

The table has been started for you.

x	0	1	2	3
$y = 2x + 1$	1			
$y = 2x + 2$		4		
$y = 2x + 3$			7	
$y = 2x + 4$				10

b Draw a coordinate grid. Number the x-axis from 0 to 3 and the y-axis from 0 to 10.

c On your grid, draw the graph for each equation in the table.

d What do you notice about each graph?

e Use what you have noticed to draw the graphs of these equations.

i $y = 2x + 5$ **ii** $y = 2x + 0.5$

3 **a** Copy and complete this table for the four equations shown.
The table has been started for you.

x	0	1	2	3
$y = x$	0			
$y = 2x$		2		
$y = 3x$			6	
$y = 4x$				12

b Draw a coordinate grid. Number the x-axis from 0 to 3 and the y-axis from 0 to 12.

c On your grid, draw the graph for each equation in the table.

d What do you notice about each graph?

e Use what you have noticed to draw the graphs of these equations.

i $y = 5x$ **ii** $y = 0.5x$

4 **a** Copy and complete this table for the four equations shown.
The table has been started for you.

x	0	1	2	3
$y = x + 4$	4			
$y = 2x + 4$		6		
$y = 3x + 4$			10	
$y = 4x + 4$				16

b Draw a coordinate grid. Number the x-axis from 0 to 3 and the y-axis from 0 to 16.

c On your grid, draw the graph for each equation in the table.

d What do you notice about each graph?

e If you draw a graph of $y = mx + c$, where m and c are any number, what does the value of m tell you about the straight-line graph?

f Use what you have noticed to draw the graphs of these equations.

i $y = 5x + 4$ **ii** $y = 0.5x + 4$

Investigation: The graph of $x + y = c$

The line $x + y = 6$ passes through points (0, 6), (1, 5), (2, 4), (3, 3), (4, 2), (5, 1), (6, 0).

A Draw a coordinate grid. Number axes for both x and y from 0 to 10.

B Plot the points listed above and join them up.

C Now find six points that the line $x + y = 5$ passes through.

Start with (0, 5) and plot these on the same graph.

D You should now be able to draw, without plotting any points, the lines $x + y = 10$ and $x + y = 3$.

11.2 Problems involving straight-line graphs

Learning objective

- To draw graphs to solve some problems

When a car is being filled with petrol, both the amount and the cost of the petrol are displayed on the pump. One litre of petrol costs £1.30. So, 2 litres cost £2.60 and 5 litres cost £6.50.

The table below shows the costs of different quantities of petrol as displayed on the pump.

Amount of petrol (litres)	5	10	15	20	25	30
Cost (£)	6.50	13.00	19.50	26.00	32.50	39.00

The information can be shown on a graph, like this.

Because of this fixed relationship, the graph is a straight line.

You can use this idea to solve a number of different types of problem.

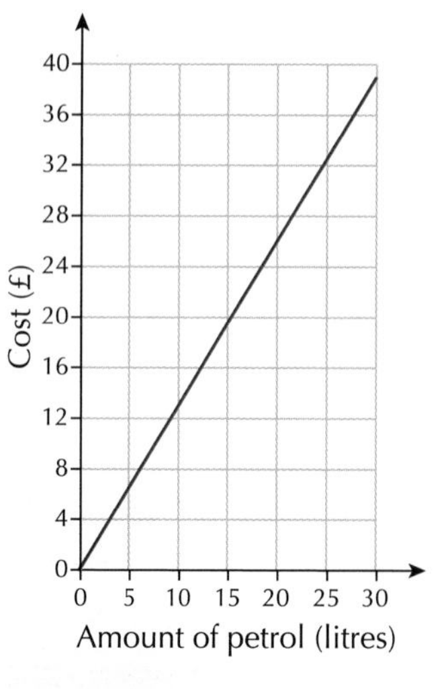

Example 2

Mr Evans set his pupils a French test. He wanted to convert all their marks to percentages. The highest possible score was 60 marks.

a Use this fact to draw a conversion graph to change the marks to percentages.

b Use the conversion graph to change the marks to percentages, for:

i Stephanie, who scored 30

ii Joe who scored 38.

a Use what you know to draw a table of values.

A mark of 0 is equivalent to 0%.

A mark of 60 is equivalent to 100%.

French marks	0	60
Percentage	0	100

Now use these facts to draw a conversion graph.

This is a straight-line graph, as shown here.

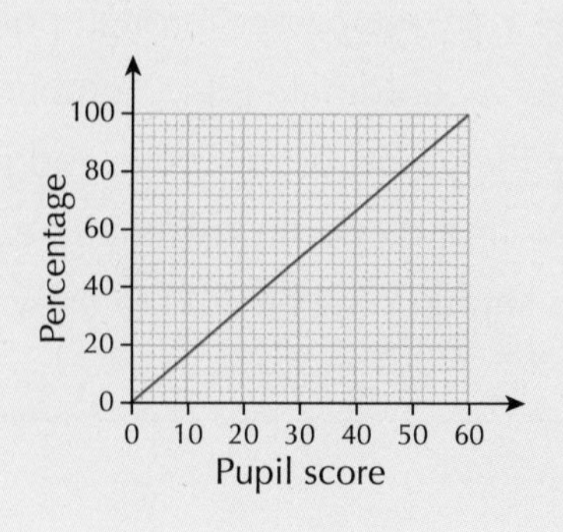

b i Stephanie scored 30. Using the graph, you can convert this to 50%.

ii Joe scored 38. Again, using the graph, you can convert this to 63%.

Exercise 11B

1 Eve was employed to sell apples on a market stall.

She was told just two prices.

Number of apples	0	12
Cost (£)	0	1.50

a Draw a coordinate grid. Use the horizontal axis for the number of apples, numbering it from 0 to 20. Use the vertical axis for the cost of apples, in pounds (£), numbering it from 0 to 3.

Plot the two points on your graph and join them with a straight line.

Continue your line to the edge of your graph.

b Use your graph to find the cost of:

i 4 apples **ii** 10 apples **iii** 20 apples.

c Use your graph to find how many apples you could buy for:

i 75p **ii** £1.75 **iii** £2.75.

2 At a Joe King concert, fans can buy posters of Joe from one of the stalls.

Benny sells the posters. He knows these facts about the costs.

Number of posters	0	20
Cost (£)	0	27

a Draw a coordinate grid. Use the horizontal axis for the number of posters, from 0 to 20, and the vertical axis for the cost of posters, going up to £27. Plot the two points on your graph and join them with a straight line.

b Use your graph to find the cost of:

i 6 posters **ii** 12 posters **iii** 16 posters.

c Use your graph to find how many posters you could buy for:

i £17.00 **ii** £12.00 **iii** £23.00.

3 Tom put different masses on the end of a spring to see how much it stretched each time. After his experiment, he knew these facts.

Mass (g)	0	900
Stretch (cm)	0	18

a Draw a coordinate grid. Use the horizontal axis for the mass added, going up to 1000 g, and the vertical axis for the stretch of the spring, going up to 20 cm. Plot the two points on your graph and join them with a straight line.

b Use your graph to find the stretch when the mass is:

i 200 g **ii** 300 g **iii** 1 kg.

c Use your graph to find the mass needed to stretch the spring by:

i 2 cm **ii** 5 cm **iii** 14 cm.

4 At a garden party, Eve looked after the hoopla stall. She knew two prices for the hoops.

Number of hoops	0	10
Cost (£)	0	2.20

a Draw a coordinate grid. Use the horizontal axis for the numbers of hoops, going up to 12, and the vertical axis for the cost of hoops, going up to £3. Plot the two points on your graph and join them with a straight line.

b Use your graph to find the cost of:

i 3 hoops **ii** 8 hoops **iii** 12 hoops.

c Use your graph to find how many hoops you get for:

i 88p **ii** £1.10 **iii** £2.42.

5 Sophia went to France for her holiday. This is what she knew about the relationship between British pounds (£) and euros (€).

British pound (£)	0	100
Euro (€)	0	120

a Draw a coordinate grid. Use the horizontal axis for British pounds, going up to £150, and the vertical axis for euros, going up to €200. Plot the two points on your graph and join them with a straight line.

b Use your graph to find the value, in euros, of:

i £20 **ii** £50 **iii** £130.

c Use your graph to find the value, in British pounds, of:

i €40 **ii** €80 **iii** €160.

Challenge: Strolling sisters

Try to solve this problem by drawing a graph.

Two sisters, Aliya and Bryana, are walking on the same long, straight road towards each other.

Aliya sets off at 9:00 am at a speed of 4 km/h.

Bryana also sets off at 9:00 am, 15 km away, at a speed of 5 km/h.

A At what time do the sisters meet?

B How far will Aliya have walked when she meets Bryana?

11.3 Solving simple quadratic equations by drawing graphs

Learning objective

- To solve a simple quadratic equation by drawing a graph

You can find a lot of information from a quadratic graph, when you know how to do it. The example will show you how to start.

Example 3

a Draw the graph of $y = x^2 + x$.

b Use the graph to find the value of y when $x = 1$.

c Use the graph to find the value of x when $y = 4$.

d Use the graph to solve the equation $x^2 + x = 5$.

e What is the lowest value y can have?

a Draw up a table of values.

x	−3	−2	−1	0	1	2
$y = x^2 + x$	6	2	0	0	2	6

Then draw the graph.

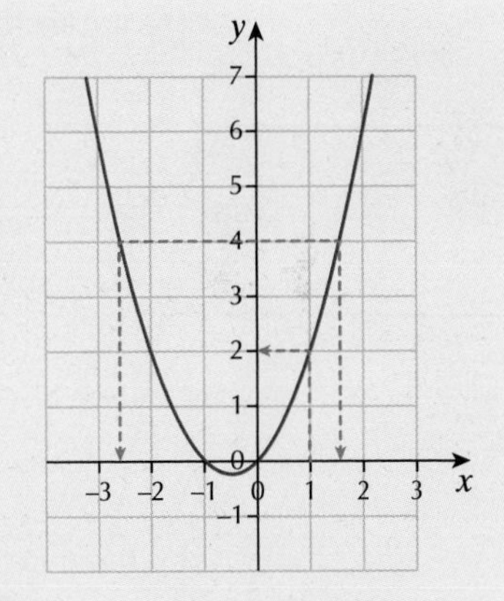

b Now draw a dotted line, from $x = 1$ on the x-axis, to the graph.

From the point where the line meets the graph, draw across to the y-axis.

When $x = 1$, $y = 2$.

c Draw a dotted line across the graph, through $y = 4$ on the y-axis.

The line meets the graph in two different places, so there will be two solutions to this equation.

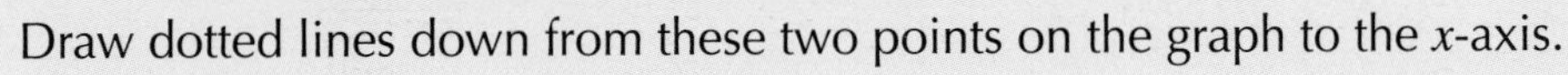

Draw dotted lines down from these two points on the graph to the x-axis.

The solutions are $x = -2.6$ and $x = 1.6$.

d This is similar to part **c**, but now $y = 5$.

Drawing a line where $y = 5$ to the graph and following the lines down to the x-axis gives the solutions as $x = 1.8$ and $x = -2.8$.

e Look at the bottom of the curve of the graph to find the lowest point.

This gives the lowest value y can have. This is $y = -0.25$.

Exercise 11C

1 **a** This is a graph of $y = x^2 + 2x$.

Copy it carefully onto graph paper.

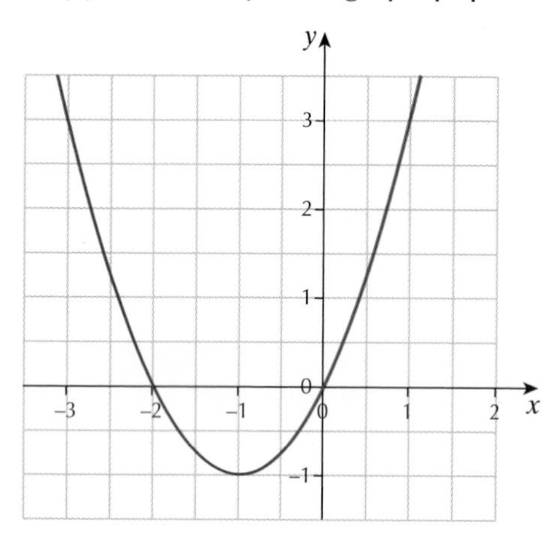

b Draw lines on your graph to find the value of y when $x = 0.5$.

c Draw lines on your graph to find the values of x when $y = 2$.

d What is the lowest value y can have?

e Use your graph to find the solutions to $x^2 + 2x = 1$.

2 **a** This is a graph of $y = x^2 + 3x$.

Copy it carefully onto graph paper.

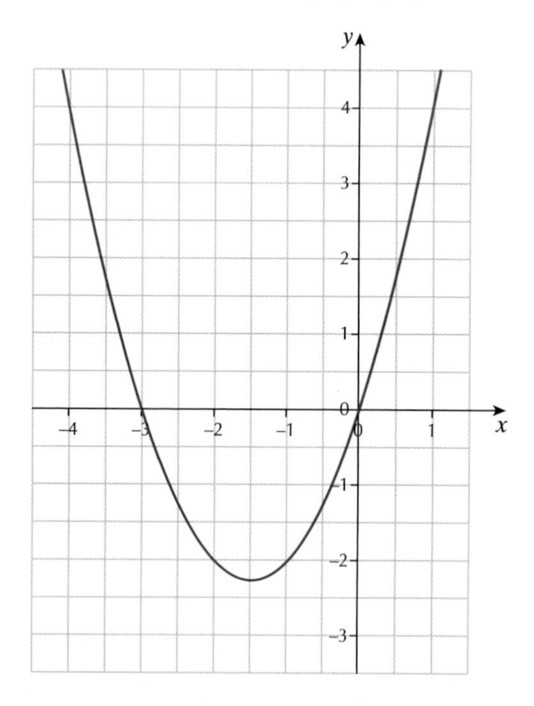

b Draw lines on your graph to find the value of y when $x = -0.5$.

c Draw lines on your graph to find the values of x when $y = 3$.

d What is the lowest value y can have?

e Use your graph to find the solutions to $x^2 + 3x = 1$.

3 **a** This is a graph of $y = x^2 - x$.

Copy it carefully onto graph paper.

b Draw lines on your graph to find the value of y when $x = 2.5$.

c Draw lines on your graph to find the values of x when $y = 1$.

d What is the lowest value y can have?

e Use your graph to find the solutions to $x^2 - x = 5$.

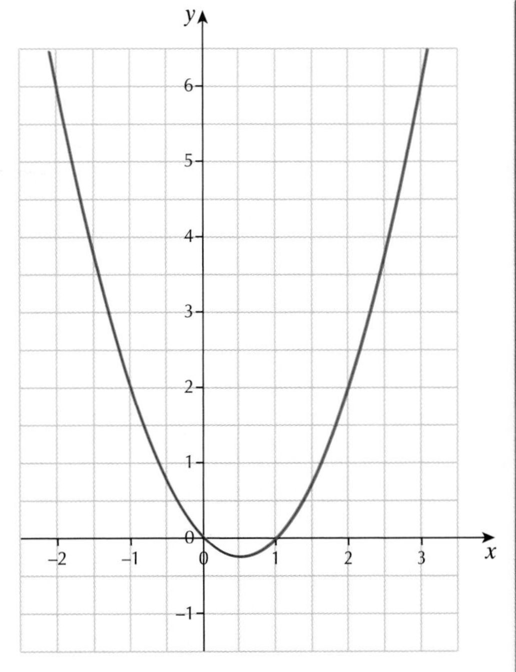

4 **a** This is a graph of $y = x^2 - 2x$.

Copy it carefully onto graph paper.

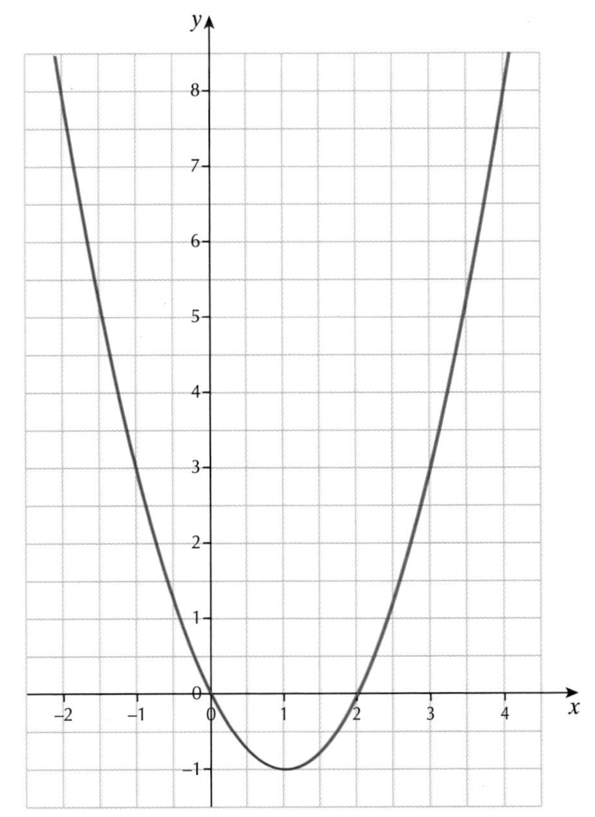

b Use your graph to find the value of y when $x = 1.5$.

c Use the graph to find the values of x when $y = 2$.

d What is the lowest value y can have?

e Find the solutions to $x^2 - 2x = 6$.

5 **a** This is a graph of $y = x^2 - 3x$.

Copy it carefully onto graph paper.

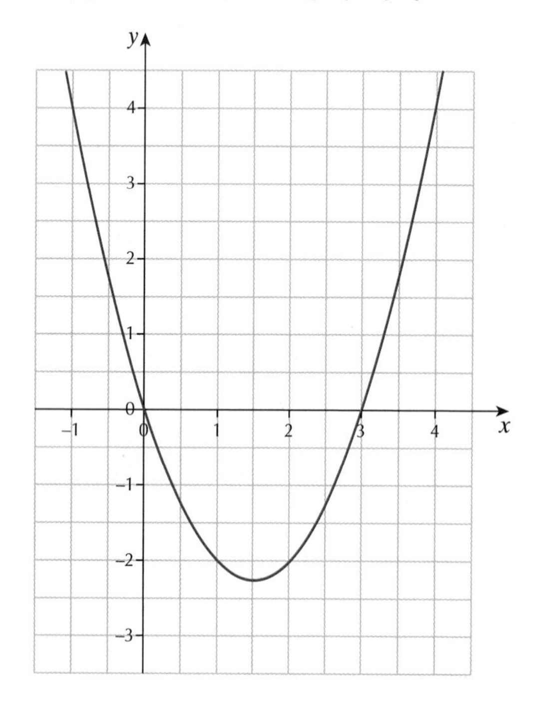

b Use your graph to find the value of y when $x = 2.5$.

c Use the graph to find the values of x when $y = 2$.

d What is the lowest value y can have?

e Find the solutions to $x^2 - 3x = 3$.

Investigation: You can't solve them all

A The table shows the points that you need to plot for the graph of $y = x^2 + x + 1$.

x	−3	−2	−1	0	1	2
$y = x^2 + x + 1$	7	3	1	1	3	7

Draw the graph of $y = x^2 + x + 1$.

B Use the graph to solve these equations.

a $x^2 + x + 1 = 5$ **b** $x^2 + x + 1 = 2$

C You cannot solve the equation $x^2 + x + 1 = 0$.

Explain why it is not possible to solve this equation.

D Write down another equation that you think you will not be able to solve.

Check your equation by drawing the graph.

11.4 Problems involving quadratic graphs

Learning objective

- To solve problems that use quadratic graphs

There are quite a lot of problems that can be solved from graphs shaped like quadratic curves. Work through the example, to find out more.

Example 4

Owen had 40 metres of fencing.

He was planning to make a rectangular sheep pen.

He wanted to know the area of the pen, depending on the length of one side of the pen.

The supplier gave him the following information.

Length of one side (metres)	0	4	8	12	16	20
Area (m^2)	0	64	96	96	64	0

a Use the data in the table to draw a graph.

b Use your graph to find out:

i the greatest area that could be made

ii the side length that gives this area.

c What lengths should he make the sides, if he wanted the area to be 80 m^2?

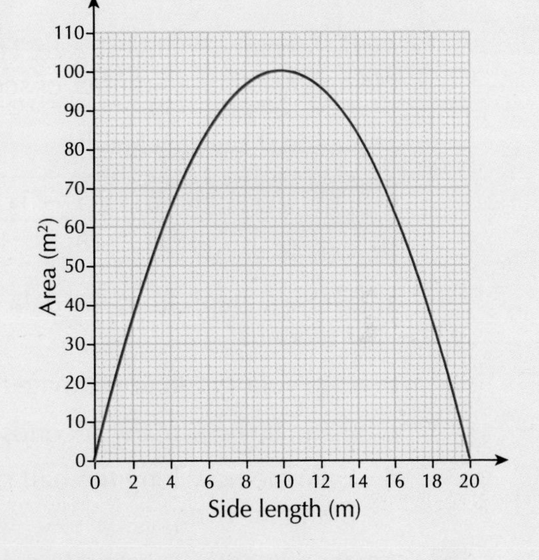

a Use the data in the table to draw the graph.

b i The greatest area is the value at the highest part of the graph.

The greatest area is 100 m^2.

ii At this point, the side length is 10 m.

c A line drawn on the graph where the area is 80 would meet it at two points.

At these points, the values of the side length are about 5.8 m and 14.2 m.

Check this by laying your ruler over the graph.

Exercise 11D

Each graph used in this exercise is a quadratic graph.

Therefore your graphs should have the usual symmetrical shape of a quadratic graph.

In each question, use the top line in the table for the horizontal axis on your graph.

1 Richie was watching his dad, who was throwing a ball up in the air.

Richie decided to investigate. He used markers on the side of his house to measure the heights after a number of seconds.

The table shows his results.

After t seconds	0.5	1	2	2.5	3
Height (metres)	8	12	13	10	5

a Draw a graph from the data in the table, to show the height of the ball after t seconds.

b Use your graph to estimate the greatest height the ball was thrown.

c Estimate how many seconds it takes for the ball to get to its greatest height.

d Estimate how long the ball stays in the air, if it always falls to the ground after being thrown.

2 An accountant was asked to calculate the costs linked with various profit margins in a company.

He did some calculations and came up with this information.

Costs (£1000)	1	3	4	7	10
Profit (£1000)	7	15	18	17	3

a Draw a graph from the data above.

b Use your graph to estimate the greatest profit possible.

c Estimate the costs needed to get the best profit.

d Estimate the costs to generate a profit of ten thousand pounds.

3 Andrew was laying a path around the outside of a rose bed in a park. He found that the costs were different for different widths of path.

He found these facts.

Path width (m)	1	5	9	11	14
Costs (£)	540	2000	2160	1760	560

a Draw a graph from the data above.

b Use your graph to estimate the largest cost for a path.

c Estimate the width of the path that gives rise to this largest cost.

d Estimate the widths of paths that would cost £1000 to build.

4 Solomon ran a commune.

Over the years, he kept a note of how many people were in the commune.

He also tried to work out how happy the commune seemed, on a scale of 1 to 10.

He decided that 0 was not very happy but 10 was very happy.

One day he gave this information to a friend.

Number in commune	4	12	26	34	40
Happiness	3	8	9	5	0

a Draw a graph from the data above.

b Use your graph to estimate the number in the commune that would give a happiness score of 10.

c Estimate how happy the commune would be with 16 people in.

d How many are needed in the commune to keep the happiness level above 6?

Investigation: Cooling pies

After it is taken out of the oven, the temperature of a cooked pie drops as it cools.

A school cook noted the temperature of her pies as they cooled.

The pies came out of the oven at 150 °C.

Number of minutes cooling	1	2	3	5	8	10
Temperature (°C)	100	60	40	25	22	20

A Explain how you could use the data to find the temperature after:

a 4 minutes **b** 6 minutes **c** 12 minutes.

B The pies are best eaten when their temperature is between 50 °C and 30 °C.

Find out when the pies can be eaten at their best.

Ready to progress?

I know that a graph from an equation in the form $y = mx + c$ is a straight line.

I can solve problems from data that gives a straight-line graph.
I can solve simple quadratic equations by drawing a graph.
I can solve problems from data that gives a quadratic graph.

Review questions

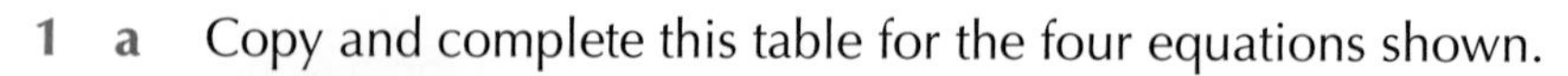

1 **a** Copy and complete this table for the four equations shown.

x	0	1	2	3
$y = 3x + 1$	1			
$y = 3x + 2$		5		
$y = 3x + 3$			9	
$y = 3x + 4$				13

b Draw a coordinate grid. Number the x-axis from 0 to 3 and the y-axis from 0 to 14.

c On your grid, draw the graph for each equation in the table.

d What do you notice about each graph?

e Use what you have noticed to draw the graphs of these equations.

i $y = 3x + 5$ **ii** $y = 3x + 0.5$

2 Look at the equation $y = 4x + 9$.

a When $x = 5$ what is the value of y?

b When $x = -5$ what is the value of y?

c Which of these equations give the same value for both $x = 5$ and $x = -5$?

$y = 4x$ $y = 4 + x$ $y = x^2$ $y = x^2 - 4$

Explain your answer.

3 Joe was asked to make a rectangle with a perimeter of 20 cm.

He drew this one.

8 cm

2 cm

a Explain why the area of this rectangle is 16 cm^2.

b Sketch another three different rectangles that Joe could have drawn. Use sides with whole-number values.

c Find the area of each of your rectangles.

d Copy and complete this table with information from your rectangles.

Longest side (cm)	8			
Area (cm^2)	16			

e Draw a graph from your table.

f What is the greatest area that a rectangle with a perimeter of 20 cm can have?

g What length of side gives this greatest area?

4 **a** Here is a graph of $y = x^2 - 4x$.

Copy it carefully onto graph paper.

b Use your graph to find the value of y when $x = 3.5$.

c Use the graph to find the values of x when $y = 2$.

d What is the lowest value y can have?

e Use your graph to find the solutions to $x^2 - 4x = 1$.

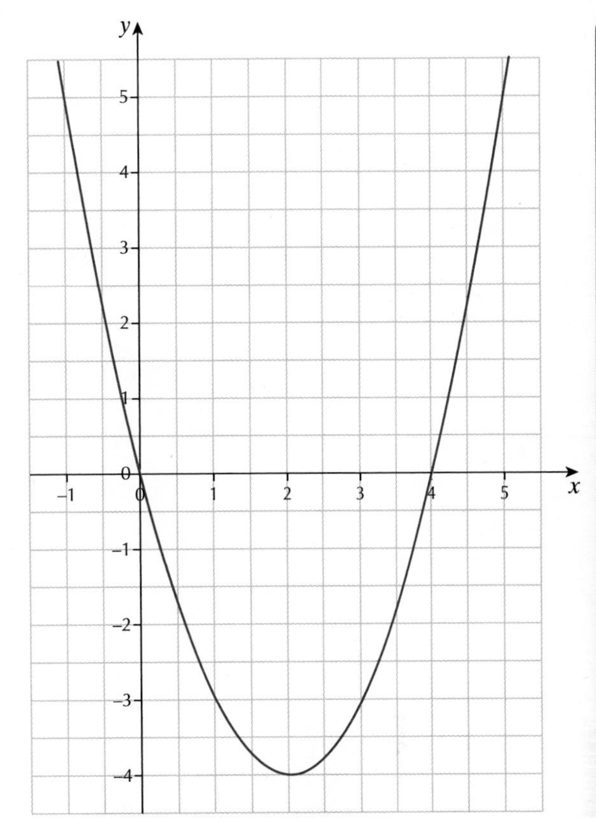

5 Nell sold oranges on a market stall. She was told just these two prices.

Number of oranges	0	15
Cost (£)	0	3.50

a Draw a coordinate grid. Use the horizontal axis for the number of oranges, numbering it from 0 to 20. Use the vertical axis for the cost of oranges, in pounds (£), numbering it from 0 to 5.

Plot the two points on your graph and join them with a straight line.

b Use your graph to estimate the cost of:

i 4 oranges **ii** 9 oranges **iii** 20 oranges.

c Use your graph to estimate how many oranges you could buy for:

i £1 **ii** £1.55 **iii** £4.50.

6 As an object falls from a height, its speed increases.

This table shows the distance, d, that an object will have fallen after t seconds.

Time, t (seconds)	1	3	5	7	10
Distance, d (metres)	5	44	122	240	490

a Draw a graph from the data in the table.

b How long will it take for an object to fall 300 metres?

c An object dropped from the top of a building took 4 seconds to get to the ground. How tall was the building?

PS **7** Todd was investigating volumes of tins in the kitchen cupboard. He noticed that all the heights were the same. The volumes, to 1 decimal place, of tins of different diameter are given in the table.

He wanted to find out the diameter of a tin of this height that would have a volume of 300 cm^3. By drawing a suitable graph, find the diameter that would give a volume of 300 cm^3.

Diameter (cm)	4	5	6	7	8
Volume (cm^3)	126	196	283	385	503

Problem solving

Squirrels

Red squirrels are native to Britain. In 1870 some North American grey squirrels were released in the North of England. The grey squirrel thrived in the conditions in Britain and slowly took over the habitats of the red squirrel, reducing their numbers dramatically.

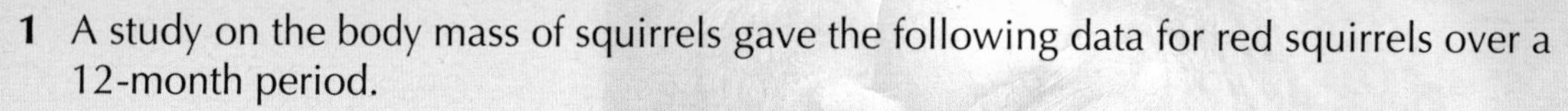

1 A study on the body mass of squirrels gave the following data for red squirrels over a 12-month period.

Month	Jan	Feb	Mar	Apr	May	Jun	Jul	Aug	Sep	Oct	Nov	Dec
Average mass (g)	270	265	275	280	285	290	310	325	345	375	330	290

a Draw a graph of the data in the table, taking the average mass as the vertical axis, starting from 250 g.

b A red squirrel was caught that had a mass of 300 g. What are the likely months for when it was caught?

This graph shows the same data for grey squirrels.

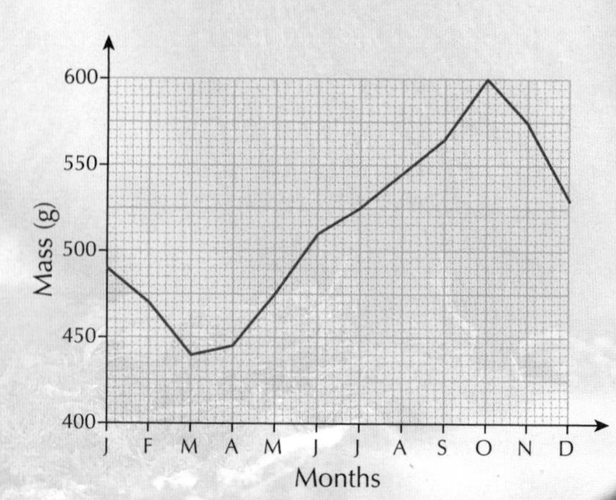

c A grey squirrel was caught that had a mass of 480 g. What are the two likely months for when it was caught?

d Why do you think the masses of the squirrels increase in the autumn?

e Comment on the differences in the masses of the red and grey squirrels over the year.

2 This scatter diagram shows the relationship between the body length and tail length of red squirrels.

a Estimate the tail length of a red squirrel with a body length of 210 mm.

b Describe the correlation between the body length and tail length of red squirrels.

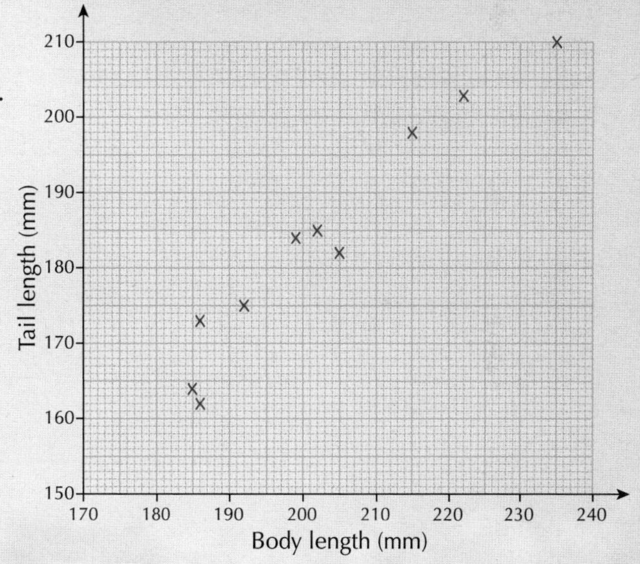

Data for grey squirrels

	A	B	C	D	E	F	G	H	I	J
Body length (mm)	272	243	278	266	269	280	251	272	278	281
Tail length (mm)	223	196	220	218	218	222	198	220	226	225

c **i** Use the data for grey squirrels to draw a scatter diagram to show the relationship between their body length and tail length. Label the horizontal axis for body length from 240 mm to 290 mm and the vertical axis for tail length from 190 mm to 230 mm.

ii Use your scatter graph to estimate the body length of a grey squirrel with a tail length of 205 mm.

iii Explain why the diagram could not be used to estimate the tail length of a young grey squirrel with a body length of 180 mm.

12 Distance, speed and time

This chapter is going to show you:

- how to solve problems involving distance, speed and time.

You should already know:

- that distance is how far you travel
- that speed is how fast you are travelling
- that time is how long you spend travelling
- how to draw graphs on a grid.

About this chapter

Voyager 1 is a 722 kg space probe, launched by NASA on 5 September 1977. Its mission was to study the outer Solar System.

As of 3 May 2014, *Voyager* 1 has been operating for 36 years, 7 months and 28 days. The spacecraft was 19 000 000 000 km (19 billion km) from the Earth. This is the farthest a human-made object has ever travelled from the Earth.

The spacecraft is moving with an average speed of about 60 000 km/h. *Voyager* 1 is expected to continue its mission until 2025, when it will no longer have enough power to operate any of its instruments.

In this chapter you will learn the basic rules that relate to travel, in terms of speed, distance and time.

12.1 Distance

Learning objectives

- To work out the distance travelled in a certain time at a given speed
- To use and interpret distance–time graphs

Key words	
average speed	distance
distance–time graph	km/h
speed	time

This speedometer in a French car shows that Raoul is travelling at 60 kilometres per hour. You would normally write this **speed** as 60 **km/h**.

This means that Raoul, in his car, travels 60 kilometres each hour.

So if he travelled at this speed for 2 hours, he would have travelled $2 \times 60 = 120$ km.

This table shows the **distance** travelled for different **times** when the speed is 60 km/h.

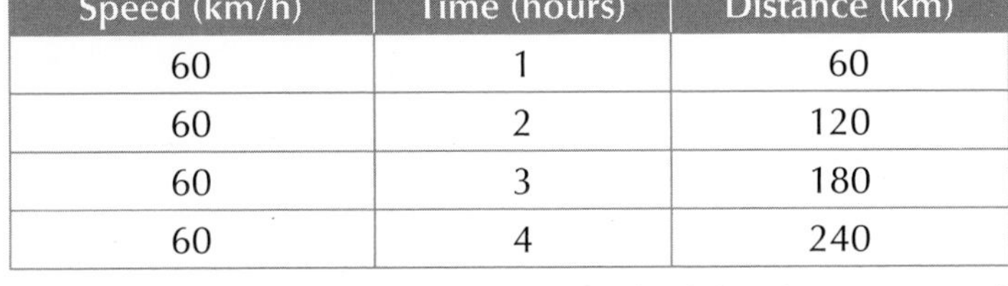

Speed (km/h)	Time (hours)	Distance (km)
60	1	60
60	2	120
60	3	180
60	4	240

From the table, you can see that you find the distance travelled by multiplying the speed by the time.

You can write this as:

distance = speed × time

You can also use the formula:

$d = s \times t$ or $d = st$

where d is the distance, s is the speed and t is the time.

Vehicles rarely move at the same speed for their whole journey, so the speed is the **average speed**.

Example 1

A train travels at an average speed of 80 km/h.

Work out the distance the train travels in:

a 1 hour **b** 3 hours **c** $4\frac{1}{2}$ hours.

Use the formula $d = st$.

a $d = 1 \times 80$

$= 80$ km

b $d = 3 \times 80$

$= 240$ km

c $d = 4\frac{1}{2} \times 80$

$= 360$ km

Example 2

This table shows the distance travelled by a cyclist moving an average speed of 10 km/h.

Time (h)	0	1	2	3	4
Distance (km)	0	10	20	30	40

This information is also shown on this **distance–time graph**.

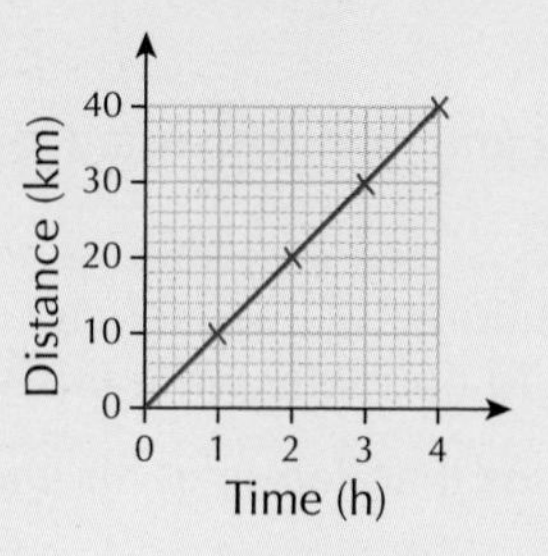

Use the graph to work out how far the cyclist travelled in $2\frac{1}{2}$ hours.

Read up from $2\frac{1}{2}$ on the time axis and read across to the answer on the distance axis, as shown by the arrows.

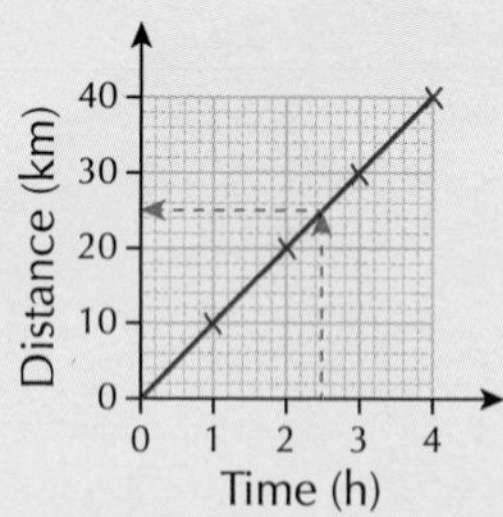

The cyclist travelled 25 km.

Exercise 12A

1 A train travels at an average speed of 120 km/h.
Work out the distance the train travels in:

a 2 hours **b** 4 hours **c** $5\frac{1}{2}$ hours.

Copy and complete each calculation.

a $d = st$
$d = \ldots \times \ldots$
$= \ldots$ km

b $d = st$
$d = \ldots \times \ldots$
$= \ldots$ km

c $d = st$
$d = \ldots \times \ldots$
$= \ldots$ km

2 Use the formula $d = st$ to work out the distance travelled for each journey.

a speed = 20 km/h, time = 4 hours

b speed = 35 km/h, time = 3 hours

c speed = 5 km/h, time = 6 hours

d speed = 64 km/h, time = $\frac{1}{2}$ hour

3 A jogger runs at an average speed of 12 km/h.
She runs for 3 hours.
How far does she run?

4 An aeroplane flies at an average speed of 640 km/h.
How far does it travel in 3 hours?

5 Tariq walks at an average speed of 5 km/h for 4 hours.
Jamil walks at an average speed of 4 km/h for 5 hours.
How far did they both walk?

6 Copy and complete the table. Use this information.

$\frac{1}{4}$ hour = 15 minutes or 0.25 hour

$\frac{1}{2}$ hour = 30 minutes or 0.5 hour

$\frac{3}{4}$ hour = 45 minutes or 0.75 hour

The first one has been done for you.

	Speed (km/h)	Time	Time as a decimal (hours)	Distance travelled (km)
a	40	$1\frac{1}{2}$ hours	1.5	60
b	60	$2\frac{1}{4}$ hours		
c	100	$\frac{3}{4}$ hour		
d	80	2 hours 15 minutes		
e	10	30 minutes		
f	4	3 hours 45 minutes		

7 Look at Bridget's working and her answers.

She has made some mistakes.

a speed = 40 km/h
time = 30 minutes
d = s × t
= 40 × 30
= 1200 km

b speed = 60 km/h
time = 1 hour 30 minutes
d = s × t
= 60 × 1.30
= 78 km

Write out her answers for her, with the correct working and answers for each problem.

8 A car travels at an average speed of 20 km/h.

a Copy and complete this table.

Time (h)	0	1	2	3	4
Distance (km)					

b Copy this grid and draw a distance–time graph to show the journey.

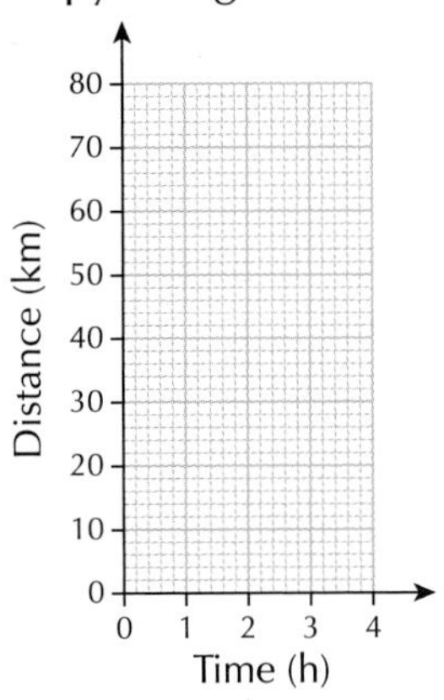

c Use your graph to work out how far the car travelled in $3\frac{1}{2}$ hours.

PS **9** This distance–time graph shows the journey of a motorcyclist.

a Work out the speed of the motorcyclist in the first part of the journey.

b Explain what is happening in the second part of the journey.

c Work out the speed of the motorcyclist in the final part of the journey.

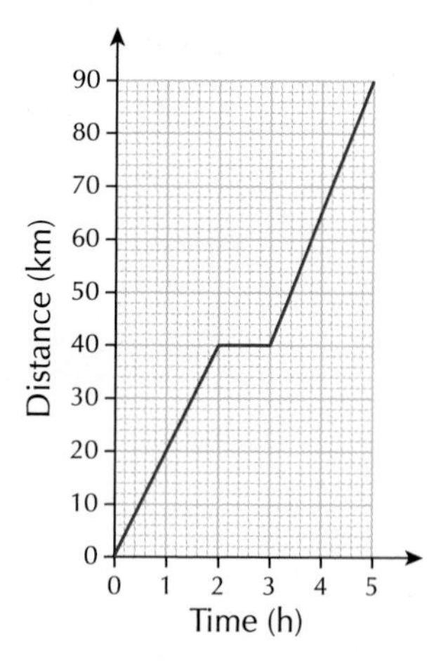

Challenge: Changing kilometres into miles

In Britain, distance is usually measured in miles and speed in miles per hour (mph).

This flow diagram shows you how to change kilometres into miles.

kilometres → ÷ 8 → × 5 → miles

For example, to change 120 km into miles:

$120 \div 8 \times 5 = 75$

So 120 km = 75 miles.

A Change these distances into miles.

a 24 km **b** 80 km **c** 100 km **d** 180 km

B This is a motorway sign that you might see in Germany.

How many miles is it to München (Munich)?

12.2 Speed

Learning objective

- To work out the speed of an object, given the distance travelled and the time taken

If a car travels 80 km in 2 hours, then in 1 hour it travels 40 km.

This means that its speed is 40 km/h.

Remember that the speed of any vehicle will not always be constant, as it will speed up and slow down from time to time – think of a typical bus journey. Because of this, what you are working with is the average speed overall.

If you know the distance an object moves in a certain time, then you can work out its average speed by dividing the distance (in kilometres) by the time (in hours).

This can be written as:

speed = distance ÷ time

You can also use the formula:

$s = d \div t$ or $s = \frac{d}{t}$

where d is the distance, s is the speed and t is the time.

Example 3

A car travels 120 km.

Work out the average speed of the car, if the time taken for it to complete the journey is:

a 2 hours **b** 4 hours **c** 5 hours.

a $s = \frac{120}{2}$

$= 60$ km/h

b $s = \frac{120}{4}$

$= 30$ km/h

c $s = \frac{120}{5}$

$= 24$ km/h

Example 4

A ship sails 96 km in 3 hours. Work out its average speed.

Use the formula:

$s = \frac{d}{t}$

$s = \frac{96}{3}$

$= 32$ km/h

Exercise 12B

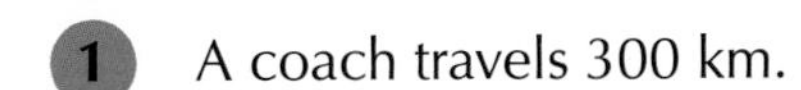

1 A coach travels 300 km.

Work out the average speed of the coach if the time taken to complete the journey is:

a 4 hours **b** 5 hours **c** 6 hours.

Copy and complete each calculation.

a $s = \frac{d}{t}$

$s = \ldots \div \ldots$

$= \ldots$ km/h

b $s = \frac{d}{t}$

$s = \ldots \div \ldots$

$= \ldots$ km/h

c $s = \frac{d}{t}$

$s = \ldots \div \ldots$

$= \ldots$ km/h

2 Use the formula $s = \frac{d}{t}$ to work out the speed for each journey.

a distance = 20 km, time = 5 hours

b distance = 60 km, time = 6 hours

c distance = 90 km, time = 3 hours

d distance = 500 km, time = 10 hours

3 A passenger jet aeroplane flies 3200 km in 4 hours.

What is the average speed of the aeroplane?

4 Nikki has to catch a ferry at 10:30 am. At 8:30 am the car's satnav shows that she still has 80 km to travel.

What average speed will she have to keep up, to catch the ferry?

5 The Earth is approximately 150 000 000 km from the Sun.

Light travelling from the Sun takes approximately 0.139 hours to reach the Earth.

Use this information to work out the approximate speed of light.

Give your answer correct to the nearest ten million km/h.

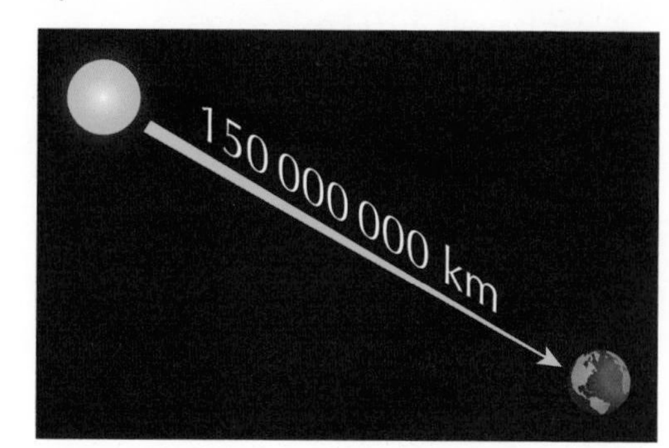

6 Jessica walks 1 km in 10 minutes.

Work out her speed, in kilometres per hour.

7 **a** A car travels 80 km in 2 hours.

Write down its average speed.

b Another car travels 80 km in 4 hours.

Write down its average speed.

c Explain what happens to your speed, if you travel the same distance but double the time it takes.

Challenge: Road signs in France

This road sign in France means that the speed limit is 40 km/h.

You can change the speed in km/h to miles per hour (mph), like this.

$40 \div 8 \times 5 = 25$

So 40 km/h = 25 mph.

For each French road sign, change the speed limits to miles per hour, giving you answer to the nearest whole number.

A

B

FRANCE

50

90

C

130

110

12.3 Time

Learning objective

- To work out the time an object will take on a journey, given its speed and the distance travelled

If a car travels 120 km at an average speed of 60 km/h, then in 1 hour it travels 60 km.

So it takes 2 hours to travel 120 km.

If you know the distance an object moves, at a certain average speed, then you can work out the time taken by dividing the distance (in kilometres) by the speed (in kilometres per hour).

You can write this as:

time = distance ÷ speed

You can also use the formula:

$t = d \div s$ or $t = \frac{d}{s}$

where d is the distance, s is the speed and t is the time.

Example 5

A motorcyclist travels 150 km.

Work out the time taken to complete the journey, if the speed of the motorcycle is:

a 30 km/h **b** 50 km/h **c** 60 km/h.

a $t = \frac{150}{30}$

$= 5$ hours

b $t = \frac{150}{50}$

$= 3$ hours

c $t = \frac{150}{60}$

$= 2\frac{1}{2}$ hours

Example 6

Work out the time it takes to walk 20 km at an average speed of 5 km/h.

Use the formula:

$t = \frac{d}{s}$

$t = \frac{20}{5}$

$= 4$ hours

Exercise 12C

1 A cyclist takes part in a 40 km race.

Work out the time the cyclist takes. if his average speed is:

a 10 km/h **b** 20 km/h **c** 16 km/h.

Copy and complete each calculation.

a $t = \frac{d}{s}$

$t = \ldots \div \ldots$

$= \ldots$ hours

b $t = \frac{d}{s}$

$t = \ldots \div \ldots$

$= \ldots$ hours

c $t = \frac{d}{s}$

$t = \ldots \div \ldots$

$= \ldots$ hours

2 Use the formula $t = \frac{d}{s}$ to work out the time for each journey.

a distance = 20 km, speed = 4 km/h

b distance = 100 km, speed = 25 km/h

c distance = 150 km, speed = 50 km/h

d distance = 420 km, speed = 120 km/h

3 How long does it take an aeroplane flying at an average speed of 720 km/h to fly:

a 2160 km **b** 3240 km?

4 The Japanese bullet train travels at an average speed of 260 km/h.

How long does it take the train to travel:

a 780 km **b** 650 km?

5 Mrs McKenzie leaves home at 10:00 am and travels 100 km to see a friend.

If her average speed is 40 km/h, at what time does she arrive at her friend's home?

6 Mia runs at an average speed of 12 km/h.

It takes her 3 hours to complete a 36 km run.

a Copy and complete this table.

Time (h)				
Distance (km)	0	12	24	36

b Copy this grid and draw a distance–time graph to show her journey.

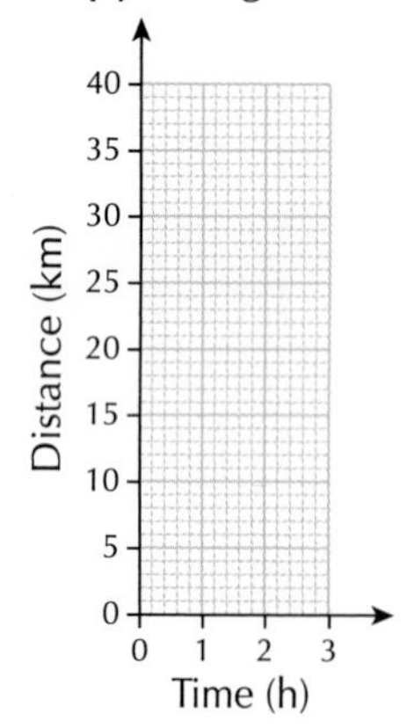

An athletic sprinter can run at an average speed of 10 metres per second (m/s). How long does it take him to run a 100 m race?

Challenge: The distance, speed and time triangle

You have been using the three formulae that connect distance, speed and time.

distance = speed × time $\quad$ time = $\frac{\text{distance}}{\text{speed}}$ $\quad$ speed = $\frac{\text{distance}}{\text{time}}$

This triangle is a useful reminder about the different formulae.

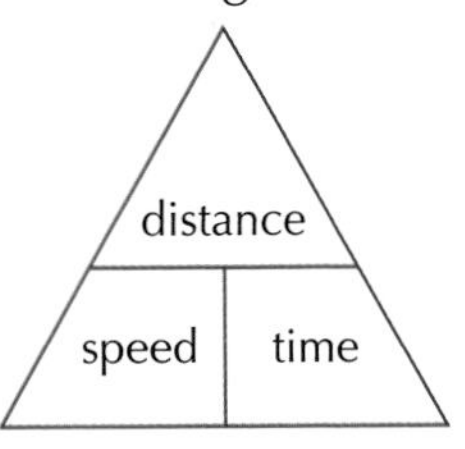

Put your finger over the word you want to use.

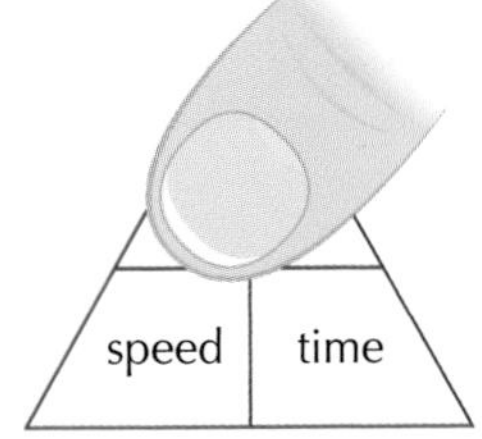

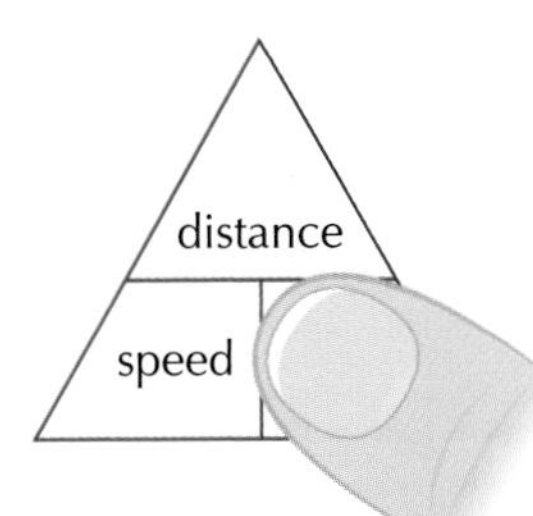

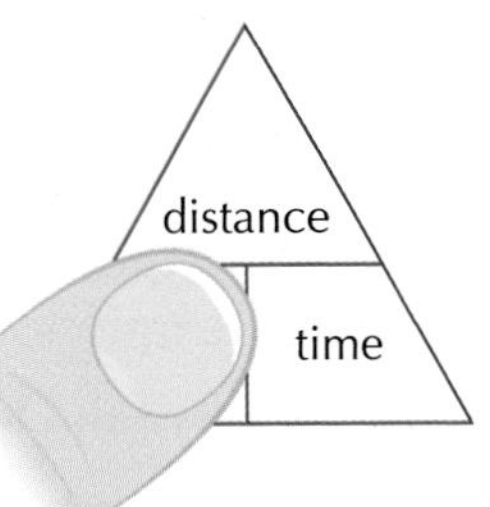

distance = speed × time $\quad$ time = $\frac{\text{distance}}{\text{speed}}$ $\quad$ speed = $\frac{\text{distance}}{\text{time}}$

Choose the correct formula to use, to work out the missing entries in this table.

	Distance (km)	Speed (km/h)	Time (hours)
A	6	3	
B	200		2
C		120	3
D	1800	400	
E	50		$\frac{1}{2}$
F		30	$4\frac{1}{2}$

Ready to progress?

I can use the formula $d = st$ to work out a distance.

I can use the formula $s = \frac{d}{t}$ to work out a speed.

I can use the formula $t = \frac{d}{s}$ to work out a time.

I can read and interpret distance–time graphs.

Review questions

1 A car travels at an average speed of 80 km/h.

Work out the distance the car travels in:

a 2 hours **b** 3 hours **c** $4\frac{1}{2}$ hours.

Copy and complete each calculation.

a $d = st$
$d = \ldots \times \ldots$
$= \ldots$ km

b $d = st$
$d = \ldots \times \ldots$
$= \ldots$ km

c $d = st$
$d = \ldots \times \ldots$
$= \ldots$ km

2 A lorry travels 400 km.

Work out the average speed of the lorry, if the time taken to complete the journey is:

a 5 hours **b** 8 hours **c** 4 hours.

Copy and complete each calculation.

a $s = \frac{d}{t}$
$s = \ldots \div \ldots$
$= \ldots$ km/h

b $s = \frac{d}{t}$
$s = \ldots \div \ldots$
$= \ldots$ km/h

c $s = \frac{d}{t}$
$s = \ldots \div \ldots$
$= \ldots$ km/h

3 A hiker walks 36 km.

Work out the time the hiker takes if his average speed is:

a 6 km/h **b** 3 km/h **c** $4\frac{1}{2}$ km/h.

Copy and complete each calculation.

a $t = \frac{d}{s}$
$t = \ldots \div \ldots$
$= \ldots$ hours

b $t = \frac{d}{s}$
$t = \ldots \div \ldots$
$= \ldots$ hours

c $t = \frac{d}{s}$
$t = \ldots \div \ldots$
$= \ldots$ hours

4 On a motorway, a car travels at an average speed of 80 km/h for 2 hours and then at an average speed of 100 km/h for $1\frac{1}{2}$ hours.

How far has the car travelled altogether?

5 Jake walks at an average speed of 5 km/h.

On a walking holiday, he walks for 5 hours.

a Copy and complete this table.

Time (h)	0	1	2	3	4	5
Distance (km)						

b Copy this grid and draw a distance–time graph to show his journey.

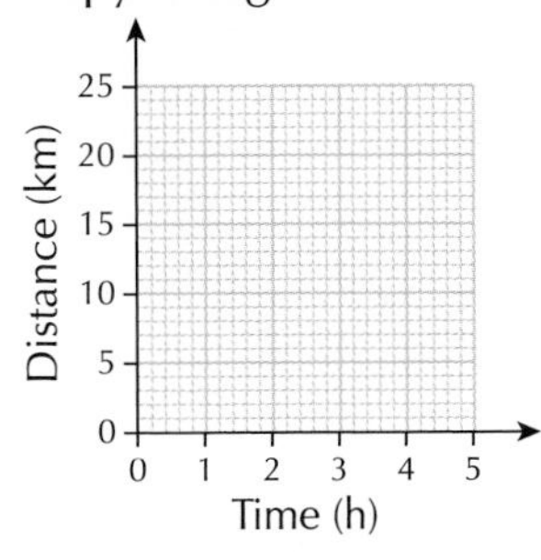

c Use the graph to work out how far he walks in $2\frac{1}{2}$ hours.

PS **6** Ewan cycles 2 km in 6 minutes.

Work out his speed, in kilometres per hour.

PS **7** The speed of sound is approximately 340 metres per second (m/s) at sea level.

Work out the speed of sound in kilometres per hour (km/h). Give your answer correct to the nearest 10 km/h.

1000 m = 1 km
3600 seconds = 1 hour

Financial skills

Shopping at the market

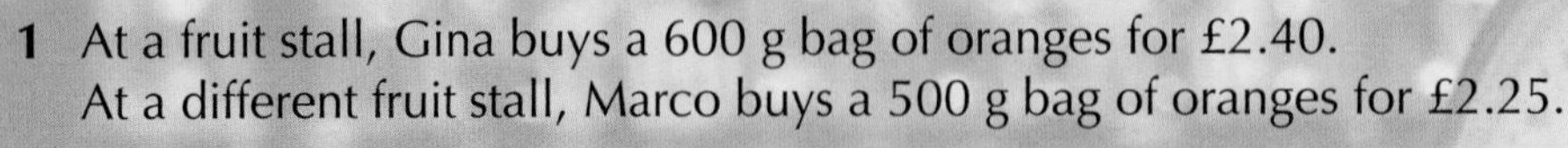

1 At a fruit stall, Gina buys a 600 g bag of oranges for £2.40.
At a different fruit stall, Marco buys a 500 g bag of oranges for £2.25.

Copy and complete these sentences.

a For Gina:

600 g costs £2.40, so 100 g costs $\frac{£2.40}{6}$ = 40p

b For Marco:

500 g costs £2.25, so 100 g costs $\frac{£2.25}{5}$ = 45p

c Gina pays 40p for 100 g and Marco pays 45p for 100g, so … gets a better deal.

2 At a cheese stall, George buys 800 g of Cheddar cheese for £6.00.
At a different cheese stall, Natasha buys 500 g of Cheddar cheese for £4.00.

Copy and complete these sentences.

a For George:

800g costs …, so 100 g costs $\frac{\ldots}{\ldots}$ = … p

b For Natasha:

500 g costs …, so 100g costs $\frac{\ldots}{\ldots}$ = … p

c George pays … p for 100 g and Natasha pays … p for 100 g, so … gets a better deal.

3 At a meat stall, Mrs Seager buys $1\frac{1}{2}$ kg of minced lamb for £12.00.
At a different meat stall, Mr Mir buys 2 kg of minced lamb for £14.00.

Copy and complete these sentences.

a For Mrs Seager:

$1\frac{1}{2}$ kg costs …, so $\frac{1}{2}$ kg costs $\frac{\ldots}{\ldots}$ = £ …

b For Mr Mir:

2 kg costs …, so $\frac{1}{2}$ kg costs $\frac{\ldots}{\ldots}$ = £ …

c Mrs Seager pays £ … for $\frac{1}{2}$ kg and Mr Mir pays £ … for $\frac{1}{2}$ kg, so … gets a better deal.

4 At a vegetable stall, Nathan buys a 5 kg bag of potatoes for £4.50.
At a different vegetable stall, Lily buys a 8 kg bag of potatoes for £6.40.

Copy and complete these sentences.

a For Nathan:

1 kg costs …p

b For Lily:

1 kg costs …p

c So … gets a better deal.

5 Here are signs at two different fish stalls.

a Work out the cost of 1 kg of cod fillet at each stall.

b Which stall gives better value?

6 These are the prices of two different-sized tins of dog food at a pet stall.

Which tin is better value for money?

Give a reason for your answer.

13 Similar triangles

This chapter is going to show you:

- what similar triangles are
- patterns you can find in similar and right-angled triangles
- how to use these patterns to solve some problems.

You should already know:

- what a right-angled triangle is
- how to use a protractor to construct triangles
- what a hypotenuse is
- how to multiply fractions.

About this chapter

How could you find the height of a very tall building, or a tree, if all you had was a tape measure, a protractor and a calculator? The answer is, you could use mathematics!

Part of mathematics looks at the relationships between the sides and angles of triangles. In this chapter, you will look at important properties of right-angled triangles and learn about similar triangles. You will discover relationships that will enable you to find the size of angles or the lengths of sides of similar triangles.

You will also be able to work out the height of that tall tree.

13.1 Similar triangles

Learning objective

- To understand what similar triangles are

Key words

denominator	numerator

Two triangles are similar if they both have the same angles.

Notice, in the two triangles shown here, that the lengths of the corresponding sides are different but the angles all pair up.

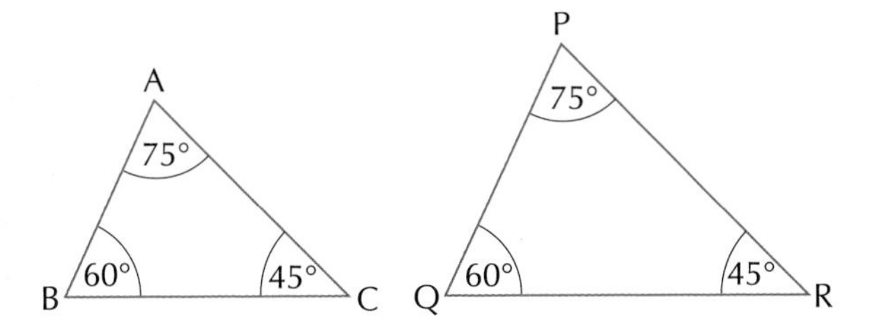

angle A = angle P

angle B = angle Q

angle C = angle R

So triangles ABC and PQR are similar.

The lengths of the sides of the triangles are not taken into account, for the test of similarity.

You only need to consider the angles.

Work through this exercise to find out more about similar triangles.

Exercise 13A

1 **a** Draw this pair of similar triangles, as accurately as you can.

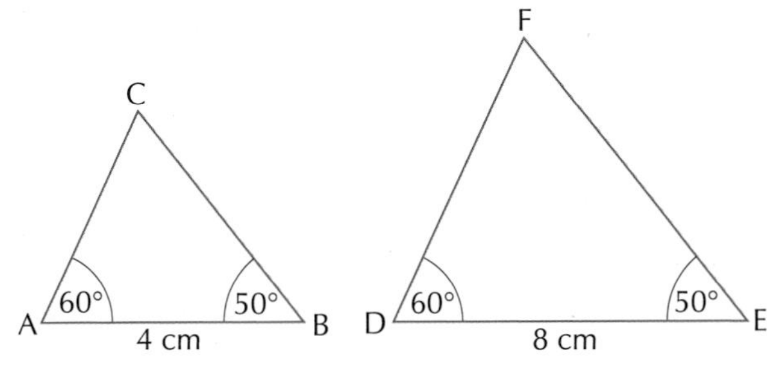

b Measure the sides of both triangles. Write the measurements on your diagrams.

c Now write down the fractions $\frac{DE}{AB}$, $\frac{EF}{BC}$ and $\frac{DF}{AC}$.

Divide the **denominator** (bottom number) into the **numerator** (top number) of each fraction. Use a calculator if you need to.

d What do you notice?

You should have found that all your answers are very close to 2.

2 **a** Draw this pair of similar triangles, as accurately as you can.

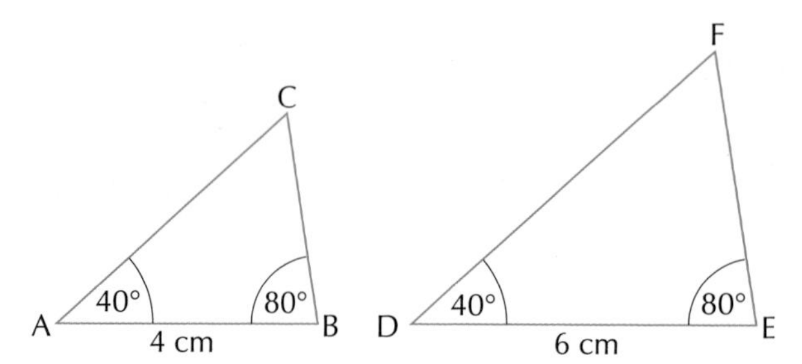

b Measure the sides of both triangles. Write the measurements on your diagrams.

c Now write down the fractions $\frac{DE}{AB}$, $\frac{EF}{BC}$ and $\frac{DF}{AC}$.

Divide the denominator into the numerator of each fraction. Use a calculator if you need to.

d What do you notice?

You should have found that all your answers are very close to 1.5.

3 **a** Draw a pair of similar triangles ABC and DEF in which:

angle A = angle D = 70°

angle B = angle E = 40°

length AB = 5 cm, length DE = 7 cm.

b Measure the sides of both triangles as accurately as you can.

Write the measurements on your diagrams.

c Now calculate $\frac{DE}{AB}$, $\frac{EF}{BC}$ and $\frac{DF}{AC}$. Use a calculator if you need to.

d What do you notice?

You should have found that all your answers are very close to 1.4.

4 **a** Draw a pair of similar triangles ABC and DEF in which:

angle A = angle D

angle B = angle E

length DE is longer than AB.

You are choosing your own starting angles and sizes for AB and DE.

b Measure the sides of both triangles as accurately as you can. Write the measurements on your diagrams.

c Now calculate $\frac{DE}{AB}$, $\frac{EF}{BC}$ and $\frac{DF}{AC}$. Use a calculator if you need to.

d What do you notice?

You should have found that all of your answers are very close to each other.

Investigation: Angles and ratios

Select an angle between 30° and 70°.

Draw six different right-angled triangles, each with the angle you have chosen at the bottom right as shown here.

If you make the base length an integer, the calculations are simpler.

Label your triangles A to F.

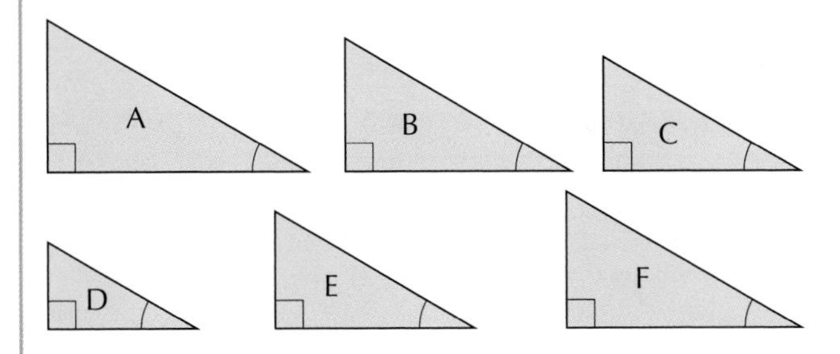

Measure, as accurately as you can, the lengths of the sides of all the triangles. Add this information to your diagram.

Now label each triangle in the same way, as shown here.

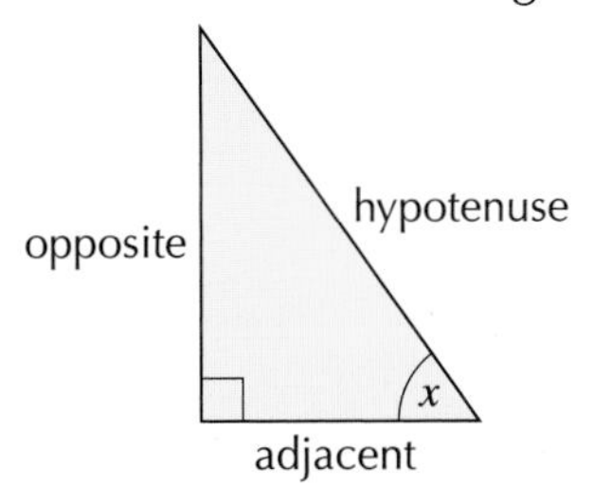

The 'hypotenuse', as you already know, is the long side opposite the right angle.

The 'opposite' is the side opposite the angle you are focused on.

The 'adjacent' is the side between the right angle and the angle you are focused on.

Copy and complete this table.

Triangle	Opposite	Adjacent	Hypotenuse	$\frac{\text{Hypotenuse}}{\text{Opposite}}$	$\frac{\text{Hypotenuse}}{\text{Adjacent}}$
A					
B					
C					
D					
E					
F					

Convert the fractions in the last two columns to decimals.

Give your answers correct to two decimal places.

What do you notice?

Do you think this will happen, whatever angle you chose?

13.2 A summary of similar triangles

For any two similar triangles ABC and DEF, the ratios of the corresponding sides are always the same.

$$\frac{DE}{AB} = \frac{EF}{BC} = \frac{DF}{AC}$$

Hint You could think of one triangle as being an enlargement of the other.

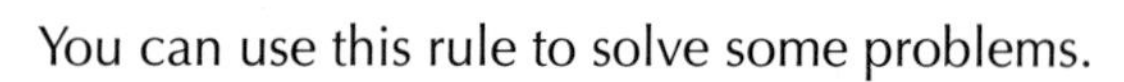

You can use this rule to solve some problems.

Example 1

The shadow of a tree is 8 m long.

At exactly the same time of day, a boy of height 1.2 m casts a shadow of 50 cm.

Find the height of the tree.

You can draw similar triangles showing the heights and the shadows.

Hint This is because the hypotenuse of each triangle is the path of the light from the Sun.

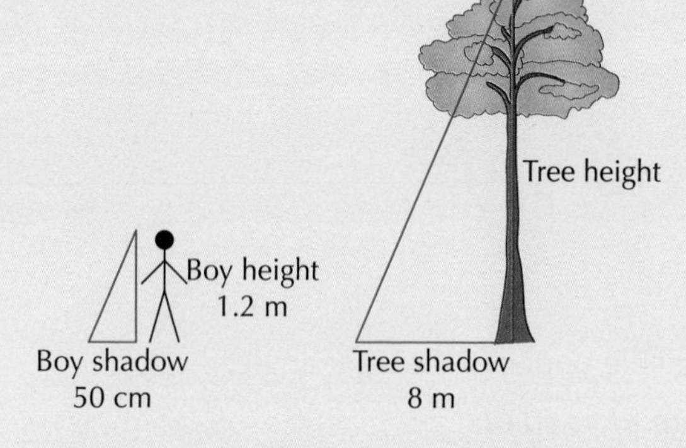

Now use the rule for similar triangles:

$$\frac{\text{tree's height}}{\text{boy's height}} = \frac{\text{length of the tree's shadow}}{\text{length of the boy's shadow}}$$

Now substitute the numbers into the equation.

$$\frac{\text{tree's height}}{1.2} = \frac{8}{0.5}$$

Then the height of the tree is:

$$\frac{8 \times 1.2}{0.5} = 19.2 \text{ metres}$$

Hint Change 50 cm to 0.5 m so that the units are the same.

Example 2

These two triangles are similar.

> **Hint** Note how to use symbols to show equal angles.

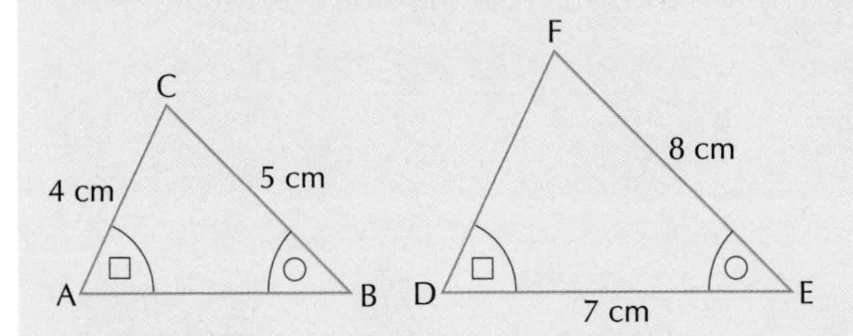

Calculate:

a the length of DF **b** the length of AB.

a Use the rule for similar triangles.

$$\frac{DF}{AC} = \frac{EF}{BC}$$

Then:

$$\frac{DF}{4} = \frac{8}{5}$$

So:

$$DF = \frac{8 \times 4}{5}$$

$$= 6.4 \text{ cm}$$

b Use the rule for similar triangles.

$$\frac{AB}{DE} = \frac{BC}{EF}$$

Then:

$$\frac{AB}{7} = \frac{5}{8}$$

So:

$$AB = \frac{5 \times 7}{8}$$

$$= 4.375 \text{ cm}$$

Exercise 13B

1 For each pair of similar triangles, write down the ratios for the sides, as fractions. The first one has been done for you.

a

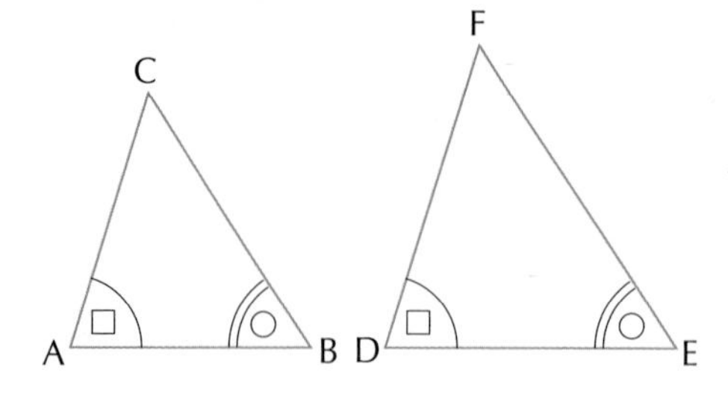

$$\frac{DE}{AB} = \frac{EF}{BC} = \frac{DF}{AC}$$

b

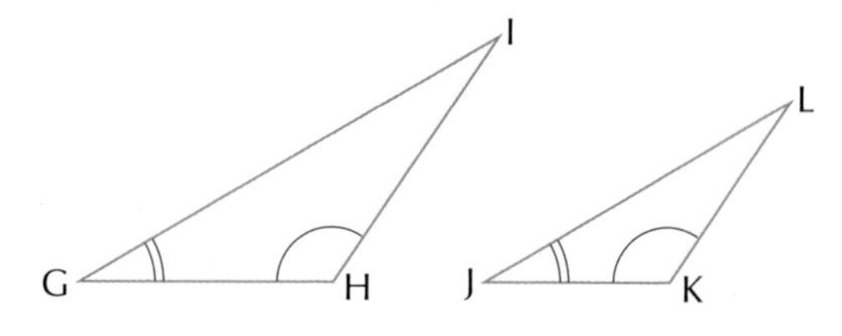

c

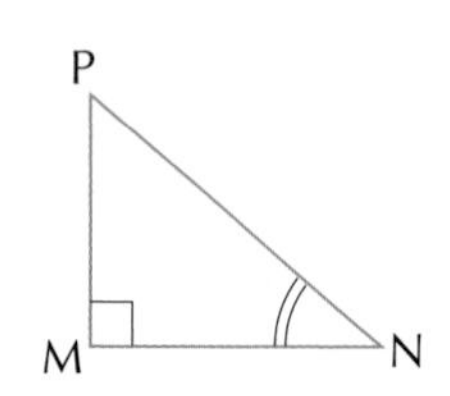

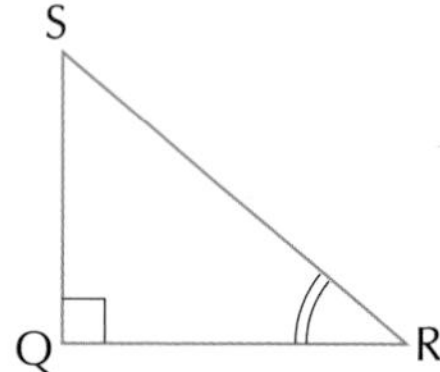

d

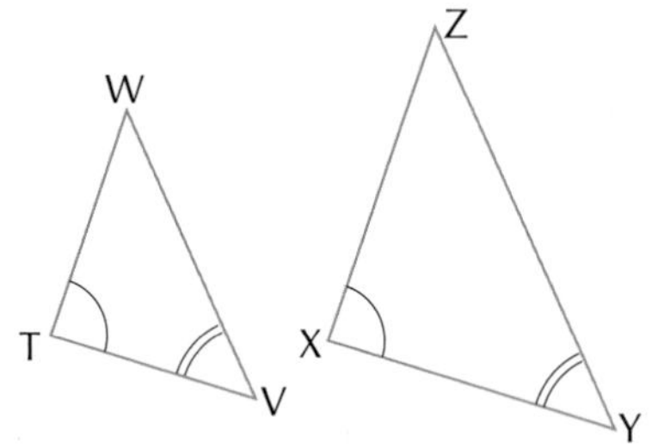

2 For each pair of similar triangles, all the measurements given are in centimetres. Calculate the missing lengths.

The first one has been done for you.

a

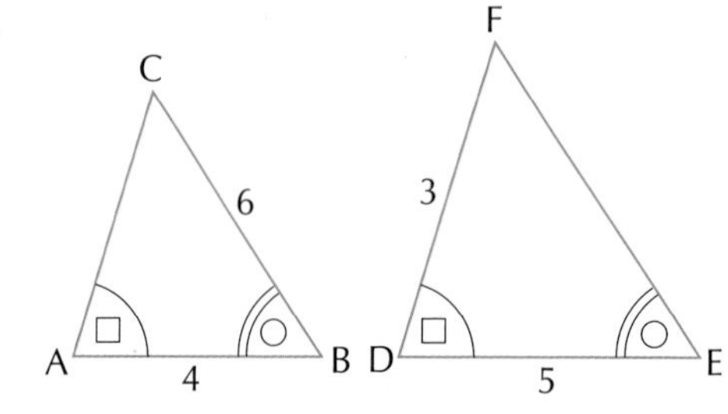

$$\frac{FE}{6} = \frac{5}{4}$$

So $FE = \dfrac{5 \times 6}{4}$

$= 7.5$

$FE = 7.5$ cm

$$\frac{AC}{3} = \frac{4}{5}$$

So $AC = \dfrac{4 \times 3}{5}$

$= 2.4$

$AC = 2.4$ cm

b

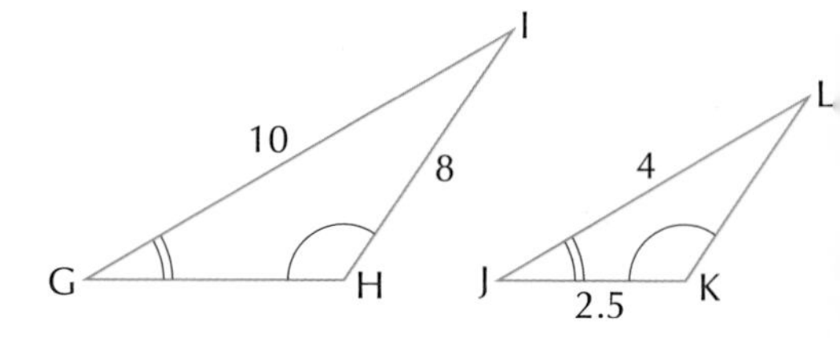

c

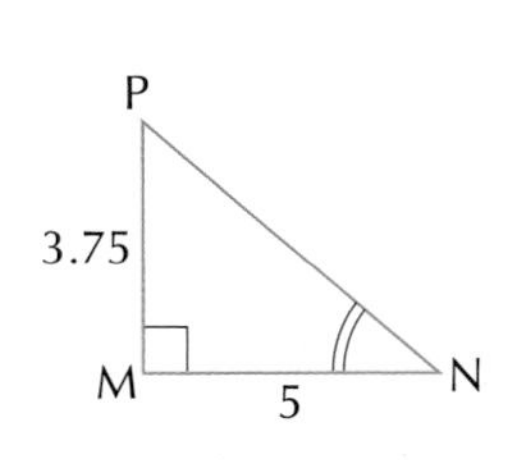

d

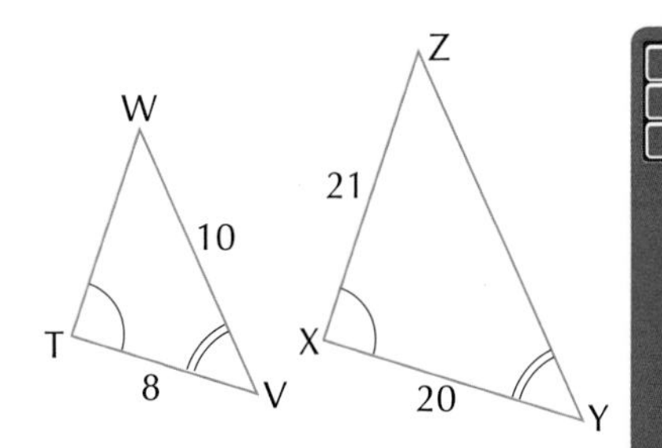

PS **3** The shadow of a tree is 6 metres long.
At the same time of day, the shadow of a pillar pox is 40 cm.
The height of the pillar box is 1.2 m.
Find the height of the tree.

PS **4** The shadow of a pole that is 1 metre high was 30 cm.
At the same time, a tall chimney cast a shadow 12 metres long.
Find the height of the chimney.

PS **5** A high building cast a shadow 15 metres long.
At the same time the shadow of a man 1.6 metres tall was 20 cm.
Find the height of the building.

Investigation: Nested triangles

Neil says he can see two similar triangles in this shape.

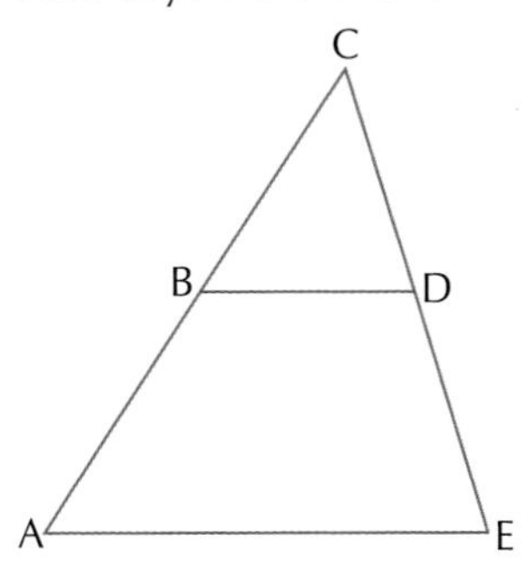

A Measure the angles and show that Neil is correct.

B You are told that AE = 10 cm, BD = 4 cm and BC = 5 cm.
Calculate the length of AB.

13.3 Using triangles to solve problems

Learning objective

- To understand that triangles can be used to solve some real problems

You may be surprised to know that we can solve many real-life problems by applying some mathematics. In this section you will see how to draw similar triangles, to model the real situation, and use them to find the solution to a problem.

Example 3

A builder has a ladder that is 4 m long.

He is told that the safest way to stand his ladder against the wall is so that it makes an angle of 75° with the floor.

Find out how far away from the wall of a building the foot of the ladder needs to be, so that it is as safe as possible.

First, draw a diagram representing the situation.

Label the triangle ABC.

Next, draw an accurate triangle with the same angles as the diagram (90° and 75°).

Label it DEF.

Measure the lengths of the relevant lines in the accurate diagram.

Now use the rules for similar triangles.

$\frac{AB}{2} = \frac{400}{7.6}$ Notice that all units are in centimetres.

This gives:

$AB = \frac{400 \times 2}{7.6}$

$= 105$ cm correct to the nearest whole number

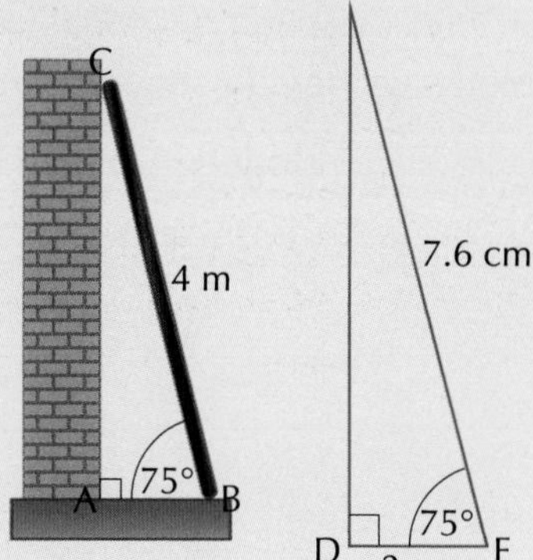

Exercise 13C

1 A ramp is built for wheelbarrow access up a small wall.

It needs to be built 50 cm high, at an angle of 10° to the horizontal.

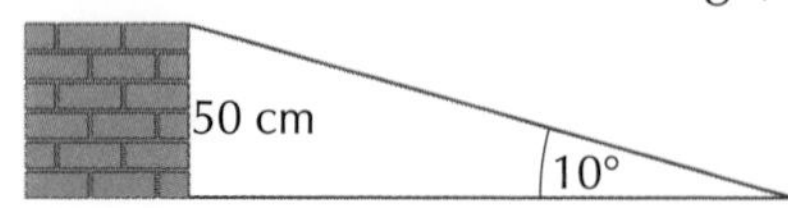

How far from the wall must the ramp start?

2 A skier skied down a 400 m slope that was at an angle of 55° to the horizontal.

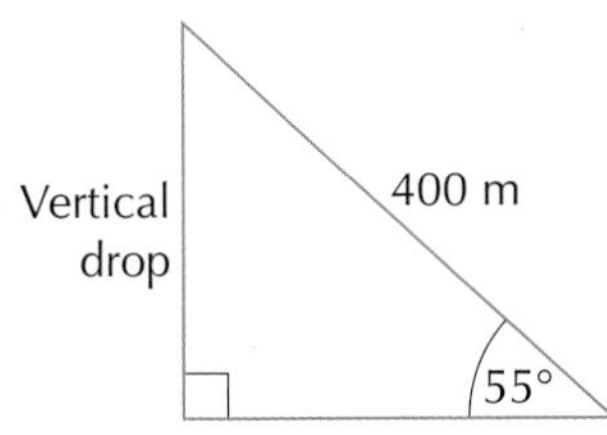

What was the vertical drop of the descent?

PS **3** A boy who was 1.5 m tall stood on level ground, looking up at the top of a tower.

He knew that he was standing 200 metres away from the foot of the tower.

He was looking up at the tower at an angle of 22° from the horizontal.

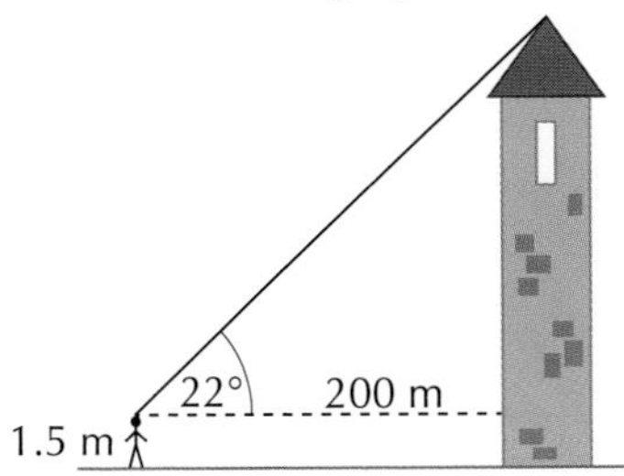

How tall was the tower?

4 A ship sails on a direction of N75°E for 150 km.

a Draw a sketch of the ship's journey. Show a north line and the direction of east.

b Find how far east the ship has sailed.

5 A plank 8 metres long is leaning against a wall. The plank is at an angle of 30° with the horizontal.

How far up the wall does the plank reach?

6 Robin flew for 300 km on a bearing of 150°.

How far: **a** east **b** south has he flown?

Investigation: Skewed triangles

Mandy said she could see two similar triangles in the shape shown.

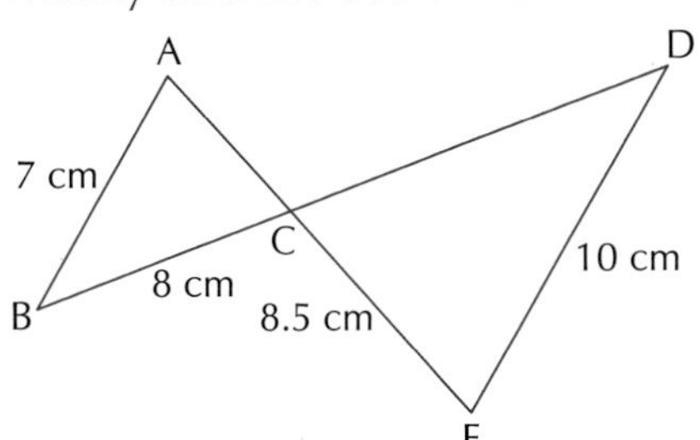

A By measuring the angles, show that Mandy is correct.

B Find the lengths of AE and BD.

Ready to progress?

I understand what similar triangles are.
I can use similar triangles to solve simple problems.

Review questions

1 For each pair of similar triangles, write down the ratios for the sides, as fractions.

The first one has been done for you.

a

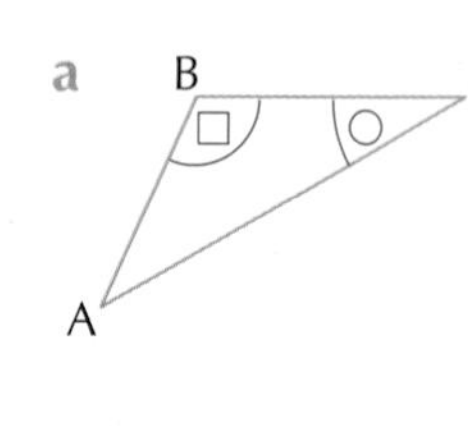

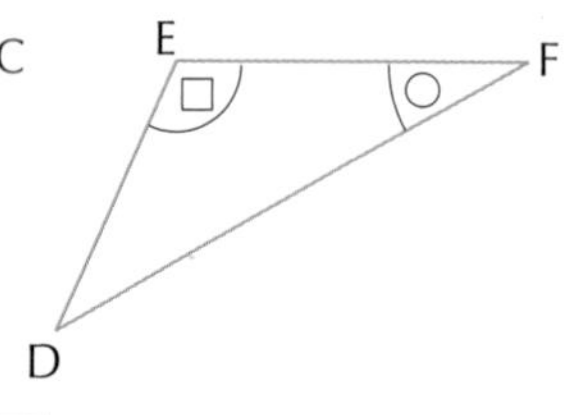

$$\frac{DE}{AB} = \frac{EF}{BC} = \frac{DF}{AC}$$

b

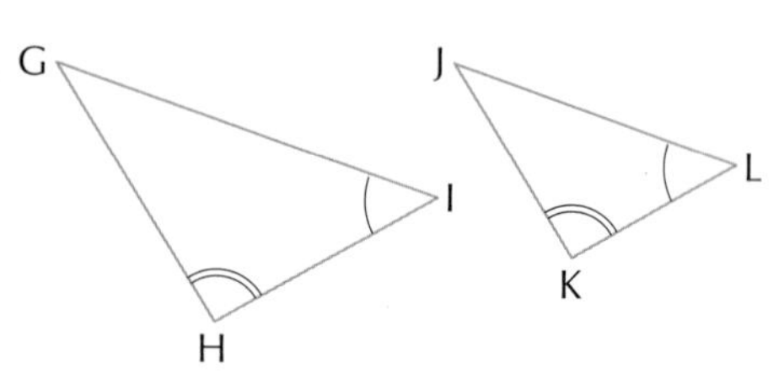

c

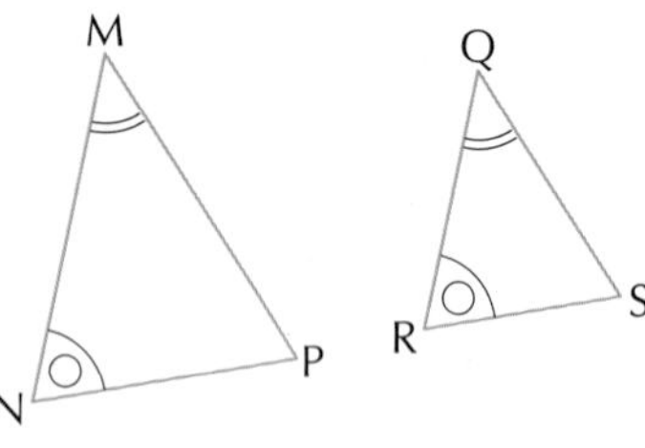

d

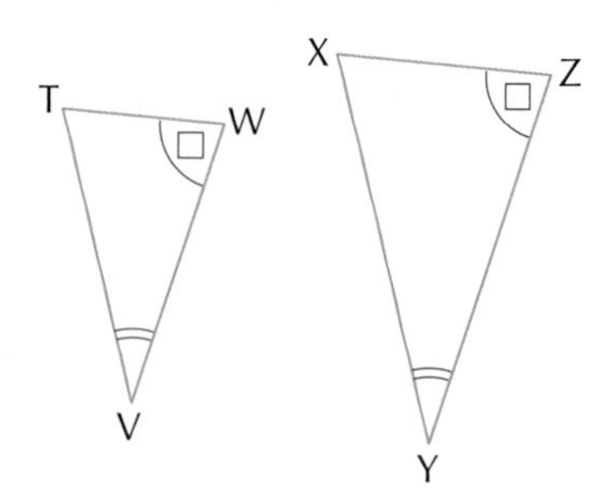

2 For each pair of similar triangles below, all the measurements are given in centimetres.

Calculate the missing lengths.

The first one has been done for you.

a

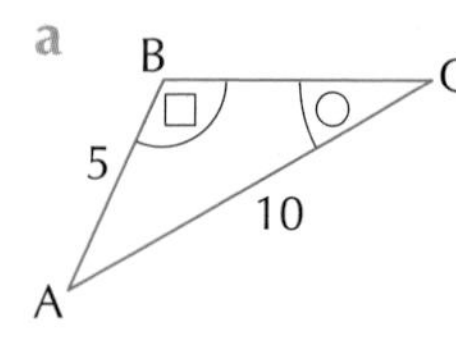

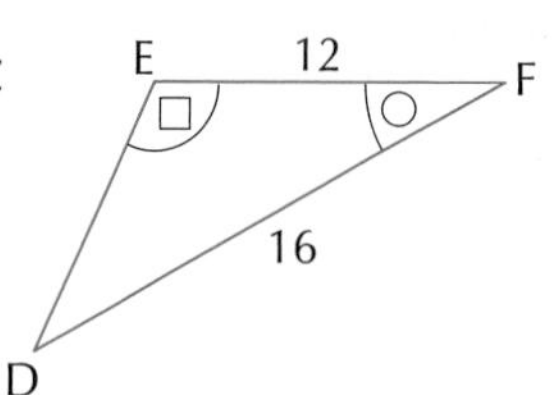

b

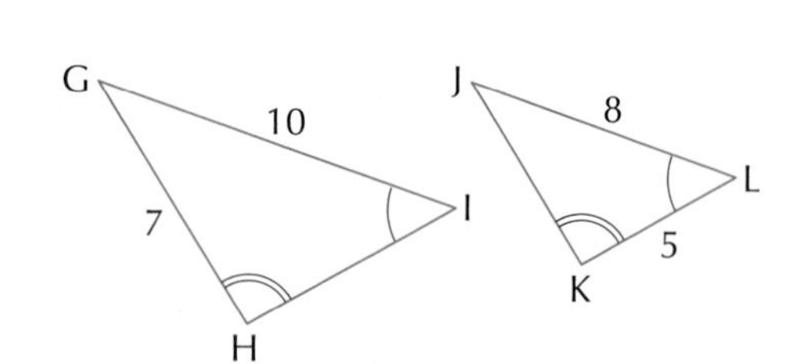

$$\frac{ED}{5} = \frac{16}{10}$$

$$\text{So ED} = \frac{16 \times 5}{10}$$

$$= 8$$

DE = 8 cm

$$\frac{BC}{12} = \frac{10}{16}$$

$$\text{So BC} = \frac{10 \times 12}{16}$$

$$= 7.5$$

BC = 7.5 cm

c

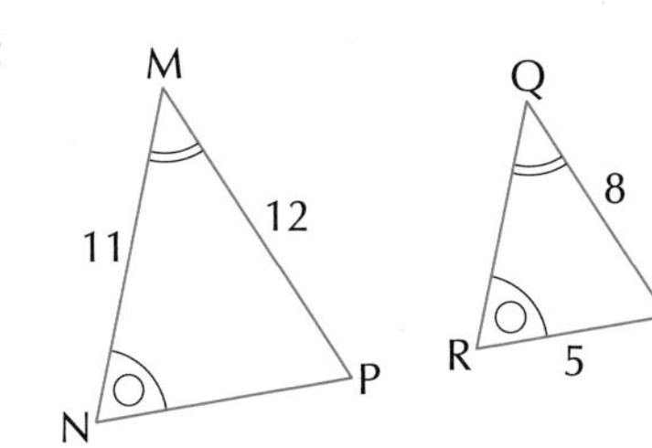

d

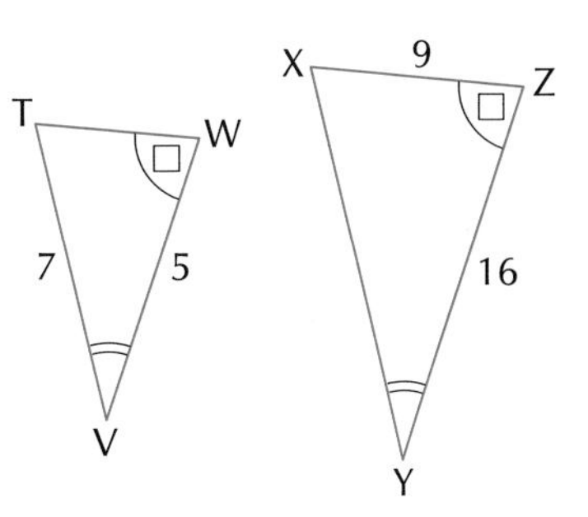

PS **3** A chimney cast a shadow 10 metres long.

At the same time the shadow of a girl 1.1 m tall was 15 cm.

Find the height of the chimney.

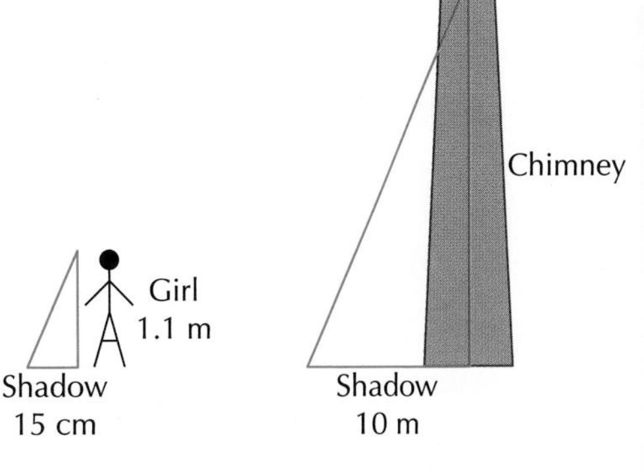

PS **4** A funicular railway goes down a 900 m track that slopes at an angle of 75° to the horizontal.

What is the vertical drop of the descent?

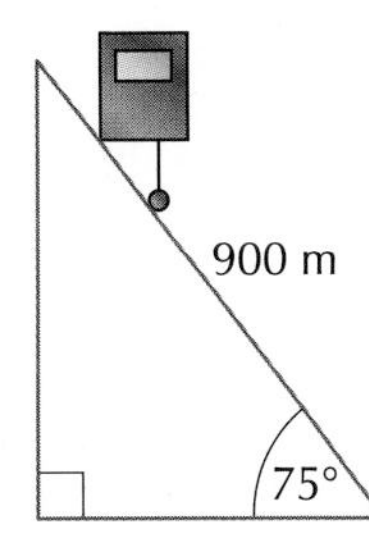

MR **5** Billy is asked to find out how wide a river is.

He places a stone at a point B directly opposite a tree at A on the riverbank on the other side of the river.

He then walks 40 m along the bank of the river to a point C, and estimates that the angle ACB is 20°.

a Explain how Billy could find the width of the river from his figures.

b Use Billy's figures to find the width of the river.

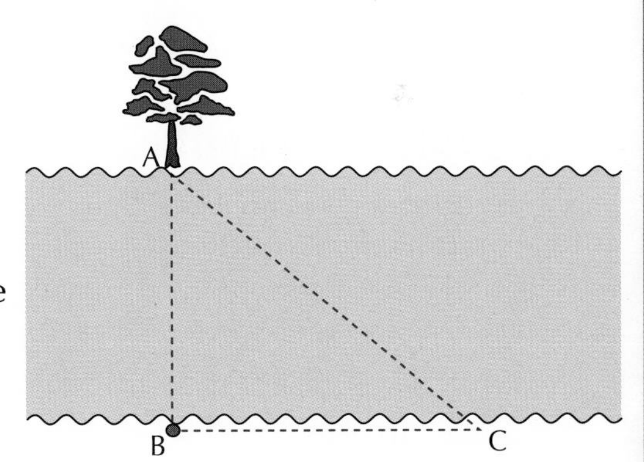

PS **6** A man, who is 160 cm tall, casts a shadow of length 50 cm.

a Work out the angle that the rays of the Sun make with the horizontal at this time.

b At the same time, the length of the shadow of a building is 23 metres. How high is the building?

Investigation

Barnes Wallis and the bouncing bomb

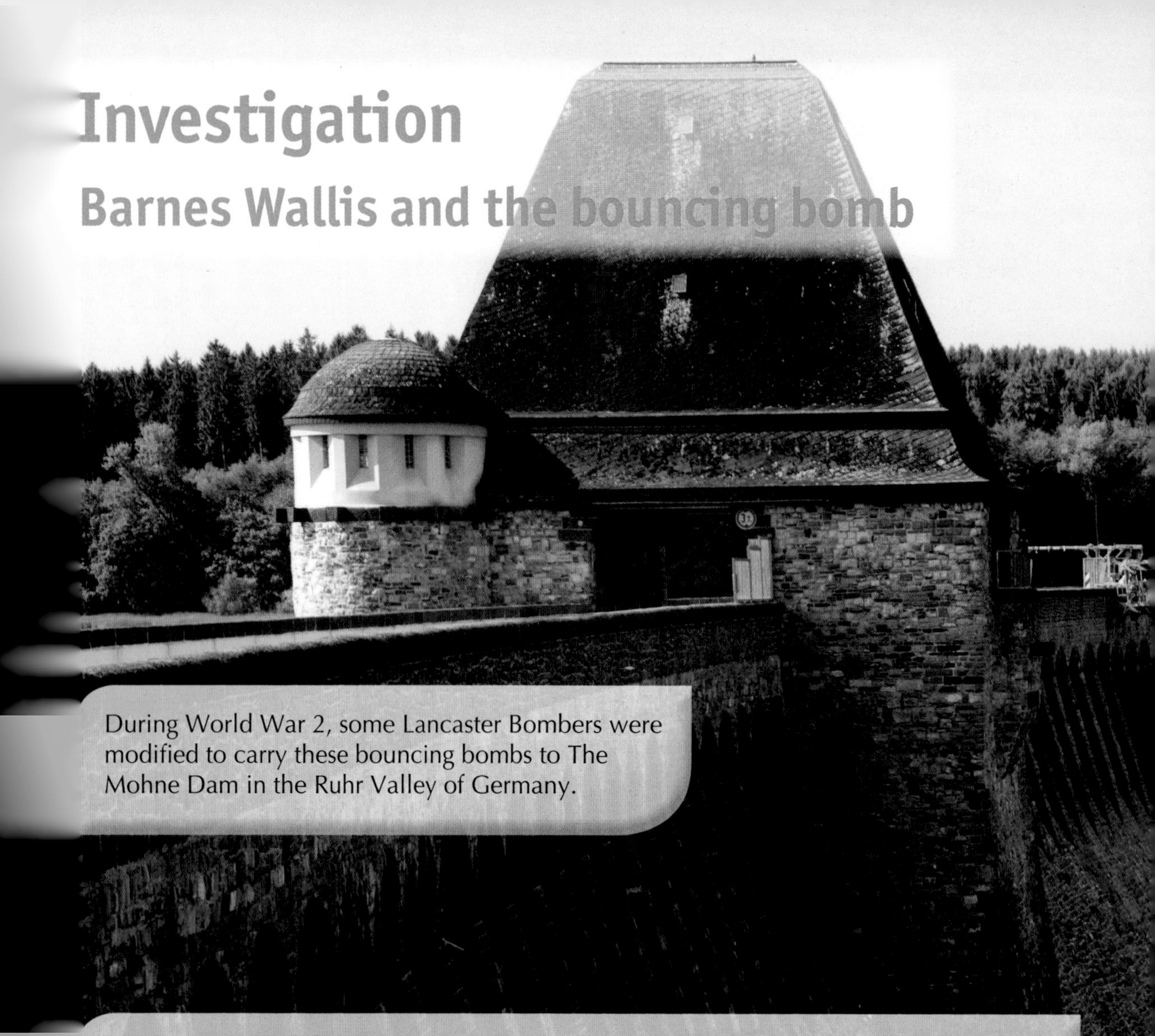

During World War 2, some Lancaster Bombers were modified to carry these bouncing bombs to The Mohne Dam in the Ruhr Valley of Germany.

The converging lights

For the bouncing bomb to work, it had to be released from the aircraft at exactly the right height (60 feet). To achieve this accuracy, two angled spotlights were mounted on the aircraft, one at the front and the other at the rear. When the two spotlights converged, the pilot knew that the aircraft was at exactly the right height.

1. Draw an accurate triangle, based on the diagram above. Make the length of the top line 102 cm and the height 6 cm.
2. Measure the angles between the beams and the underneath of the aircraft. These are the angles at which the lights would have been set.
3. If a different aircraft had been used, that was 80 feet long, at what angle would the lights need to be positioned to identify the same height of 60 ft?
4. Another aircraft was 90 feet long. It had a similar system of lights. The pilot wished to know when the aircraft was exactly 90 ft above a target. At what angle would he need to set the lights, to achieve this?

The bomb sights

Not only had the bomb to be dropped from the aircraft at the correct height, it also had to be at the right distance from the wall of the dam (390 metres). To enable the pilot to get this measurement right, two nails were mounted on a sight guide. The bomb had to be released at the precise moment when the two nails were seen to be exactly in line with the twin turrets of the dam.

5 The top diagram represents the dam turrets at A and B.

a Draw a triangle, to scale, in which the length of AB is 2.2 cm and the distance of C from AB is 3.9 cm.

b Measure the length AC on the triangle you have drawn.

c What was the actual length of AC on the top diagram?

6 The bottom diagram represents the nails at D and E on a sight guide, to help the pilot to line up the aircraft and show him when he is exactly 390 metres away from the wall of the dam.

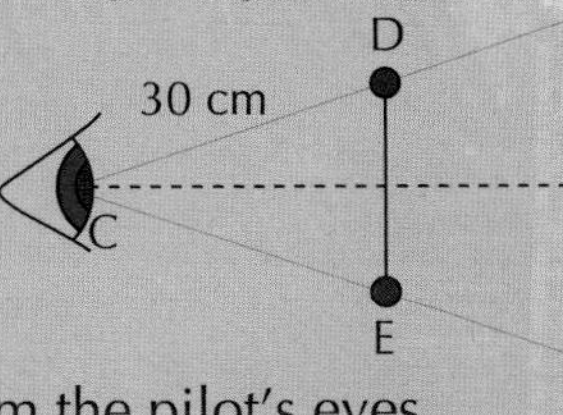

The nails on the sight are 30 cm away from the pilot's eyes.
Use what you know about similar triangles to find how far apart the nails must be set in the sight guide.

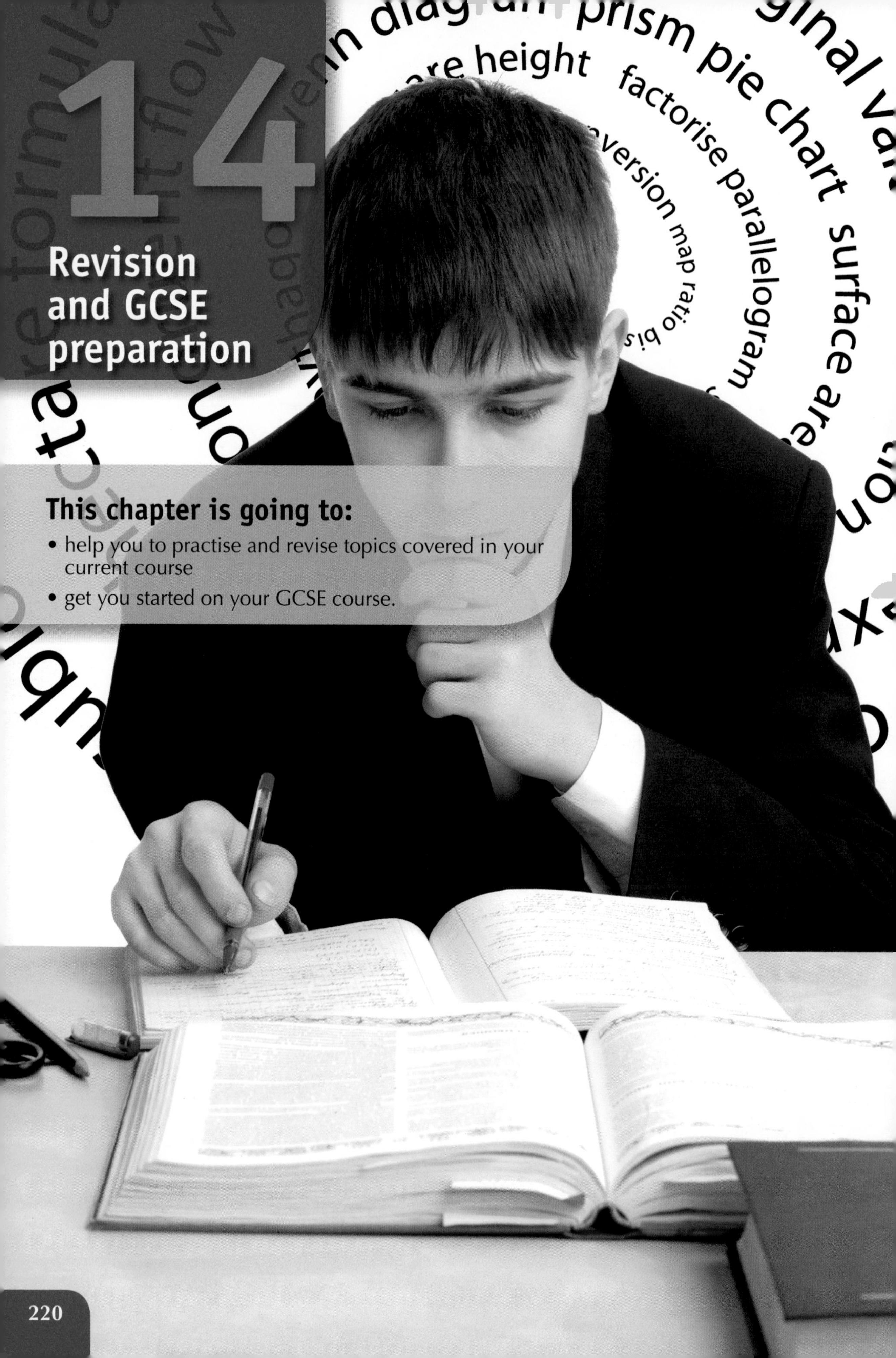

14 Revision and GCSE preparation

This chapter is going to:

- help you to practise and revise topics covered in your current course
- get you started on your GCSE course.

Practice

Practice in fractions, decimals and percentages

Exercise 14A

Do not use a calculator for the first eight questions.

1 How much of each grid is shaded?

Choose the correct answer from the box.

a

b

c

more than half	more than a third	more than a quarter
a half	a third	a quarter
less than half	less than a third	less than a quarter

2 **a** Add 356 to half of 422.

b Take a quarter of 156 from 200.

3 **a** A Scots pine tree is 4.35 m tall. A larch pine is 84 cm taller.
How tall is the larch pine?

b From Barnsley to Sheffield, travelling by the motorway, is 26.45 km. If you use the ordinary roads it is 3.8 km shorter.

How far is it from Barnsley to Sheffield on the ordinary roads?

4 If $\frac{5}{12}$ of the members of a youth club are girls, what fraction are boys?

FS **5** This is the sign at an airport's long-stay car park.
How much would it cost to park at the airport for 9 days?

FLYPARK

£6.50 per day or
£42.50 for a full week.

6 This is a method for working out 12% of 320.

10% of 320 = 32
1% of 320 = 3.2
1% of 320 = 3.2
12% of 320 = 38.4

Use a similar method, or a method of your own, to work out 28% of 480.

7 **a** About 33% of this rectangle is dotted.
About what percentage is:

i striped **ii** plain?

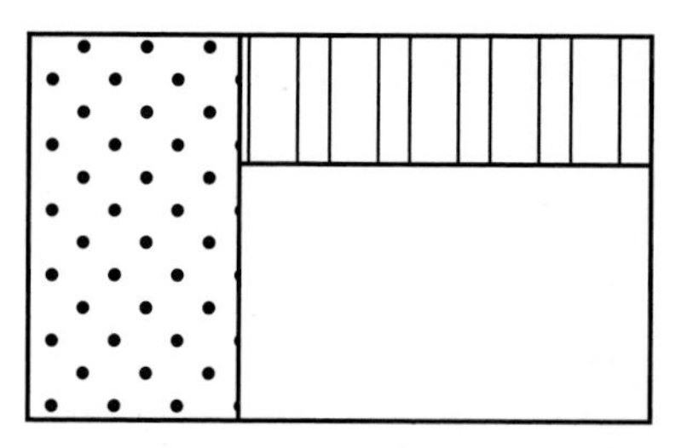

b About $\frac{1}{8}$ of this rectangle is red.
About what percentage is:

i blue **ii** white?

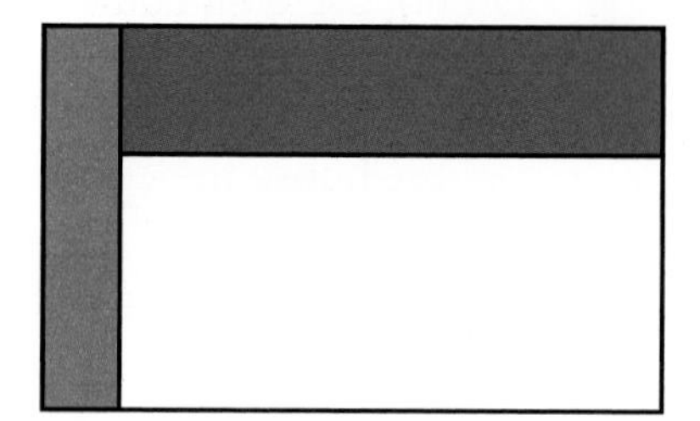

PS **8** Some bathroom scales measure in stones and pounds.
Others measure in kilograms.
The flow diagram shows how to find an approximate equivalent, in kilograms, to a mass in stones and pounds.

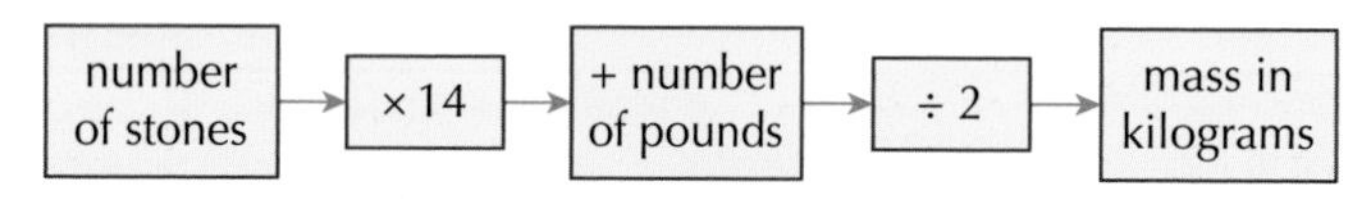

Convert 11 stones 10 pounds to kilograms.

You may use a calculator for the rest of the exercise.

FS **9** The train fare for an adult from Sheffield to London is £164.
A child's fare is 35% less than this. How much is a child's fare?

MR **10** Identify which four of these numbers are equivalent.

0.06 60% 0.6 $\frac{6}{100}$ $\frac{3}{5}$ 6% $\frac{6}{10}$

11 Complete these calculations, giving your answers as fractions.

a $\frac{3}{5} + \frac{1}{3}$ **b** $\frac{5}{9} - \frac{1}{6}$ **c** $2\frac{3}{4} + 1\frac{2}{5}$

FS **12** Jack's Jackets is having a sale.
Calculate the sale price of a jacket that is normally priced at £40.

Practice in the four rules, ratios and directed numbers

Exercise 14B

 Do not use a calculator for the first seven questions.

1 **a** Add together 143 and 328. **b** Subtract 183 from 562.
c Multiply 66 by 4. **d** Divide 132 by 6.

2 **a** Copy these number lines and fill in the missing numbers.

i

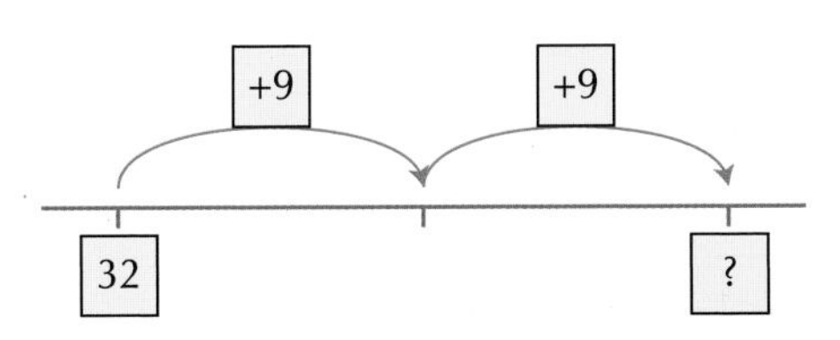

ii

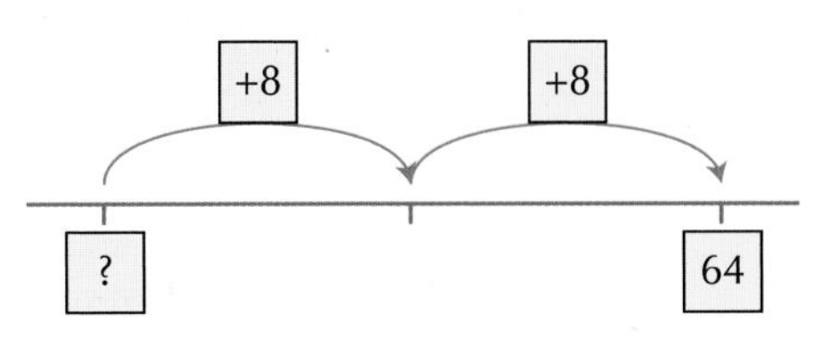

b On this number line, both steps are the same size. How big is each step?

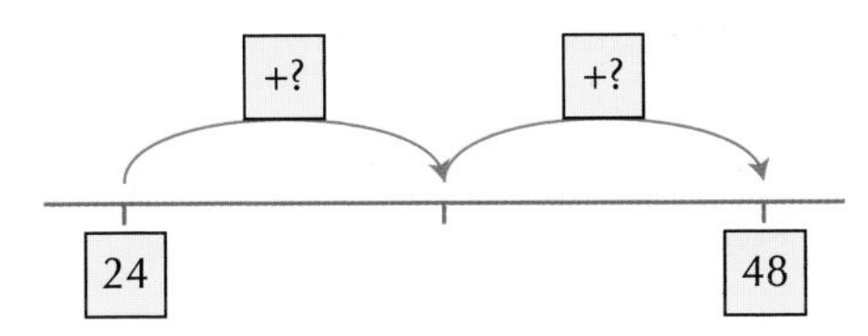

3 Copy each calculation and fill in the missing numbers.

a 780 – ... = 340 **b** ... – 235 = 625 **c** 34 × ... = 272
d 462 = ... + 61 **e** ... ÷ 33 = 9 **f** 568 = 879 – ...

FS **4** **a** Brenda buys fish, chips and mushy peas.

i How much does she pay?
ii How much change does she get from a £10 note?

MR **b** Abdul has £3.25. He wants a burger and chips in a bread bun.
Does he have enough money? Explain your answer.

5 Use +, –, × or ÷ to make each calculation correct.
For example, for 3 ... 7 = 2 ... 5, you could insert '+' and '×' to give 3 + 7 = 2 × 5.

a 9 ... 6 = 20 ... 5 **b** 15 ... 3 = 4 ... 3
c 5 ... 2 = 15 ... 5 **d** 8 ... 4 = 4 ... 2

FS **6** A teacher has 32 pupils in her class. She decides to buy each pupil a pen for Christmas. Each pen costs 98p.

How much will it cost her altogether?

7 Use numbers from the box to answer the questions.

–8 –6 –4 –2 0 1 3 5

a Choose two numbers that have a total of –1.

b What is the total of all the numbers in the box?

c Choose two different numbers from the box to make the lowest possible answer when they are written in these boxes.

□ – □ = …

You may use a calculator for the rest of the exercise.

8 Work out the missing number from each number chain.

a (36) → [+ 5] → [× 12.4] → ( )

b (36) → [– 5] → [÷ 12.4] → ( )

c (36) → [×] → (450)

d (364) → [÷] → (35)

FS **9** Litter bins cost £29 each. A school has a budget of £500 to spend on bins.

How many bins can the school afford?

FS **10** Alf and Bert are paid £48 for doing a job.

They decide to share the money in the ratio 3 : 5.

How much does Alf get?

FS **11** A car company wants to move 700 cars by rail.

Each train can carry 48 cars.

a How many trains will be needed to move the 700 cars?

b It costs £3745 to hire each train. What is the total cost of hiring the trains?

c What is the cost, per car, of transporting them by train?

PS **12** **a** A bus travels 234 miles in 4 hours and 30 minutes.

What is the average speed of the bus?

b A car travelled 270 miles, at an average speed of 60 miles per hour.

How long was the car travelling for? Give your answer in hours and minutes.

Practice in basic rules of algebra and solving linear equations

Exercise 14C

This business card shows how much a plumber charges.

a How much would Ivor charge for a job that lasted 2 hours?

b If Ivor charged £110 for a job, how long did it take him?

Ivor Wrench
Emergency plumber
£30 callout charge
plus £20 per hour

2 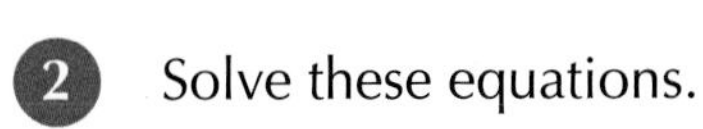
Solve these equations.

a $x + 5 = 7$ **b** $3x = 12$ **c** $x - 6 = 10$

3 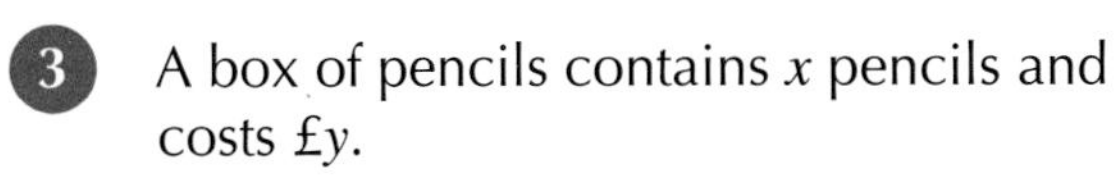
A box of pencils contains x pencils and costs £y.

a How many pencils are there in 6 boxes?

b How much do 5 boxes cost?

c Which expression represents the cost of x boxes of pencils?

i £$(x + y)$ **ii** £xy

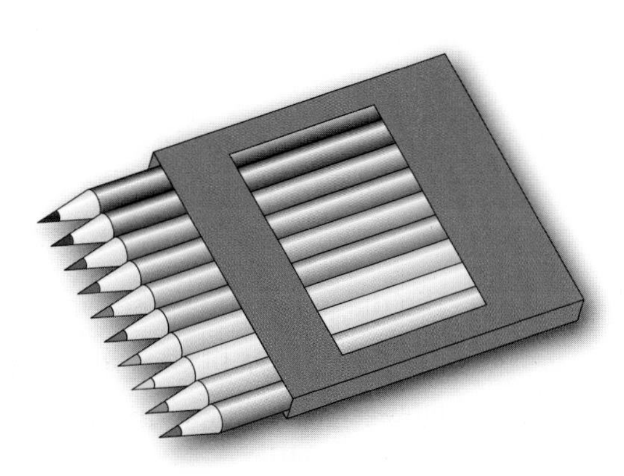

4 **a** What is the next set of coordinates in this sequence?

(2, 1), (4, 3), (6, 5), (8, 7), …

b Explain why the coordinates (29, 28) could not be part of this sequence.

5 **a** Sania has a bag of beads.

Call the number of beads that Sania has in her bag x.

She puts four more beads into the bag.

Write an expression to show the total number of beads in Sania's bag now.

b Dorin has a bag of marbles.

Call the number of marbles that Dorin has in his bag y.

He drops two of the marbles out of his bag and loses them.

Write an expression to show the total number of marbles in Dorin's bag now.

6 Three children, Ali, Billie and Charlie have masses, in kilograms, represented by a, b and c.

a Match each algebraic expression with one of the statements below.

$a = 30$ $\quad b = 2a$ $\quad b + c = 75$ $\quad \frac{a + b + c}{3} = 35$

Statement 1: Billie weighs twice as much as Ali.

Statement 2: The mean weight of all three children is 35 kg.

Statement 3: Ali weighs 30 kg.

Statement 4: Billie and Charlie weigh 75 kg together.

b Use the information to work out Billie's weight and Charlie's weight.

7 The diagram shows a square with sides of length $(n + 4)$ cm.

The square has been split into four parts. The area of one part is shown.

a Copy the diagram and write a number or an algebraic expression in each part.

b Write down a simplified expression for the total area of the square.

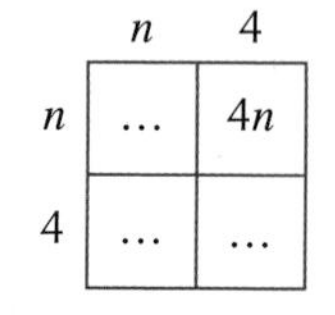

8 Expand the brackets and simplify each expression, if possible.

a $4(x - 5)$ **b** $3(2x + 1) + 5x$ **c** $3(x - 2) + 2(x + 4)$

d $5(3x + 4) + 2(x - 2)$ **e** $4(2x + 1) - 3(x - 6)$

9 **a** Work out the value of each expression, when $x = 4$ and $y = 6$.

i $3x + 9$ **ii** $4x - y$ **iii** $2(3x + 2y + 1)$

b Solve each equation to find the value of z.

i $5z + 9 = 24$ **ii** $\frac{z - 8}{2} = 7$ **iii** $5z + 9 = 3z + 7$

10 Selma is revising algebra.

Selma says: 'I am thinking of a number. If you multiply it by 6 and add 3 you get an answer of 12.'

Call Selma's number x and form an equation. Then solve the equation.

Practice in graphs

Exercise 14D

You will need graph paper or centimetre-squared paper for this exercise.

When you need to draw a graph, number both axes from 0 to 6.

1 Emma and Shehab are playing a game.

Emma has to make a line of four crosses, like this, to win.

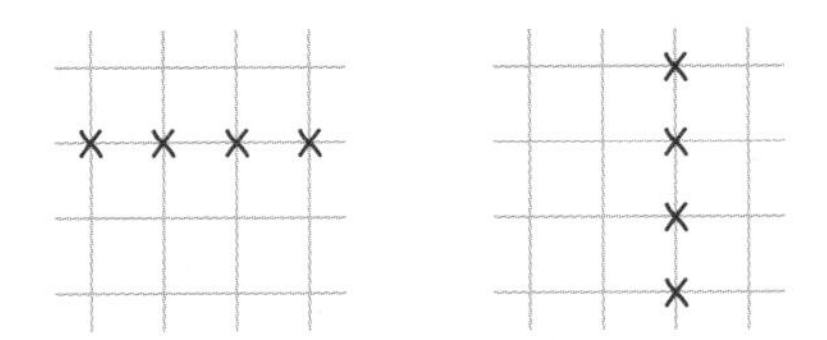

a Copy this grid and place one more cross to make a winning line for Emma.

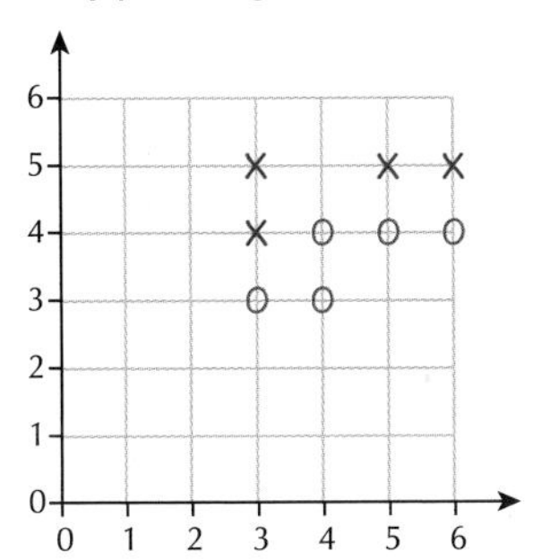

b Write down the coordinates of the four crosses in Emma's winning line.

c Look at the numbers in the coordinates. What do you notice?

2 **a** The point M is halfway between points A and C.

What are the coordinates of M?

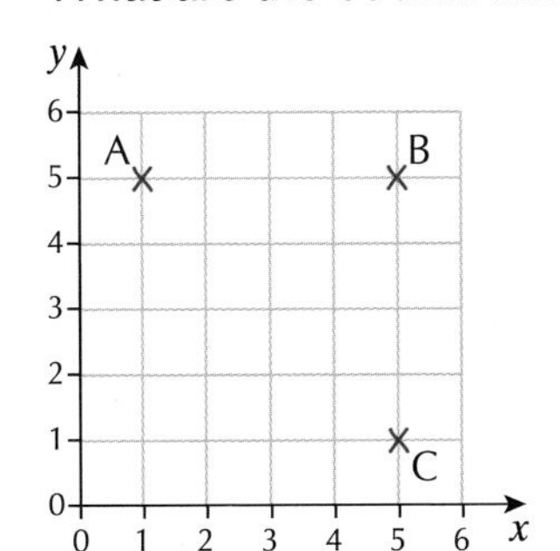

b Shape ABCD is a square.

What are the coordinates of the point D?

3 The graph shows points A, B and C.

a What are the coordinates of A and B?

b ABCD is a rectangle. What are the coordinates of D?

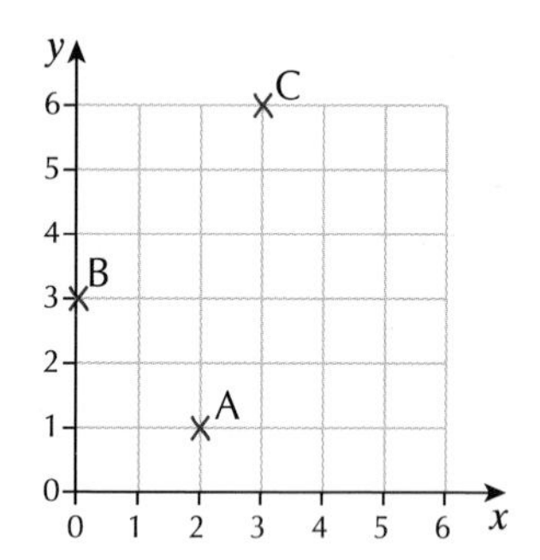

4 The graph shows the line $y = 3$.

Copy the diagram. Add these lines to your diagram.

a $y = 5$ **b** $x = 4$

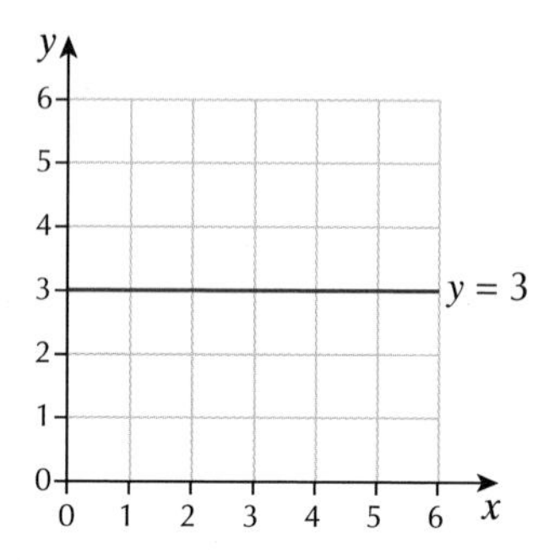

5 Each of the lines labelled l_1, l_2, l_3 and l_4 represents one of the equations in the list.

Match each line to its equation.

a $y = 2$ **b** $y = x$ **c** $x = -3$ **d** $y = -x$

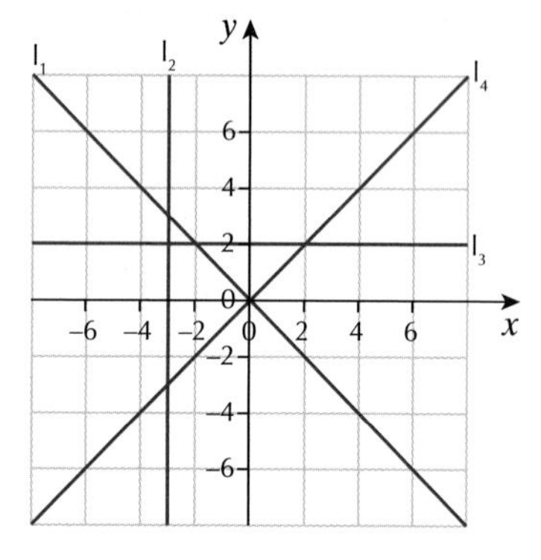

6 The distance–time graph shows the journey of a jogger on a 5-mile run. At one point she stopped to admire the view and at another point she ran up a steep hill.

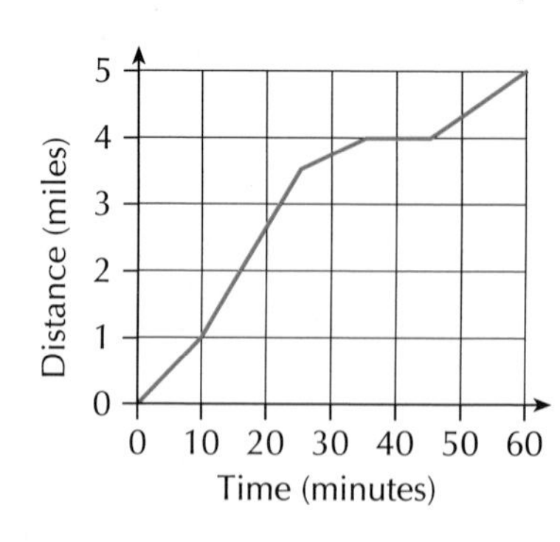

a For how long did she stop to admire the view?

b What distance into her run was the start of the hill?

7 Draw and label a graph for each equation.

Use x-values from –3 to 3.

a $y = 2x + 1$ **b** $y = x - 1$ **c** $x + y = 3$

8 Does the point (20, 30) lie on the line $y = 2x - 10$?

Explain your answer.

9 For every point on the graph of $x + y = 6$, the x-coordinate and the y-coordinate add up to 6.

a Which of these points lie on the line?

i (3, –3) **ii** (6, 0) **iii** (–7, –1) **iv** (–1, 7)

b On a grid, draw the graph of $x + y = 6$. Take x-values from 0 to 6.

Practice in geometry and measures

Exercise 14E

1 For each of these shapes, write down:

i the number of lines of symmetry.

ii the order of rotational symmetry.

a

b

c

2 **a** Write down the name of the quadrilateral being described.

i It has four right angles. It has two lines of symmetry.

ii It has one pair of equal angles. It has two pairs of equal sides.

iii It has no lines of symmetry. It has rotational symmetry of order 2.

b Complete these sentences to describe a rhombus.

i It has … equal sides.

ii It has … lines of symmetry.

iii It has rotational symmetry of order … .

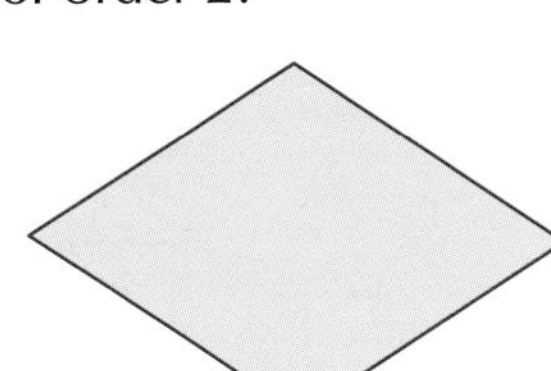

3 **a** Tebor drew this rectangle on a centimetre grid.

What was the area of Tebor's rectangle?

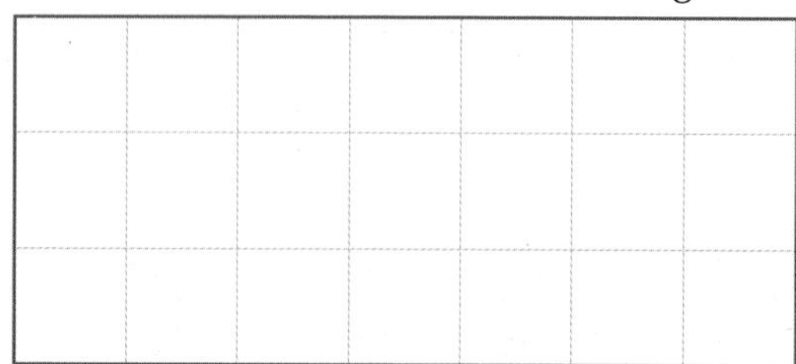

b Tebor cut the rectangle into four triangles, as shown.

What is the area of one of the larger triangles?

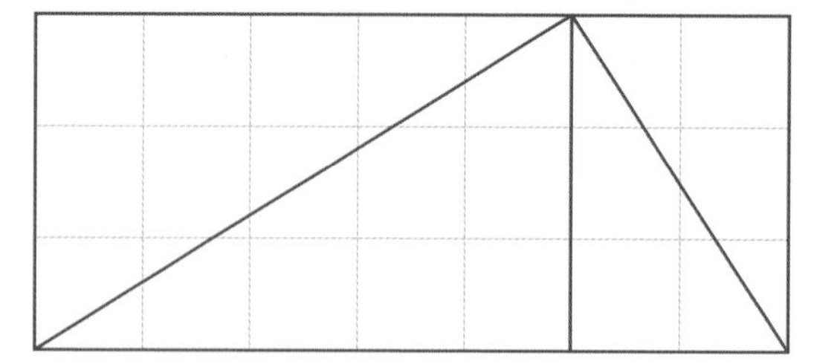

c He put the four triangles together to form a kite.

Explain how to find the area of the kite.

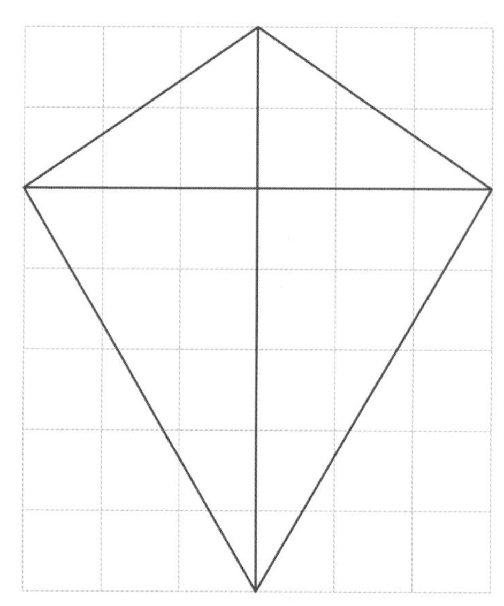

4 **a** Describe the angles labelled *a*–*e* in the diagram, choosing the correct words from the box.

acute	obtuse	reflex	right angle

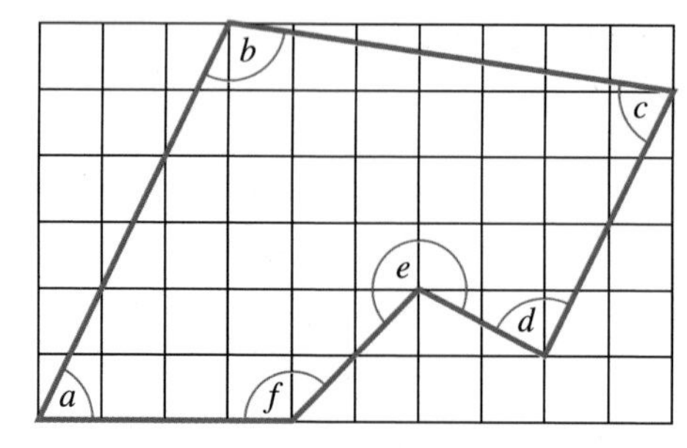

b Is the angle labelled *a* bigger, smaller or the same size as the angle labelled *c*? Explain your answer.

5 **a** Copy and complete the two-way table to show the symmetries of each of these shapes. Shape A has been done for you.

		Number of lines of symmetry				
		0	1	2	3	4
Order of rotational symmetry	1		A			
	2					
	3					
	4					

b Name a quadrilateral that has two lines of symmetry and rotational symmetry of order 2.

6 **a** Construct this triangle accurately.

b Measure the angle at A.

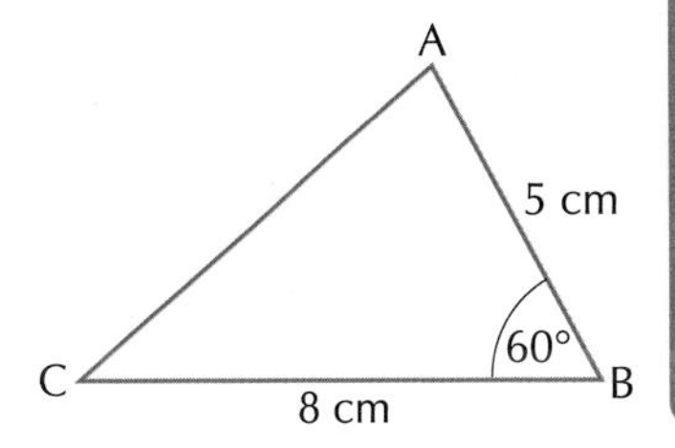

7 Find the values of angles a, b and c in this diagram.

The lines marked with arrows are parallel.

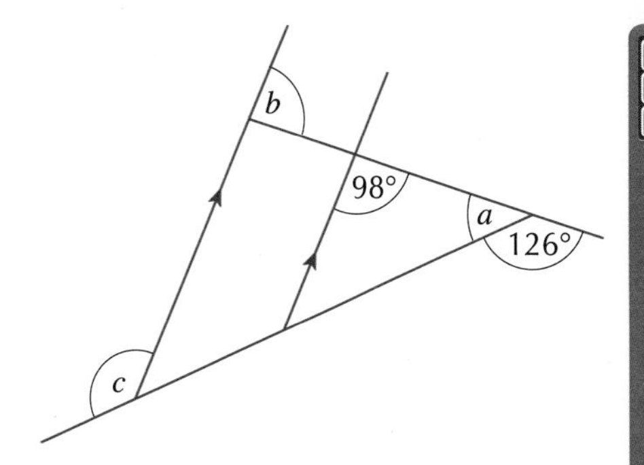

MR **8** This car speedometer shows speed in both miles per hour (mph) and kilometres per hour (km/h). Use it to answer the questions.

a How many kilometres are equivalent to 50 miles?

b Is someone travelling at 100 kph breaking the speed limit of 70 mph? Justify your answer.

c About how many miles is 150 km? Explain your answer.

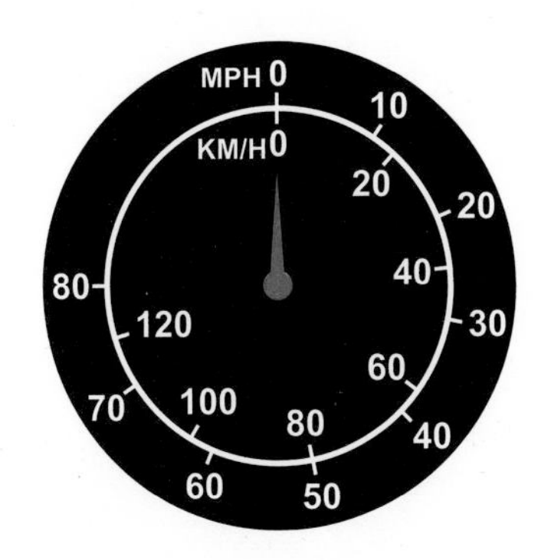

a A rectangle measures 24 cm by 12 cm.

What is its area?

b The rectangle is folded in half several times until it measures 6 cm by 3 cm.

How many times was it folded?

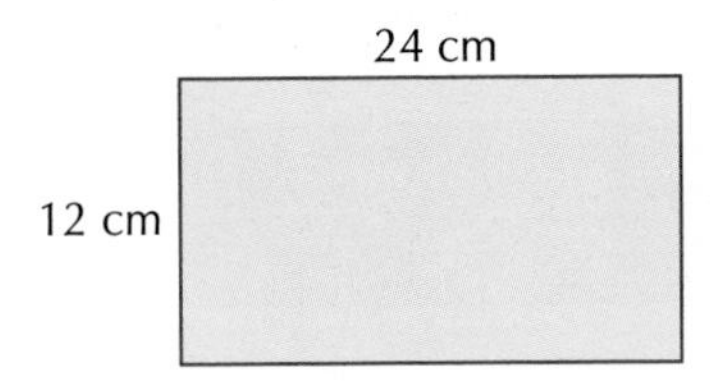

c What is the ratio of the area of the original rectangle to the area of the smaller rectangle?

Give your answer in its simplest form.

Practice in statistics and probability

Exercise 14F

1 This bar chart shows the favourite pets of 80 pupils.

a How many pupils chose a rabbit as their favourite pet?

b How many more pupils preferred a cat to a horse?

c What is the difference between the number of pupils who chose the most popular pet and the number who chose the least popular?

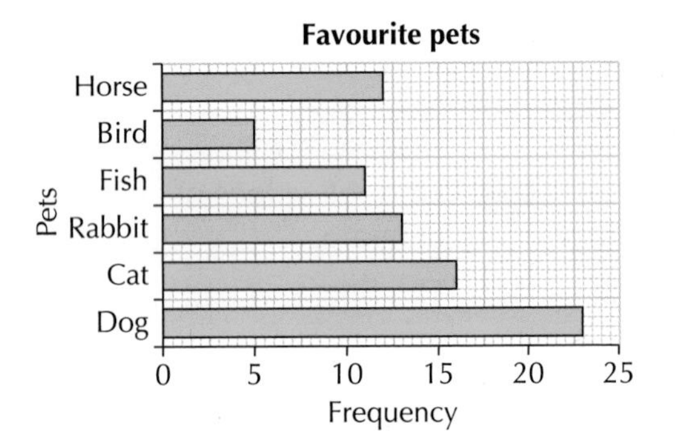

2 This table shows the types and colours of vehicles passing a school between 9:00 am and 10:00 am.

	Red	Black	White	Blue
Lorries	2	6	0	3
Vans	3	1	7	2
Cars	6	5	9	8

a How many white vans passed the school?

b How many lorries passed the school altogether?

c How many more blue vehicles than red vehicles passed the school?

3 **a** Zeenat rolls an ordinary six-sided dice.

What is the probability that the dice shows an even number?

b Zeenat now rolls the dice and tosses a coin. One way that the dice and the coin could land is to show a head and a score of 1. This can be written as (H, 1).

Copy and complete the list below to show all the possible outcomes.

(H, 1), (H, 2), …

c Zeenat rolls the dice and it shows a score of 6. She rolls the dice again. What is the probability that the dice shows a score of 6 this time?

4 Anakin has five cards.

a What is the mode of the numbers on the cards?

b What is the median of the numbers on the cards?

c What is the mean of the numbers on the cards?

d What is the range of the numbers on the cards?

5 Look at the three different spinners, P, Q and R.

P

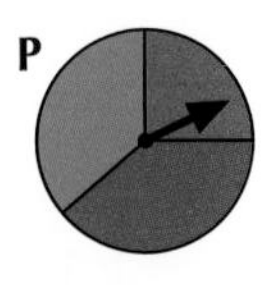

Q

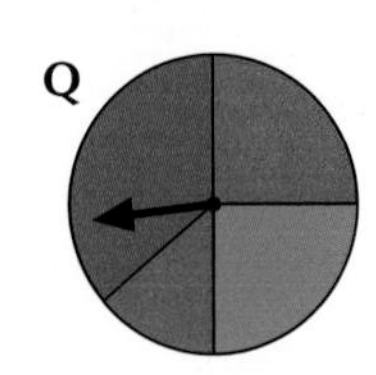

R

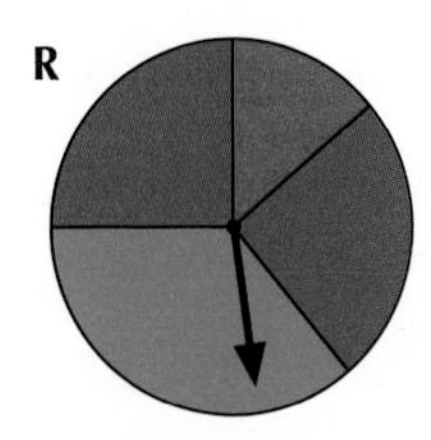

a Which spinner has the greatest chance of landing on red?

b Which spinner has an evens chance of landing on blue?

c Explain how you know which two spinners have an equal chance of landing on green?

6 **a** At a leisure centre, people take part in one of five different sports.

The table shows the percentages of people who played badminton, five-a-side and squash on Saturday.

Copy the pie chart below and label the two sections for badminton and five-a-side.

Sport	Percentage
Badminton	20
Five-a-side	30
Squash	10
Swimming	
Running	

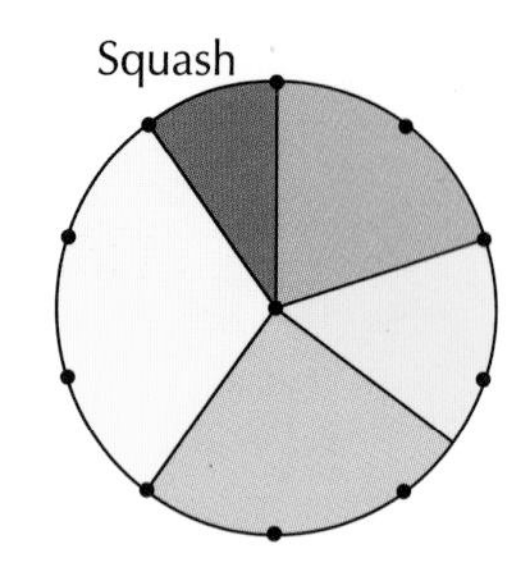

b On this Saturday, more people went swimming than went running.

Use the chart to estimate the percentage of people who:

i went swimming **ii** went running.

c Altogether, 180 people played the different sports on Saturday.

Use this table to find out how many played:

i five-a-side **ii** badminton.

Sport	Percentage	Number of people
Badminton	20%	
Five-a-side	30%	
Squash	10%	18

d Altogether, 180 people played the different sports on Saturday and 280 people played the different sports on Sunday.

30% of the people played five-a-side on Saturday but only 20% of the people played five-a-side on Sunday.

Conrad said: '30% is more than 20%, so more people played five-a-side on Saturday.'

Explain why Conrad is wrong.

7 Omar throws two four-sided dice, each numbered 1, 2, 3, 4. The table shows all his possible total scores.

Score on second dice \ Score on first dice	1	2	3	4
1	2	3	4	5
2	3	4	5	6
3	4	5	6	7
4	5	6	7	8

a When the two dice are thrown, what is the probability that the total score is a square number?

b When the two dice are thrown, what is the probability that the total score is greater than 5?

c **i** Draw a table to show all the possible products if the numbers on each of the dice are multiplied together.

ii What is the probability that the product is a number less than 17?

8 A bag contains only red and blue marbles. A marble is to be taken from the bag at random.

It is twice as likely that the marble will be red as blue. Give a possible number of red and blue marbles in the bag.

PS **9** A ball is taken from a bag at random. The probability that it is black is 0.7. What is the probability that a ball taken at random from the same bag is not black?

Revision

Revision of BIDMAS

You have been using BIDMAS since Year 7. It gives the order in which mathematical operations are carried out in calculations.

Remember that if a calculation is made up of a string of additions and subtractions, multiplications and divisions, you work from left to right, following BIDMAS.

B Brackets

I Indices

DM Division and Multiplication

AS Addition and Subtraction

Example 1

Work out each calculation, using the order of operations given by BIDMAS.

Show each step of your working.

a $10 \div 2 + 3 \times 3$ **b** $10 \div (2 + 3) \times 3$

a First, work out the division and multiplication. $10 \div 2 + 3 \times 3 = 5 + 9$

Then work out the addition. $5 + 9 = 14$

b First, work out the sum inside the brackets. $10 \div (2 + 3) \times 3 = 10 \div 5 \times 3$

Then work from left to right.

Work out the left-hand operation first. $10 \div 5 \times 3 = 2 \times 3$

Then work out the remaining operation. $2 \times 3 = 6$

Example 2

Work these out. **a** $30 - 4 \times 2^2$ **b** $(30 - 4) \times 2^2$

Show each step of the calculation.

a First, work out the power. $30 - 4 \times 2^2 = 30 - 4 \times 4$

Second, work out the multiplication. $30 - 4 \times 4 = 30 - 16$

Finally, do the subtraction. $30 - 16 = 14$

b First, work out what is inside the brackets. $(30 - 4) \times 2^2 = 26 \times 2^2$

Second, the power. $26 \times 2^2 = 26 \times 4$

Finally, the multiplication. $26 \times 4 = 104$

Exercise 14G

1 Use BIDMAS to work these out.

a $3 \times 6 + 7$ **b** $8 \div 4 + 8$ **c** $6 + 9 - 3$

d $15 \div 3 + 7$ **e** $4 \times 6 \div 2$ **f** $3^2 \times 4 + 1$

2 Use BIDMAS to work these out. Remember to work out the expression in brackets first.

a $3 \times (3 + 7)$ **b** $12 \div (3 + 1)$ **c** $(9 + 4) - 4$

d $4 \times (6 \div 2)$ **e** $20 \div (2 + 3)$ **f** $3 + (2 + 1)^2$

3 Use BIDMAS to work these out.

a $16 - 4 \times 2$ **b** $7 \times (4 + 3)$ **c** $12 \div 4 + 8$

d $(18 - 6) \div 4$ **e** $15 \div (4 + 1)$ **f** $12 + 4 \times 5$

g $(24 \div 4) + 7$ **h** $5 + 3^2 \times 2$ **i** $5 \times 4 - 4^2$

j $(3^2 + 1) \times 5$ **k** $4^2 \times (4 - 1)$ **l** $(6 - 1)^2 - 5$

4 Copy each of these calculations and then put in brackets to make them true.

a $4 \times 3 + 7 = 40$ **b** $10 \div 2 + 3 = 2$ **c** $18 \div 3 + 3 = 3$

d $5 - 2 \times 4 = 12$ **e** $20 - 5 \times 2 = 30$ **f** $5 \times 12 - 8 = 20$

g $10 - 2^2 \times 2 = 12$ **h** $10 - 2^2 \times 2 = 2$ **i** $20 - 4^2 \times 5 = 20$

5 Three dice are thrown. They give scores of 2, 4 and 5.

A class makes up these questions with the numbers. Work out the answers.

a $(2 + 4) \times 5 =$ **b** $2 + 4 \times 5 =$ **c** $4^2 + 5 =$

d $4 \times (5 - 2) =$ **e** $4 + 5 - 2 =$ **f** $(4 + 5)^2 =$

Revision of adding and subtracting negative numbers

Negative numbers are used to describe many situations, such as temperatures, distances above and below ground or how much money you have in your bank account.

Example 3

John is overdrawn by £42.56 at the bank. He gets his wages of £189.50 paid in and takes out £30 in cash. How much has he got in the bank now?

An overdrawn amount is negative, so the calculation is:

$-42.56 + 189.50 - 30$

$= 189.50 - 72.56$

$= £116.94$

Example 4

Find the missing number to make each calculation true.

a $10 + \ldots = 7$ **b** $-8 + \ldots = 12$ **c** $-9 - \ldots = 6$

You should be able to work out the answers to these using your knowledge of number facts. If you find this difficult, try visualising a number line or, for more difficult questions, rearrange the equation to find the unknown.

a $\ldots = 7 - 10$
$= -3$

b $\ldots = 12 + 8$
$= 20$

c $-\ldots = 6 + 9$
$= 15$
So $\ldots = -15$

Exercise 14H

1 Copy and complete the balance column in this statement table.

Transaction	Amount paid in (£)	Amount paid out (£)	Balance (£)
Starting balance			64.37
Standing order		53.20	11.17
Cheque	32.00		
Direct debit		65.50	
Cash	20.00		
Wages	124.80		
Loan		169.38	

2 The thermometer shows temperatures in degrees Celsius (°C).
Five temperatures are marked on it.

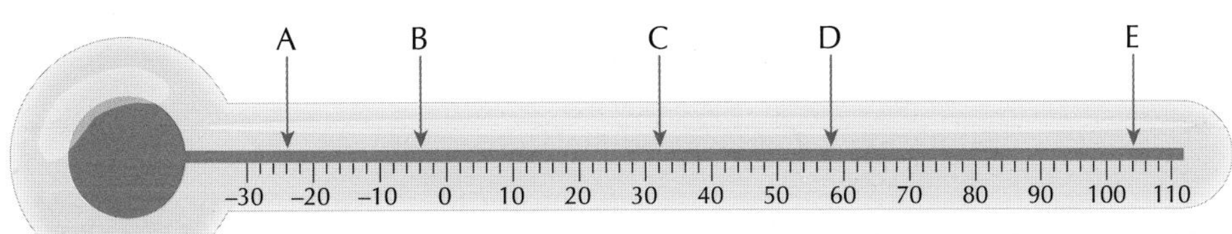

Remember to give your answer as a number of degrees.

Calculate the difference between:

a A and B **b** A and D **c** A and E **d** C and E
e B and E **f** B and D **g** A and C **h** D and E.

3 Calculate the answers.

Hint **Remember to use BIDMAS.**

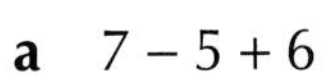

a $7 - 5 + 6$ **b** $6 - 8 - 3$ **c** $-4 - 3 - 6$ **d** $-1 + 3 + 6$
e $2 - (-5)$ **f** $-2 + (-3)$ **g** $-2 + (-4)$ **h** $+5 - (+7)$
i $-3 - -8 + 7$ **j** $+8 - +8 + -2$ **k** $-6 + -6 + +3$ **l** $-8 - -8 + -1$
m $-9 - +2 - -1$ **n** $-45 + 89 - 27$ **o** $+7 - -6 + -1$ **p** $-6 - +5 + -5$

4 Copy these number lines. Fill in the missing numbers.

a

b

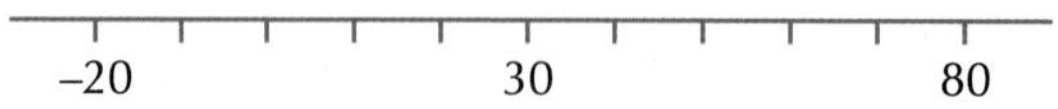

c

d

5 Copy these number sentences. Write in the correct numbers to make them true.

a $3 + -5 = \ldots$ **b** $5 + \ldots = 9$ **c** $5 + \ldots = 2$

d $\ldots - -6 = 4$ **e** $-6 - \ldots = 3$ **f** $+7 - \ldots = 4$

g $-8 + -7 = \ldots$ **h** $\ldots - +4 = 0$ **i** $3 - 4 + \ldots = 6$

Revision of multiples, factors and prime numbers

Example 5

Find the largest number less than 100 that is:

a a multiple of 3 **b** a multiple of 3 and 5.

a This will be a number in the 3 times table that is close to 100.

$30 \times 3 = 90$

$31 \times 3 = 93$

$32 \times 3 = 96$

$33 \times 3 = 99$

$34 \times 3 = 102$

So, the largest multiple of 3 that is less that 100 is 99.

b Because 3 and 5 have no common factors, multiples common to 3 and 5 are multiples of 15.

15, 30, 45, 60, 75, 90, 105, …

So, the largest number under 100 that is a multiple of both 3 and 5 is 90.

Example 6

Find the factors of: **a** 35 **b** 40.

a Find all the products that make 35.

$$1 \times 35 = 35$$
$$5 \times 7 = 35$$

So, the factors of 35 are {1, 5, 7, 35}.

b Find all the products that make 40.

$$1 \times 40 = 40$$
$$2 \times 20 = 40$$
$$4 \times 10 = 40$$
$$5 \times 8 = 40$$

So, the factors are {1, 2, 4, 5, 8, 10, 20, 40}.

Exercise 14I

1 Write down the first five multiples of each number.

a 4 **b** 9 **c** 12 **d** 25

2 Look at this list of of numbers.

3	7	8	13	14	15	18	24	36	39	45	48	64	69	90	120

Write down the numbers from the box that are:

a multiples of 3 **b** multiples of 5

c multiples of 4 **d** multiples of 12.

3 Find the largest number less than 50 that is:

a a multiple of 3 **b** a multiple of 8

c a multiple of 5 and 9 **d** a multiple of 3 and 5.

4 **a** Which of the numbers from 2 to 20 have only two factors?

b What are these numbers called?

5 Write down all the factors of:

a 48 **b** 52 **c** 60

d 75 **e** 100 **f** 130.

6 Copy this grid.

1	2	3	4	5	6
7	8	9	10	11	12
13	14	15	16	17	18
19	20	21	22	23	24
25	26	27	28	29	30
31	32	33	34	35	36
37	38	39	40	41	42
43	44	45	46	47	48
49	50	51	52	53	54
55	56	57	58	59	60

a Shade in, or cross out, the number 1.

b Leave the number 2 blank and then shade in, or cross out, the rest of the multiples of 2.

c Leave the number 3 blank and then shade in, or cross out, the rest of the multiples of 3.

Some of them will have already been shaded in or crossed out.

d Leave the number 5 blank and then shade in, or cross out, the rest of the multiples of 5.

All but three of them will have already been shaded in or crossed out.

e Leave the number 7 blank and then shade in, or cross out, the rest of the multiples of 7.

All but one of them will have already been shaded in or crossed out.

f Explain what type of numbers have been left blank.

Revision of squares, square roots and powers

Example 7

Calculate: **a** 22^2 **b** $\sqrt{289}$ **c** $\sqrt{600}$.

a You can either use the square button on your calculator or calculate 22×22.

$22^2 = 22 \times 22$

$= 484$

b Use the square root button on your calculator.

$\sqrt{289} = 17$

c Use the square root button on your calculator.

$\sqrt{600} = 24.5$, correct to one decimal place.

Example 8

Calculate 7^4.

Use the power button on your calculator.

$7^4 = 2401$

Remember that $7^4 = 7 \times 7 \times 7 \times 7$.

Exercise 14J

1 Write down the value of each number.

a 7^2 **b** 9^2 **c** 11^2 **d** 13^2 **e** 15^2

f 19^2 **g** 24^2 **h** 25^2 **i** 32^2 **j** 53^2

2 Calculate each square root. Round your answers to two decimal places where necessary.

a $\sqrt{36}$ **b** $\sqrt{64}$ **c** $\sqrt{100}$ **d** $\sqrt{144}$ **e** $\sqrt{196}$

f $\sqrt{40}$ **g** $\sqrt{80}$ **h** $\sqrt{120}$ **i** $\sqrt{500}$ **j** $\sqrt{800}$

3 Calculate the value of each expression.

a 4^5 **b** 12^3 **c** 13^4 **d** 21^3

4 $\sqrt{2} = 1.4142136$ $\sqrt{20} = 4.472136$ $\sqrt{200} = 14.142136$ $\sqrt{2000} = 44.72136$

Use this pattern to write down the value of:

a $\sqrt{20000}$ **b** $\sqrt{200000}$ **c** $\sqrt{2000000}$.

5 **a** Work out the value of:

i 1^4 **ii** 1^9 **iii** $(-1)^3$ **iv** $(-1)^5$.

b Explain how you know what $(-1)^{55}$ is.

Revision of decimals in context: addition and subtraction

Think where you may have seen decimal numbers recently.

The most obvious example of decimals in context is money. Prices in shops are usually given to two decimal places.

Speedometers in cars and weighing scales usually have digital displays involving decimal numbers.

Nutritional information on food packets, such as cereals, often involves decimal numbers.

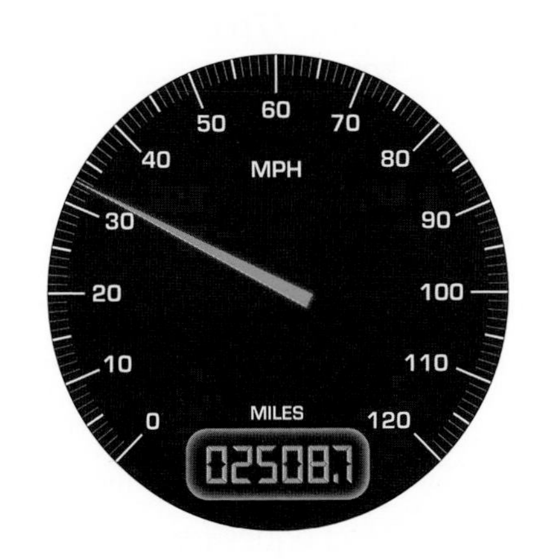

Nutritional information per 100 g		
ENERGY	16 kj	280 kcal
PROTEIN		4 g
CARBOHYDRATES		90 g
of which sugars		40 g
starch		50 g
FAT		0.7 g
of which saturates		0.2g
FIBRE		0.9 g
SODIUM		0.45 g
VITAMINS:		(% RDA)
VITAMIN D µg	4.2	(85)
TIAMIN (B_1) mg	1.2	(85)
RIBOFLAVIN (B_2) mg	1.3	(85)
NIACIN mg	15.0	(85)

Example 9

Over the course of the year, Mr Key's gas bills were £125.23, £98.07, £68.45 and £102.67. What was the total cost of Mr Key's gas for the year?

This is a straightforward addition problem.

$$\begin{array}{r} £125.23 \\ £98.07 \\ £68.45 \\ +\ £102.67 \\ \hline £394.42 \end{array}$$

Example 10

Asif earns £2457.82 in a month. From this, £324.78 is deducted for tax, £128.03 for national insurance and £76.54 for other deductions. How much does Asif take home each month?

This is a subtraction problem. The easiest method to solve it is to add up all the deductions and then subtract from his total pay.

Deductions

$$\begin{array}{r} £324.78 \\ £128.03 \\ +\ £76.54 \\ \hline £529.35 \end{array}$$

Take-home pay

$$\begin{array}{r} £2457.82 \\ -\ £529.35 \\ \hline £1928.47 \end{array}$$

Exercise 14K

1 A businesswoman pays five cheques into her bank account. The cheques are for £1456.08, £256.78, £1905.00, £46.89 and £694.58.

How much did she deposit, in total?

2 At the local shop Mary bought two tins of soup costing 77p each, a packet of sugar costing 88p, a loaf of bread costing £1.15, a packet of bacon costing £2.66 and a box of chocolates costing £4.99.

What was her total bill?

3 Five books are placed on top of one another.

The books are 2.3 cm, 15 mm, 3.9 cm, 3.4 cm and 18 mm thick.

What is the total thickness of the pile of books? Give your answer in centimetres.

4 The ingredients of a cake were 132 g of butter, 0.362 kg of flour and 96 g of sugar. What is the total mass of these ingredients? Give your answer in kilograms.

5 Misha's bank account had £467.92 in it. She wrote cheques for £67.50, £42.35 and £105.99.

How much money is left in Misha's account after these cheques have been cashed?

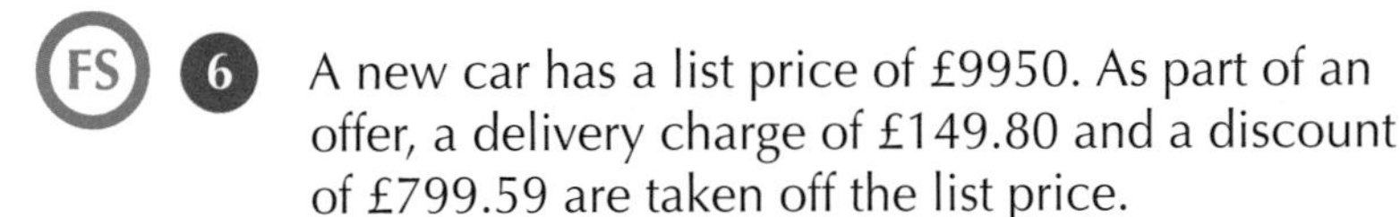

FS **6** A new car has a list price of £9950. As part of an offer, a delivery charge of £149.80 and a discount of £799.59 are taken off the list price.

How much will a customer pay for the car?

PS **7** A quadrilateral has a perimeter of 32 cm. The lengths of three of the sides are 8.2 cm, 3.5 cm and 12.8 cm.

What is the length of the fourth side?

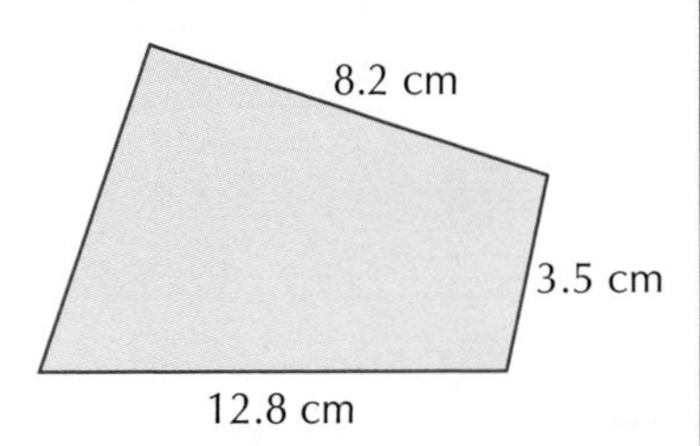

FS **8** Mr Ball's payslip shows that he is paid a basic wage of £456.78 each week. In addition to his basic wage, he gets a bonus of £152.45.

He has £77.82 tax, £38.80 national insurance and £7.78 health insurance deducted from his pay.

How much does Mr Ball take home each week?

Revision of decimals in context: multiplication and division

Example 11

One chair costs £45.76 and a table costs £123.47.

How much is a dining suite consisting of six chairs and a table?

This is a multiplication and addition problem.

Chairs cost $\begin{array}{r} 45.76 \\ \times\ 6 \\ \hline 274.56 \end{array}$ Total cost $\begin{array}{r} 123.47 \\ +\ 274.56 \\ \hline 398.03 \end{array}$

Hence the total cost is £398.03.

Example 12

Four litres of petrol and a can of oil together cost £8.95.

If the can of oil costs £2.59, how much does one litre of petrol cost?

This is a subtraction and division problem.

Petrol costs $\begin{array}{r} 8.{}^{8}\not{9}{}^{1}5 \\ -\ 2.59 \\ \hline 6.36 \end{array}$ $\begin{array}{r} 1.59 \\ 4\overline{)6.{}^{2}3{}^{3}6} \end{array}$

Hence one litre of petrol costs £1.59.

Exercise 14L

1 Work these out.

a	**b**	**c**	**d**
£17.80 × 6 = ☐	£6.07 × 12 = ☐	£76.32 × 25 = ☐	£18.95 × 54 = ☐

You may use your calculator for the rest of the exercise.

2 A packet of four AA batteries costs £4.15.

How much money would you need to buy nine packets of four AA batteries?

3 John bought five tins of cocoa, costing £1.12 each, and seven jars of coffee costing £2.09 each.

What was his total bill?

4 To make some shelves, Uncle George orders seven pieces of wood each 53.4 cm in length and two pieces of wood each 178.5 cm in length.

What is the total length of wood that Uncle George ordered?

5 A man earns £27 746.40 a year.

How much does he earn each month?

6 A crystal decanter costs £56.32 and a crystal wine glass costs £11.58.

How much will a decanter and a set of six wine glasses cost?

7 A table and four chairs are advertised for £385.

If the table costs £106, how much does each chair cost?

8 A holiday for two adults and three children costs £967.80 in total.

If the cost per child is £158.20, what is the cost for each adult?

Revision of long multiplication and long division in real-life problems

Example 13

When he checks his running diary, Paul finds that he has run an average of 65 miles a week during the last year.

How many miles did he run in the year altogether?

You need to identify that this is a multiplication problem, then recall that there are 52 weeks in a year. Finally, you must decide which method you are going to use.

The multiplication has been done here.

$$\begin{array}{r} 65 \\ \times\ 52 \\ \hline 130 \\ 3250 \\ \hline 3380 \\ \hline \end{array}$$

So Paul has run a total of 3380 miles.

Example 14

Mr Swingler buys a car for £36480. He agrees to pay for it in 24 equal, monthly instalments.

How much does he pay each month?

First you need to identify that this is a division problem.

The calculation is done here.

```
   1520
24)36480
   24
   124
   120
     48
     48
```

The answer is £1520 per month.

Exercise 14M

Show your working for each of your answers.

1 A typist can type 54 words per minute, on average.
How many words can he type in 15 minutes?

2 There are 972 pupils in a school. Each tutor group has 27 pupils in it.
How many tutor groups are there?

3 In a road-race, there were 2200 entrants.

- **a** To get them to the start, the organisers used a fleet of 52-seater buses.
  How many buses were needed?
- **b** The race was 15 miles long and all the entrants completed the course.
  How many miles, in total, did all the runners cover?

4

- **a** A cinema has 37 rows of seats. Each row contains 22 seats.
  How many people can sit in the cinema altogether?
- **b** Tuesday is 'all seats one price' night. There were 220 customers who paid a total of £1232.
  What was the cost of one seat?

5 A library gets 700 books to distribute equally among 12 local schools.

- **a** How many books will each school get?
- **b** The library keeps any books left over. How many books is this?

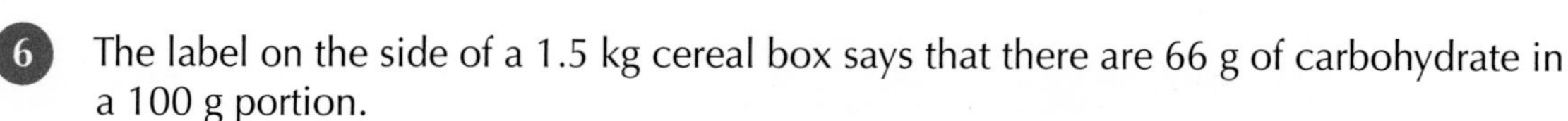

6 The label on the side of a 1.5 kg cereal box says that there are 66 g of carbohydrate in a 100 g portion.

How many grams of carbohydrate will Dan consume, if he eats the whole box at once?

7 Twelve members of a running club hire a minivan to do the Three Peaks race (climbing the highest mountains in England, Scotland and Wales).

The van costs £25 per day plus 12p per mile. The van uses a litre of petrol for every 6 miles travelled. Petrol costs £1.39 per litre. The van is hired for 3 days and the total mileage covered is 1500.

a How much does it cost to hire the van?

b How many litres of petrol are used?

c If they share the total cost equally, how much does each member pay?

Revision of geometry

You should be familiar with all the formulae and terms that you have met so far.

This section provides a checklist.

Perimeter and area

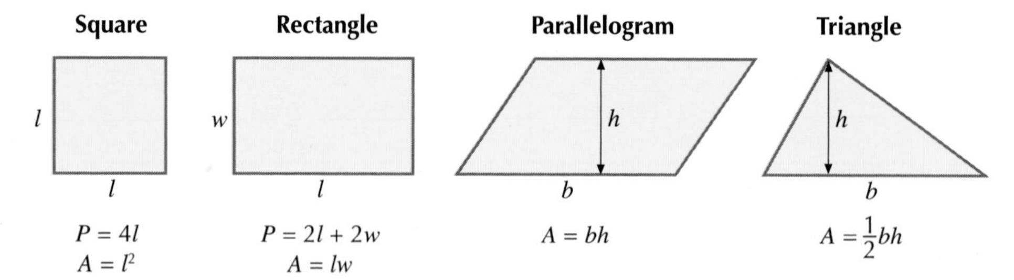

Remember that the metric units for perimeter are the same as for length:

- millimetres (mm), centimetres (cm) and metres (m).

Remember that the metric units for area are:

- square millimetres (mm^2), square centimetres (cm^2) and square metres (m^2).

Volume and surface area

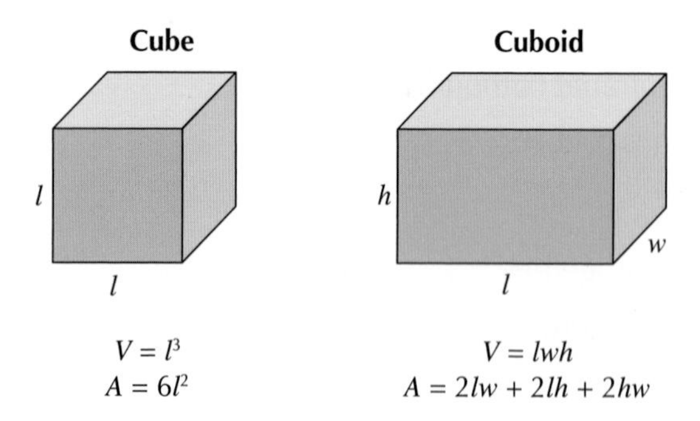

Remember that the metric units for volume are:

- cubic millimetres (mm^3)
- cubic centimetres (cm^3)
- cubic metres (m^3).

Exercise 14N

1 For each rectangle, work out:

i the perimeter **ii** the area.

a

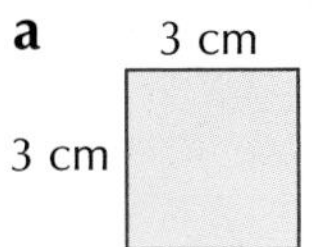

b

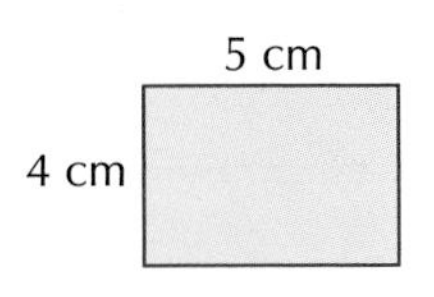

c

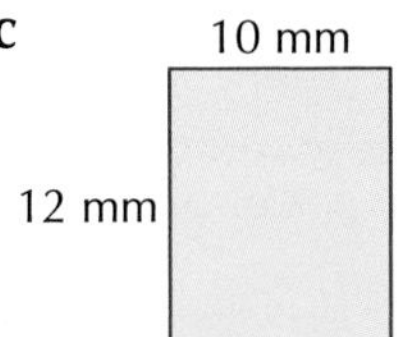

d

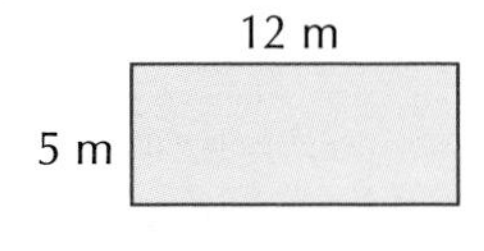

2 Work out the area of each triangle.

a

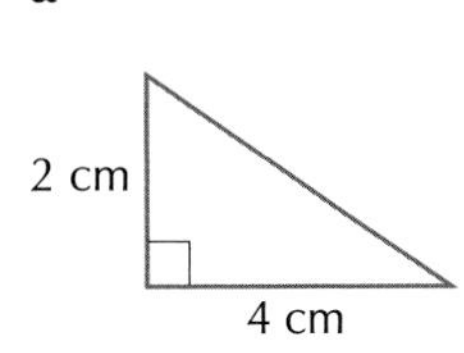

b

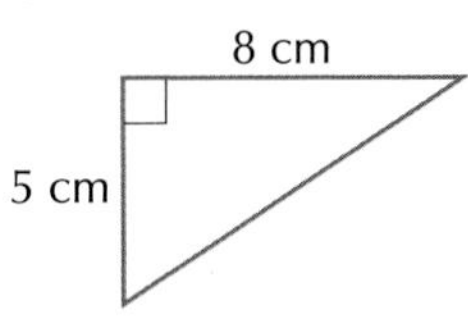

c

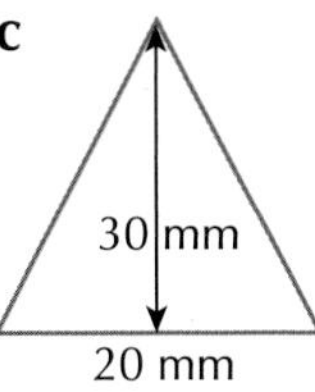

d

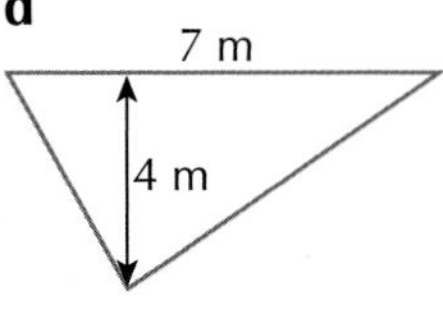

3 Work out the area of each parallelogram.

a

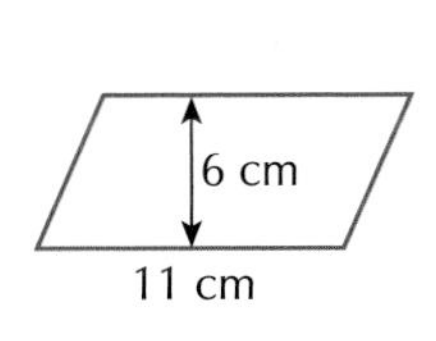

b

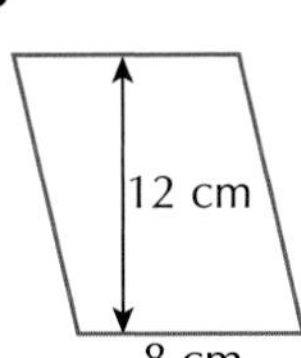

c

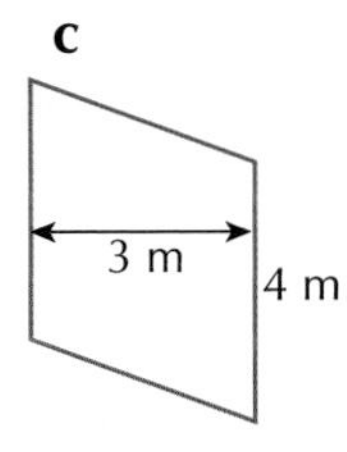

d

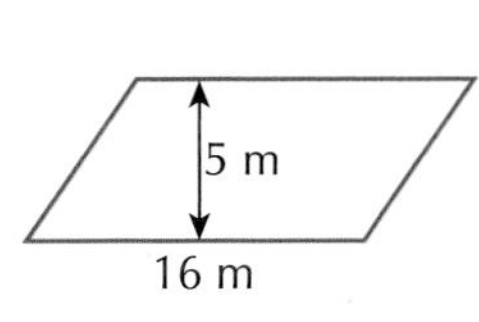

4 For each cuboid, work out:

i the surface area **ii** the volume.

a

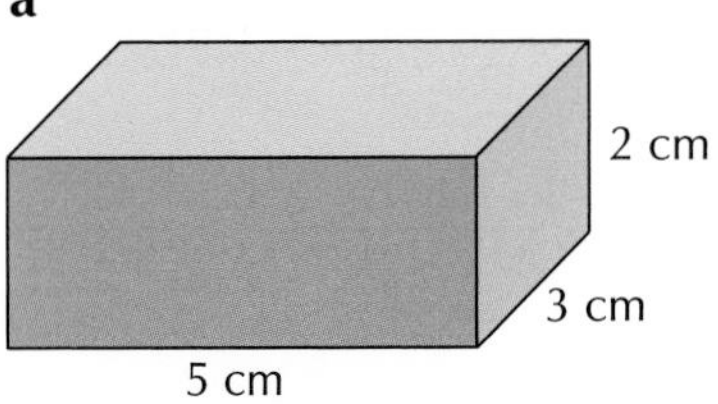

b

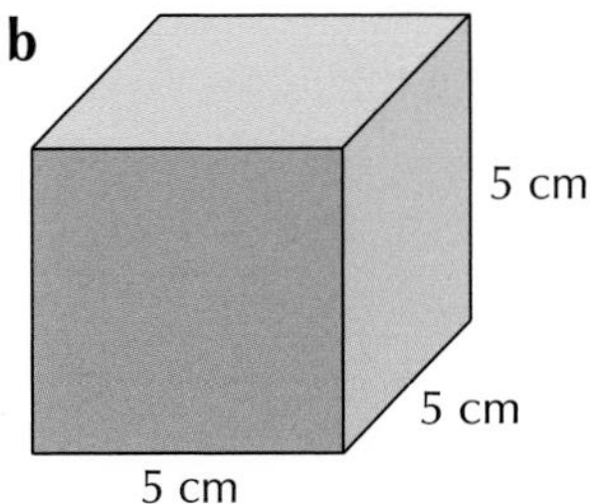

c

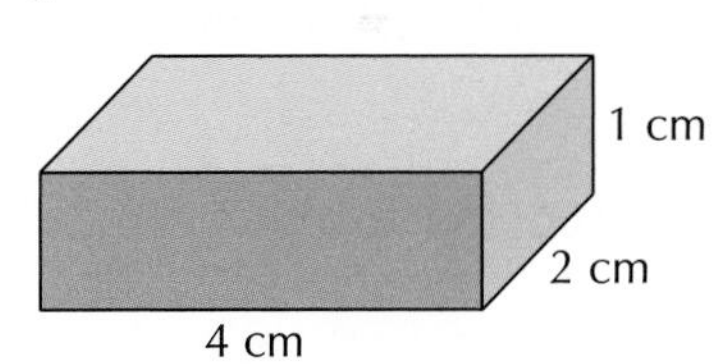

PS **5** Calculate the area of the square drawn on the centimetre grid.

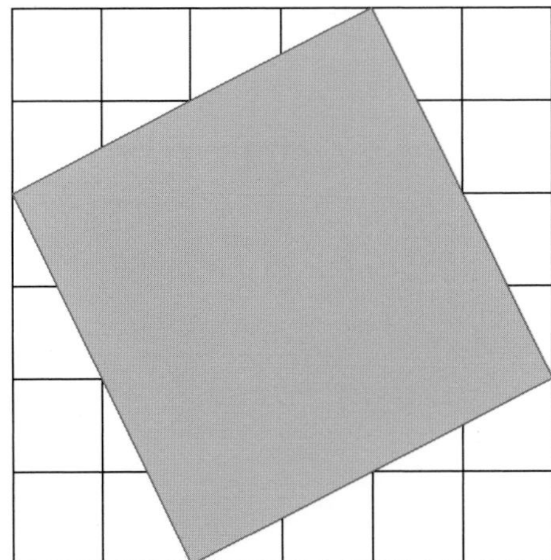

Revision of symmetry

There are two types of symmetry: reflective symmetry and rotational symmetry.

Some 2D shapes have both types of symmetry, while some have only one type.

Reflective symmetry

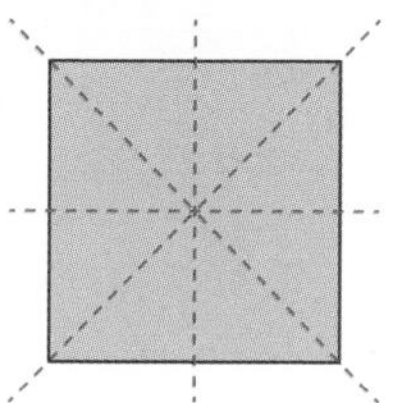

A square has four lines of symmetry.

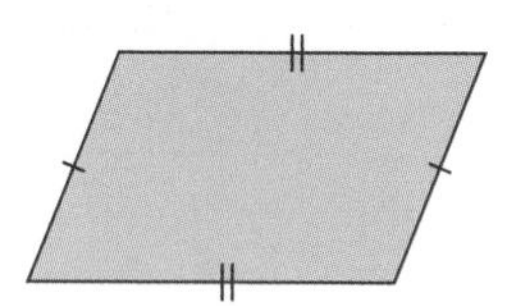

A parallelogram has no lines of symmetry.

Remember that you can use tracing paper or a mirror to find the lines of symmetry of a shape.

Rotational symmetry

A 2D shape has rotational symmetry when it can be rotated about a point to look exactly the same in its new position.

The order of rotational symmetry is the number of different positions in which the shape looks the same when rotated about the point.

All 2D shapes have rotational symmetry of order 1 or more.

A square has rotational symmetry of order 4.

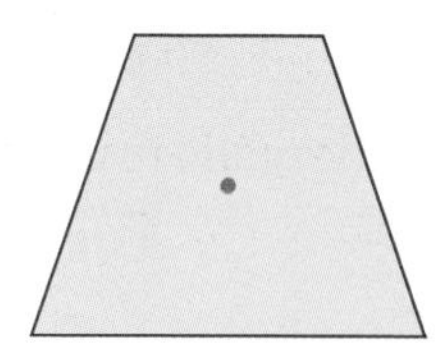

This trapezium has rotational symmetry of order 1.

Remember that you can use tracing paper to find the order of rotational symmetry of a shape.

Exercise 140

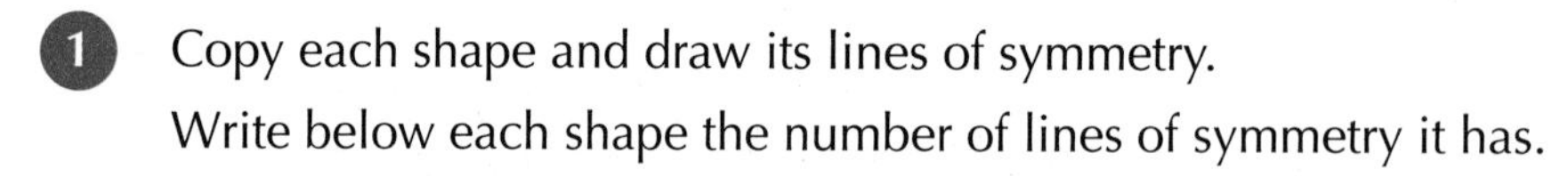

1 Copy each shape and draw its lines of symmetry.
Write below each shape the number of lines of symmetry it has.

a 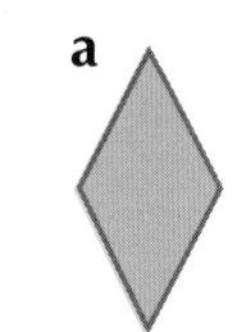

b 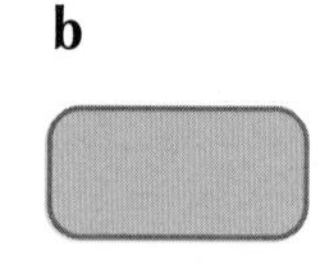

c 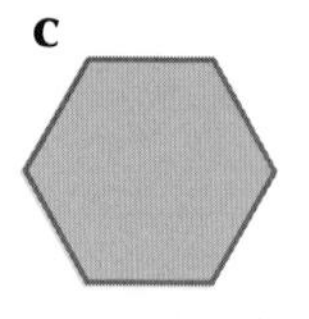

d 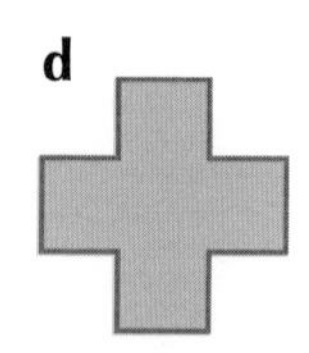

e 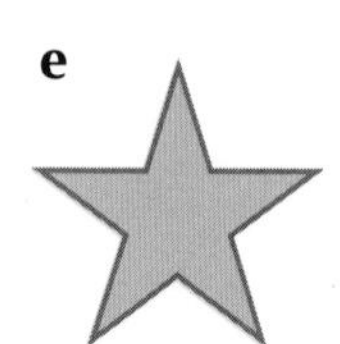

2 Write down the number of lines of symmetry for each shape.

a 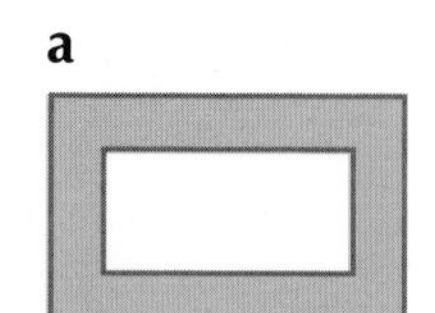

b 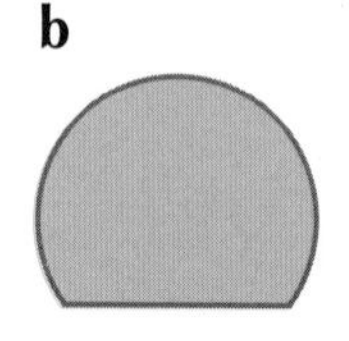

c 

d 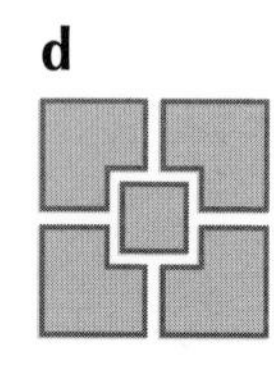

3 Copy each diagram. Write down its order of rotational symmetry.

a
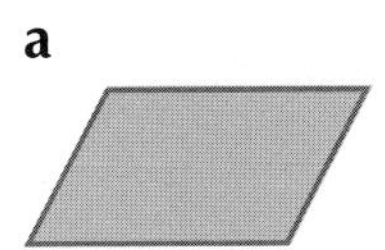
b
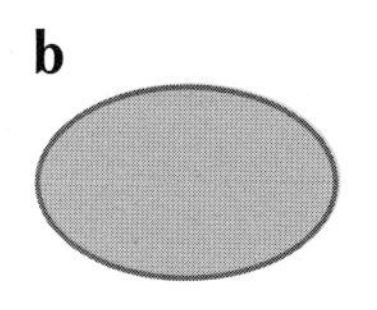
c
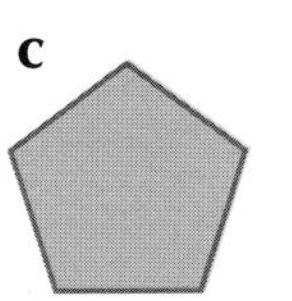
d
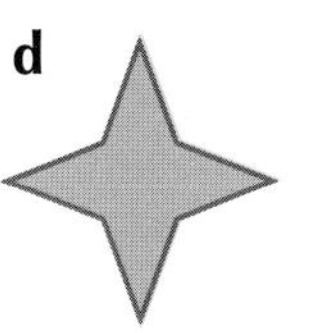
e
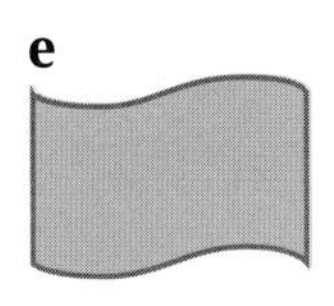

4 Write down the order of rotational symmetry for each shape.

a
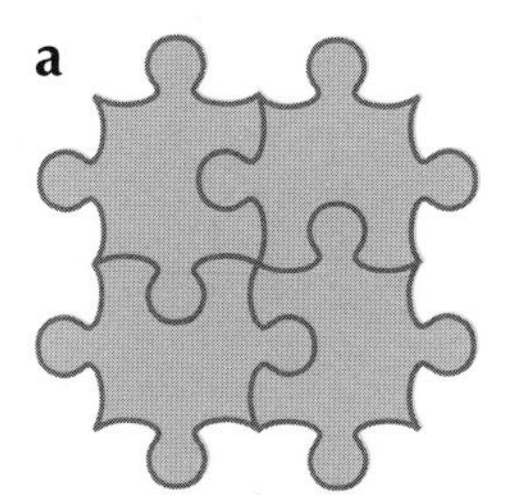
b
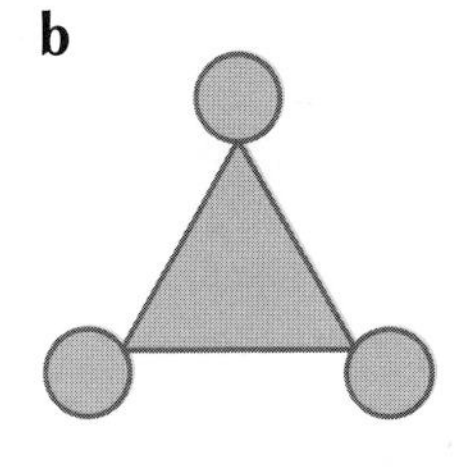
c
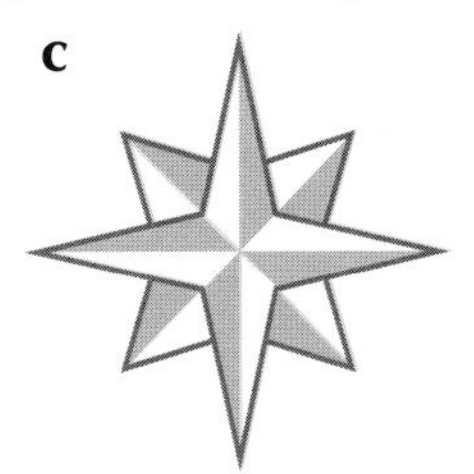
d
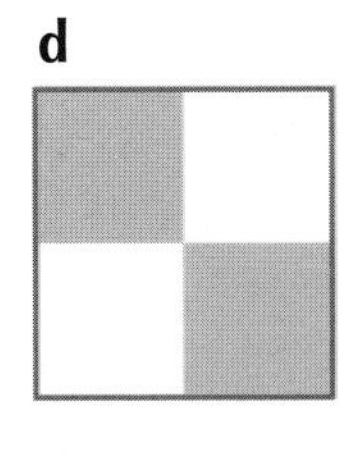

5 Three squares are shaded on this 3 by 3 tile.

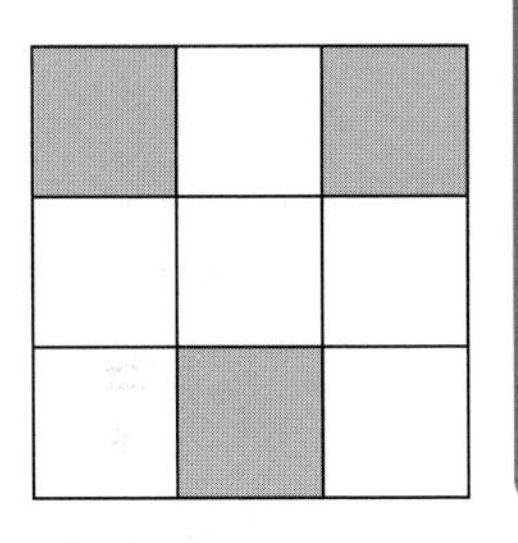

a How many lines of symmetry does the tile have?

b Copy the tile and shade in three more squares to give it rotational symmetry of order 2.

Revision of statistics and statistical techniques

This section will remind you of the statistical techniques that you have met before.

Collecting data

<table>
<tr><th></th><th>Definition</th><th>Example</th></tr>
<tr><td>Data collection sheet or observation sheet</td><td>A form for recording results</td><td rowspan="2">Favourite colours of a group of 42 pupils
<table>
<tr><td>Blue</td><td>卌 III</td></tr>
<tr><td>Red</td><td>卌 卌 卌</td></tr>
<tr><td>Green</td><td>卌 卌 II</td></tr>
<tr><td>Other</td><td>卌 II</td></tr>
</table></td></tr>
<tr><td>Tally</td><td>A means of recording data quickly</td></tr>
<tr><td>Two-way table</td><td>A table for combining two sets of data</td><td>
<table>
<tr><th></th><th>Ford</th><th>Vauxhall</th><th>Peugeot</th></tr>
<tr><td>Red</td><td>3</td><td>5</td><td>2</td></tr>
<tr><td>Blue</td><td>1</td><td>0</td><td>4</td></tr>
<tr><td>Green</td><td>2</td><td>0</td><td>1</td></tr>
</table></td></tr>
<tr><td>Frequency table</td><td>A table showing the quantities of different items or values</td><td>
<table>
<tr><th>Mass of parcels, M (kg)</th><th>Number of parcels (frequency)</th></tr>
<tr><td>$0 < M \leqslant 1$</td><td>5</td></tr>
<tr><td>$1 < M \leqslant 2$</td><td>7</td></tr>
<tr><td>$M > 2$</td><td>3</td></tr>
</table></td></tr>
</table>

(Continued)

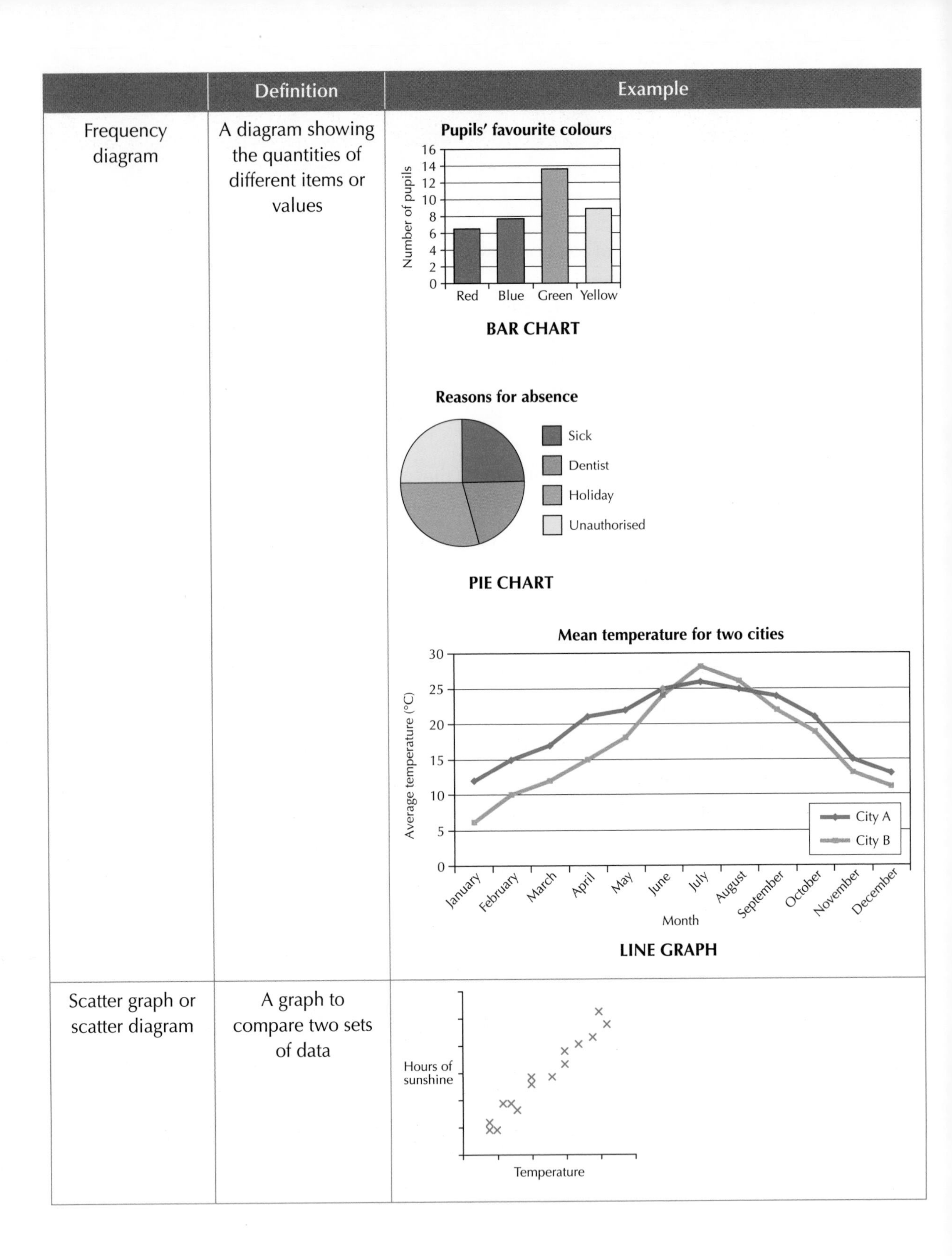

	Definition	Example
Frequency diagram	A diagram showing the quantities of different items or values	BAR CHART PIE CHART LINE GRAPH
Scatter graph or scatter diagram	A graph to compare two sets of data	

Processing data

	Definition	Example
Mode	The value that occurs most often	**Example 15** Find the mode, median, mean and range of this set of data. 23, 17, 25, 19, 17, 23, 21, 23 Sorting the data into order, smallest first, gives: 17, 17, 19, 21, 23, 23, 23, 25 Mode $= 23$ Median $= \frac{21 + 23}{2}$ $= 22$ Mean $= \frac{17 + 17 + 19 + 21 + 23 + 23 + 23 + 25}{8}$ $= 21$ Range $= 25 - 17$ $= 8$
Median	The middle value when the data is written in numerical order (or the average of the middle two values)	
Mean	The sum of all the values divided by the number of items of data	
Range	The difference between the largest and smallest values	

Exercise 14P

1 Calculate:

i the mode **ii** the median **iii** the mean

for each set of data.

a 1, 1, 1, 4, 8, 17, 24

b 2, 5, 11, 5, 8, 7, 6, 1, 9

c £2.50, £4.50, £2, £3, £4.50, £2.50, £3, £4.50, £3.50, £4

d 18, 18, 19, 21, 24, 26

2 These are the times taken (T seconds) by 20 pupils to run 100 m.

Boys	13.1	14.0	17.9	15.2	15.9	17.5	13.9	21.3	15.5	17.6
Girls	15.3	17.8	16.3	18.1	19.2	21.4	13.5	18.2	18.4	13.6

a Copy and complete the two-way table to show the frequencies.

	Boys		Girls	
	Tally	Frequency	Tally	Frequency
$12 \leqslant T < 14$	| |	2	| |	2
$14 \leqslant T < 16$				
$16 \leqslant T < 18$				
$18 \leqslant T < 20$				
$20 \leqslant T < 22$				

b Which is the modal class for the boys?

c Which is the modal class for the girls?

3. A marathon is held in Millhouses one year.

The scatter diagram shows the age of runners and the times they took to finish.

Explain what the scatter diagram tells you.

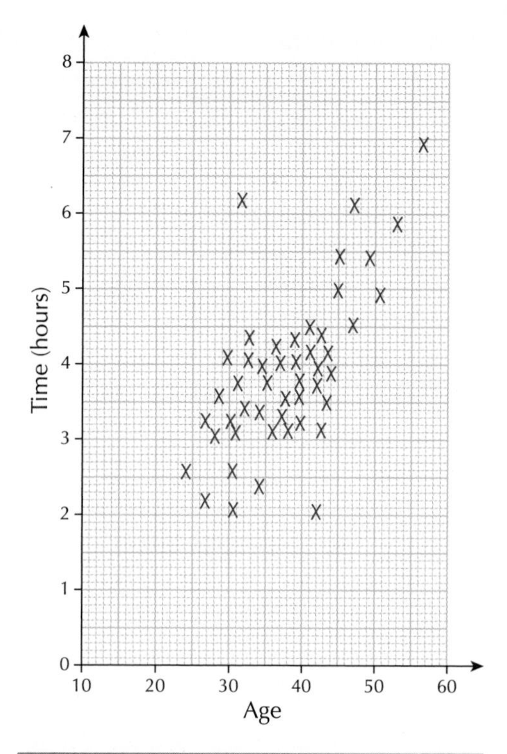

4. A school quiz team is made up of pupils from four different classes. The table shows the numbers of pupils in the team from each class.

Class	Number of pupils
A	4
B	3
C	8
D	5

a Represent this information in a pie chart.

b Holly says: 'The percentage of pupils chosen from class C is double the percentage chosen from class A.' Explain why this might not be true.

Revision of probability

Make sure that you are familiar with the vocabulary to do with probability, which is listed on the probability scale below.

Probability scale

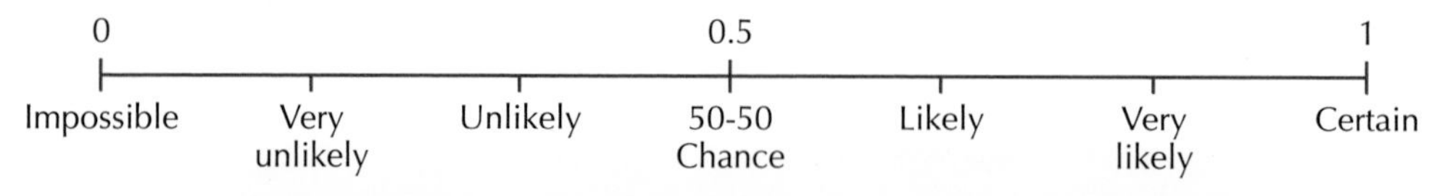

Example 16

A fair spinner is numbered 1, 2, 3.

a The spinner is spun twice.

List all the possible outcomes.

b How many possible outcomes are there if the spinner is spun three times?

a 1, 1; 1, 2; 1, 3; 2, 1; 2, 2; 2, 3; 3, 1; 3, 2; 3, 3

b $3 \times 3 \times 3 = 27$

Example 17

A six-sided dice is rolled 60 times.

It lands on a 6 fifteen times.

a What is the experimental probability of landing on a 6?

b Do you think the dice is fair?

a $\frac{15}{60} = \frac{1}{4}$

b No, because the experimental probability is $\frac{1}{4}$ and the theoretical probability is $\frac{1}{6}$ so two very different probabilities.

Example 18

A coin is thrown and a dice is rolled at the same time.

a Draw a sample space diagram.

b Write down the probability of getting a head and a 6.

a

		Dice					
		1	2	3	4	5	6
Coin	Head	H, 1	H, 2	H, 3	H, 4	H, 5	H, 6
	Tail	T, 1	T, 2	T, 3	T, 4	T, 5	T, 6

b $\frac{1}{12}$

Example 19

In a raffle there are 100 tickets, coloured blue, green or yellow.

The table shows the numbers of tickets of each colour.

Ticket colour	Number of tickets
Blue	50
Green	20
Yellow	30

a What is the probability of picking a blue ticket?

b What is the probability of picking a yellow ticket?

c What is the probability of picking a blue or green ticket?

d What is the probability of picking a ticket that is not green?

a $\frac{1}{2}$ **b** $\frac{3}{10}$ **c** $\frac{7}{10}$ **d** $1 - \frac{1}{5} = \frac{4}{5}$

Exercise 14Q

In this exercise, give each answer as a fraction in its simplest form.

1 Ten cards are numbered 1 to 10. A card is picked at random.

Work out the probability of picking:

a the number 5 **b** an even number

c a number greater than 8 **d** a number less than or equal to 4.

2 Two coins are thrown.

a How many different possible outcomes are there?

b Work out the probability of getting no heads.

c Work out the probability of getting two heads.

d Work out the probability of getting exactly one head.

3 A five-sided spinner is spun 50 times. These are the results.

Number on spinner	1	2	3	4	5
Frequency	8	11	10	6	15

a Write down the experimental probability of the spinner landing on the number 4.

b Write down the theoretical probability of a fair, five-sided spinner landing on the number 4.

c Compare the experimental and theoretical probabilities and say whether you think the spinner is fair. Give your reasons.

GCSE-type questions

1 Look at these numbers.

15	27	37	42	55	69	72	73

From the numbers in the box, write down:

a two numbers that have a total of 100

b two numbers that have a difference of 40

c the numbers that are multiples of 9

d the number that is the product of 6 and 7.

2 Andrew went out to a museum with his parents.

a In the café, his dad bought two cups of coffee and a cola.

Mum had a scone and jam and Andrew had a chocolate sundae.

They paid with a £20 note.

Calculate their change.

b Picture postcards of the polar bear cost 44p.

How many postcards of the polar bear could Andrew buy if he had £2.50?

DINOCAFÉ

Tea	£2.10
Coffee	£3.25
Cola	£1.99
Cheesecake	£2.80
Scone & Jam	£3.45
Chocolate Sundae	£4.99

PS **3** Sophia went to a wildlife park with Grandpa.

They saw a sign that said the length of the walk all round the park lake was 4.7 km.

Grandpa said that he wouldn't be able to walk more than 3 miles.

1 mile is about 1.6 km.

Would Grandpa be able to walk around the park lake?

Show your working.

4 Work out 378×52.

5 Look at these eight numbers.

10	14	17	27	65	69	81	89

a Write down two numbers from the box that have a sum of 98.

b Write down a number from the box that is:

i a multiple of 13 **ii** a square number.

6 Use words from the box to complete the sentences correctly.

cube	square	multiple	factor	square root

a 48 is a … of 12.

b 7 is the … of 49.

c 27 is the … of 3.

d 9 is a … of 72.

e 64 is the … of 8.

7 **a** Write down all the factors of 18.

b Write down a multiple of 10.

c Work out $\frac{5}{6}$ of 18.

d Write 10 out of 25 as a fraction in its simplest form.

8 The table shows the temperatures in three cities at noon one day.

Delhi	Rome	London
14 °C	–2 °C	–6 °C

What is the difference in temperature between:

a Delhi and Rome **b** Rome and London?

9 Write the numbers in each set in order of size.

Start with the smallest number.

a 77, 104, 14, 140, 147 **b** –4, 4, 0, –8, –2

c 0.91, 0.9, 0.091, 0.09, 0.901 **d** 60%, $\frac{1}{2}$, 0.55, $\frac{5}{8}$

10 Tom says 6^2 is 12.

Is Tom correct?

Explain your answer.

11 A group of 90 pupils went on an end-of-term treat.

They all went either to a theme park or Blackpool.

25 boys and 17 girls went to the theme park.

18 boys went to Blackpool.

a Copy and complete the two-way table, using the given data.

	Theme park	Blackpool	Total
Boys			
Girls			
Total			

b After school one day, the headteacher was informed that one of the pupils had lost their mobile phone.

What is the probability that this pupil:

i was a girl **ii** had been to Blackpool?

Give each answer as a fraction in its simplest form.

12 This diagram is wrong.

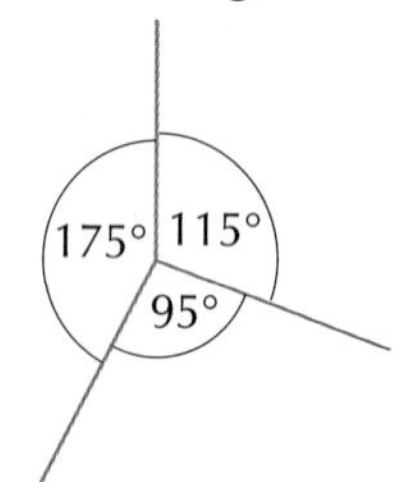

Explain how you know.

13 **a** What is the value of $3w + 4q$, given that $w = 7$ and $q = 3$?

b If n is an even number, what type of number is $n + 1$?

14 A tin of dog food cost 75p.

A supermarket had an offer on dog food one week.

Joy wants 12 tins of dog food.

a How much will Joy pay for the dog food?

Chris tells Joy that another supermarket was selling the dog food with one-third taken off the price.

b Explain if Joy has got the best deal on her dog food.

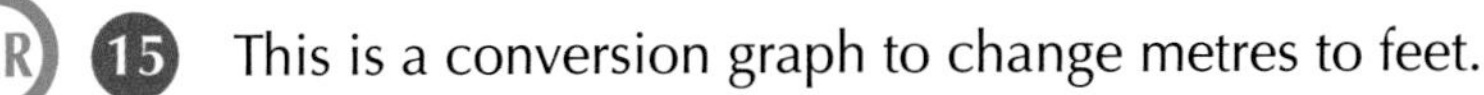

MR **15** This is a conversion graph to change metres to feet.

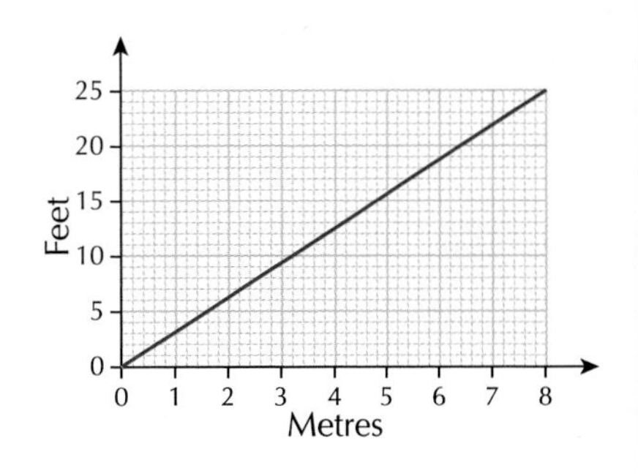

a Approximately how many feet are the same as 3 metres?

b Approximately how many metres are the same as 15 feet?

In the pole vault, Kathy's record is 5 metres and Helen's record is 16 feet.

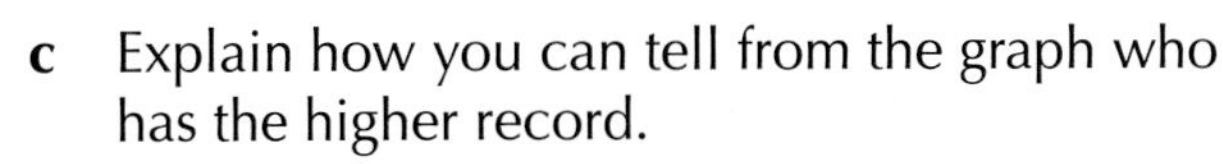

c Explain how you can tell from the graph who has the higher record.

16 A youth club had 36 members.

$\frac{1}{4}$ of the members preferred to play games.

$\frac{2}{9}$ of the members wanted to just chat.

How many members had no preference for what they did?

17 Construct an equilateral triangle with side lengths of 5 cm.

You must show all your construction lines.

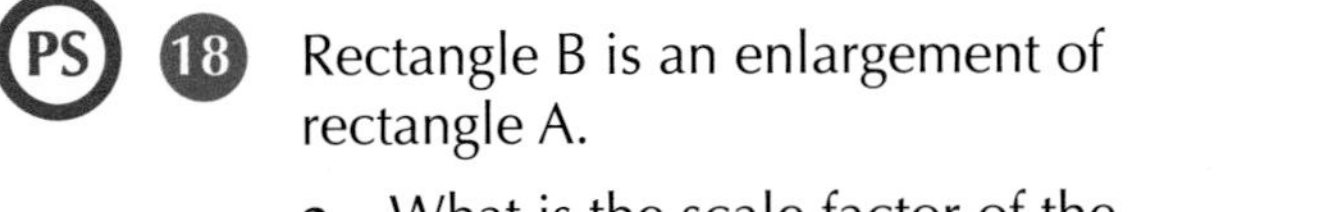

PS **18** Rectangle B is an enlargement of rectangle A.

4 cm, 3 cm, A; 20 cm, B

a What is the scale factor of the enlargement?

b What is the area of rectangle B?

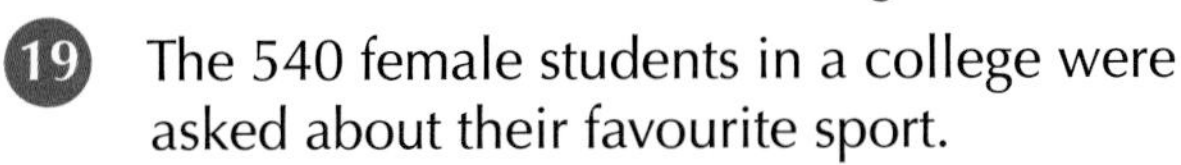

19 The 540 female students in a college were asked about their favourite sport.

The pie chart shows the results.

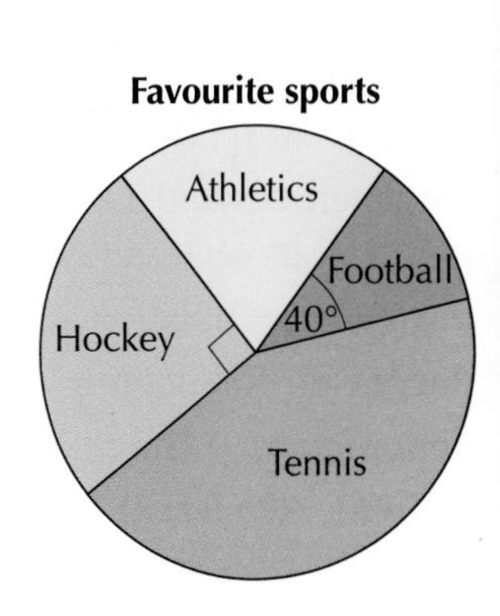

a How many chose: **i** hockey **ii** football?

240 of the students chose tennis.

b Calculate the angle in the pie chart that represents tennis.

c How many chose athletics?

Glossary

π The result of dividing the circumference of a circle by its diameter, represented by the Greek letter π (pi).

average speed The result of dividing the total distance travelled by the total time taken for a journey.

capacity The amount of space inside a container; how much fluid a container can hold.

centre of enlargement The point, inside or outside the object, on which the enlargement is centred; the point from which the enlargement of an object is measured.

concave polygon A polygon in which one or more interior angle is greater than 180°; it has a diagonal that is outside the polygon.

convex polygon A polygon in which all the interior angles are less than 180° and all the diagonals are inside the polygon.

cube **1** In geometry, a 3D shape with six square faces, eight vertices and 12 edges.

2 To multiply a number by itself twice; for example, $x^3 = x \times x \times x$.

cuboid A 3D shape with six rectangular faces, eight vertices and 12 edges; opposite faces are identical to each other.

decagon A polygon with 10 sides.

decrease Reduction; make smaller.

denominator The number below the line in a fraction, which says how many parts there are in the whole; for example, a denominator of 6 tells you that you are dealing with sixths.

distance How far something moves or travels.

distance–time graph A graph showing the distance travelled (vertical axis) against the time taken (horizontal axis), for a journey.

enlarge To make a shape bigger.

enlargement A transformation in which the object is enlarged to form an image.

expand To expand a term with brackets, you multiply everything inside the brackets by the value in front of the brackets.

factorise Take out a common factor and write an expression as a bracketed term with a multiplying factor in front.

formula A mathematical rule, using numbers and letters, that shows a relationship between variables; for example, the conversion formula from temperatures in Fahrenheit to temperatures in Celsius is: $C = \frac{5}{9}(F - 32)$.

height The vertical distance, from bottom to top, of a 2D or 3D shape.

heptagon A polygon with seven sides.

hexagon A polygon with six sides.

increase Make larger.

interior angle The angle between two adjacent sides of a 2D shape, at a vertex.

invert Turn upside down.

irregular polygon A polygon in which the sides are not all equal, and the angles are not all equal.

km/h The distance, in kilometres, that an object moves in an hour; the rate of movement.

lender Someone who allows someone else (the borrower) to borrow money from them.

linear equation An equation such as $y = 4x - 7$, that will produce a straight-line graph.

litre A measure of capacity, equal to 1000 cm^3.

multiplier A number that is used to find the result of increasing or decreasing an amount by a percentage.

negative correlation A relationship between two sets of data, in which the value of one variable increases as the value of the other variable decreases.

negative power The reciprocal of a positive power; for example, $10^{-1} = \frac{1}{10}$, $10^{-2} = \frac{1}{10^2} = \frac{1}{100}$.

no correlation A lack of any sort of relationship between two sets of data.

nonagon A polygon with nine sides.

numerator The number above the line in a fraction, which tells you haw many of the equal parts of the whole you have; for example, $\frac{3}{5}$ of a whole is made up of three of the five equal parts. The number of equal parts is the denominator.

octagon A polygon with eight sides.

original value The value before a change (increase or decrease).

pentagon A polygon with five sides.

polygon A closed 2D shape with straight sides.

positive correlation A relationship between two sets of data, in which the value of one variable increases as the value of the other variable increases.

ray A straight line drawn through two points.

regular polygon A polygon in which the sides are all equal, and the angles are all equal.

scale factor The number of times the image is bigger than the object under an enlargement.

scatter graph A diagram showing corresponding values between two sets of data.

similar Exactly the same shape but a different size; an enlargement.

similar triangles Two or more triangles that have exactly the same angles as each other but sides of different lengths.

simple interest Money that a borrower pays a lender, for allowing them to borrow money.

speed The rate at which an object is travelling.

subject In a formula, the subject is on its own, on the left-hand side of the equals sign.

suitable degree of accuracy A degree of accuracy that is suitable for the way the number is being used.

surface area The total area of all of the surfaces of a 3D shape.

time The duration of a journey.

triangular prism A prism that has a triangular cross-section; its end faces are congruent triangles.

two-way table A table that records values that depend on two sets of criteria.

variable **1** A quantity that may take many values.

2 A letter that stands for a quantity that can take various values.

volume The amount of space taken up by a 3D object.

Index

N

O

P

Q

R

S

T

U

V

W

X

Y

Z

William Collins' dream of knowledge for all began with the publication of his first book in 1819. A self-educated mill worker, he not only enriched millions of lives, but also founded a flourishing publishing house. Today, staying true to this spirit, Collins books are packed with inspiration, innovation and practical expertise. They place you at the centre of a world of possibility and give you exactly what you need to explore it.

Collins. Freedom to teach.

Published by Collins
An imprint of HarperCollins*Publishers*
77–85 Fulham Palace Road
Hammersmith
London
W6 8JB

Browse the complete Collins catalogue at
www.collins.co.uk

10 9 8 7 6 5 4 3 2 1

ISBN-13 978-0-00-753777-8

The authors Kevin Evans, Keith Gordon, Chris Pearce, Trevor Senior and Brian Speed assert their moral rights to be identified as the authors of this work.

British Library Cataloguing in Publication Data
A catalogue record for this publication is available from the British Library.

Commissioned by Katie Sergeant
Project managed by Elektra Media Ltd
Developed and copy-edited by Joan Miller
Proofread by Amanda Dickson
Edited by Helen Marsden
Illustrations by Ann Paganuzzi, Nigel Jordan and Tony Wilkins
Typeset by Jouve India Private Limited
Cover design by Angela English
Index by Indexing Specialists (UK) Ltd

Printed and bound by L.E.G.O. S.p.A, Italy

Acknowledgements
The publishers wish to thank the following for permission to reproduce photographs. Every effort has been made to trace copyright holders and to obtain their permission for the use of copyright materials. The publishers will gladly receive any information enabling them to rectify any error or omission at the first opportunity.

(t = top, c = centre, b = bottom, r = right, l = left)

Cover Hupeng/Dreamstime, p 6 godrick/Shutterstock, p 8 mangostock/Shutterstock, p 12 Stephaniellen/Shutterstock, p 14 Christian De Araujo/Shutterstock, p 16 Mila Atkovska/Shutterstock, p 22 tratong/Shutterstock, p 23 LifePhotoStudio/Shutterstock, p 24–25 r.nagy/Shutterstock, p 26 SSPL/Getty Images, p 42–43 Sofia Andreevna/Shutterstock, p 44 Horia Bogdan/Shutterstock, p 56–57 solkanar/Shutterstock, p 58 Iakov Kalinin/Shutterstock, p 65 Zeljko Radojko/Shutterstock, p 66l DDCoral/Shutterstock, p 66r Posmetukhov Andrey/Shutterstock, p 67 Jari Hindstroem/Shutterstock, p 69 Oleg M./Shutterstock, p 70 Joinmepic/Shutterstock, p 80–81 guentermanaus/Shutterstock, p 82 AlexanderZam/Shutterstock, p 85 Claudio Divizia/Shutterstock, p 94–95 ArtThailand/Shutterstock, p 96 Universal Studios/Photoshot, p 112–113 Yeko Photo Studio/Shutterstock, p 114 Nils Z/Shutterstock, p 120 Gravicapa/Shutterstock, p 126–127 Paul Bonugli/Demotix/Corbis, p 128 Nicholas Rjabow/Shutterstock, p 138–139 Hein Nouwens/Shutterstock, p 140 leungchopan/Shutterstock, p 156–157 severija/Shutterstock, p 158 Andrea Danti/Shutterstock, p 172–173 tavi/Shutterstock, p 174 MrOK/Shutterstock, p 190–191 Neil Burton/Shutterstock, p 192 NASA/Getty Images, p 204–205 Christian Mueller/Shutterstock, p 206 Jeffrey M. Frank/Shutterstock, p 218–219 chrisdorney/Shutterstock, p 219 Lebrecht Music and Arts Photo Library/Alamy, p 220 Sabphoto/Shutterstock.